38

Springer Series in Solid-State Sciences

Edited by Peter Fulde

Springer Series in Solid-State Sciences

Editors: M. Cardona P. Fulde H.-J. Queisser

1 **Principles of Magnetic Resonance** 2nd Edition 2nd Printing By C. P. Slichter

2 **Introduction to Solid-State Theory** 2nd Printing. By O. Madelung

3 **Dynamical Scattering of X-Rays in Crystals** By Z. G. Pinsker

4 **Inelastic Electron Tunneling Spectroscopy** Editor: T. Wolfram

5 **Fundamentals of Crystal Growth I** Macroscopic Equilibrium and Transport Concepts. 2nd Printing By F. Rosenberger

6 **Magnetic Flux Structures in Superconductors** By R. P. Huebener

7 **Green's Functions in Quantum Physics** By E. N. Economou

8 **Solitons and Condensed Matter Physics** 2nd Printing Editors: A. R. Bishop and T. Schneider

9 **Photoferroelectrics** By. V. M. Fridkin

10 **Phonon Dispersion Relations in Insulators** By H. Bilz and W. Kress

11 **Electron Transport in Compound Semiconductors** By B. R. Nag

12 **The Physics of Elementary Excitations** By S. Nakajima, Y. Toyozawa, and R. Abe

13 **The Physics of Selenium and Tellurium** Editors: E. Gerlach and P. Grosse

14 **Magnetic Bubble Technology** 2nd Edition By A. H. Eschenfelder

15 **Modern Crystallography I** Symmetry of Crystals, Methods of Structural Crystallography By B. K. Vainshtein

16 **Organic Molecular Crystals** Their Electronic States. By E. Silinsh

17 **The Theory of Magnetism I** Statics and Dynamics By D. C. Mattis

18 **Relaxation of Elementary Excitations** Editors: R. Kubo and E. Hanamura

19 **Solitons,** Mathematical Methods for Physicists By G. Eilenberger

20 **Theory of Nonlinear Lattices** By M. Toda

21 **Modern Crystallography II** Structure of Crystals By B. K. Vainshtein, V. M. Fridkin, and V. L. Indenbom

22 **Point Defects in Semiconductors I** Theoretical Aspects By M. Lannoo and J. Bourgoin

23 **Physics in One Dimension** Editors: J. Bernasconi, T. Schneider

24 **Physics in High Magnetic Fields** Editors: S. Chikazumi and N. Miura

25 **Fundamental Physics of Amorphous Semiconductors** Editor: F. Yonezawa

26 **Elastic Media with Microstructure I** One-Dimensional Models By I.A. Kunin

27 **Superconductivity in Transition Metals and Compounds** By S. Vonsovsky, Yu. A. Isyumov, and E. Z. Kurmaev

28 **The Structure and Properties of Matter** Editor: T. Matsubara

29 **Electron Correlation and Magnetism in Narrow-Band Systems** Editor: T. Moriya

30 **Statistical Physics I** By M. Toda and R. Kubo

31 **Statistical Physics II** By R. Kubo and M. Toda

32 **Quantum Theory of Magnetism** By R. M. White

33 **Mixed Crystals** By A. I. Kitaigorodsky

34 **Phonons: Theory and Experiments I** Lattice Dynamics and Models of Interatomic Forces By P. Brüesch

35 **Point Defects in Semiconductors II** Experimental Aspects By M. Lannoo and J. Bourgoin

36 **Modern Crystallography III** Formation of Crystals By A. A. Chernov et al.

37 **Modern Crystallography IV** Physical Properties of Crystals By L. A. Shuvalov et al.

38 **Physics of Intercalation Compounds** Editors: L. Pietronero and E. Tosatti

Physics of Intercalation Compounds

Proceedings of an International Conference
Trieste, Italy, July 6–10, 1981

Editors:
L. Pietronero and E. Tosatti

With 167 Figures

Springer-Verlag Berlin Heidelberg New York 1981

Dr. *Luciano Pietronero*
Brown Boveri Research Center,
CH-5405 Baden-Dättwil, Switzerland

Professor Dr. *Erio Tosatti*
International School for Advanced Studies, Strada Costiera 11,
I-34014 Trieste, Italy

Series Editors:

Professor Dr. Manuel Cardona
Professor Dr. Peter Fulde
Professor Dr. Hans-Joachim Queisser

Max-Planck-Institut für Festkörperforschung, Heisenbergstrasse 1
D-7000 Stuttgart 80, Fed. Rep. of Germany

ISBN-13:978-3-642-81776-2 e-ISBN-13:978-3-642-81774-8
DOI: 10.1007/978-3-642-81774-8

2153/3130-543210

Preface

When an area of research is in fast growth, it often happens that no one single journal is to be found where most of the relevant publications are contained. Such is the case of the physics of intercalation compounds, a field which, by sitting at a corner point between materials science, solid-state physics, and chemistry, finds its contributions largely scattered about in the literature. Given these circumstances it is of crucial interest to find a place where the most recent contributions and up-to-date references can be found at once. For intercalated graphite and other similar compounds this role has been played so far by proceedings of international conferences, such as La Napoule (1977), Nijmegen (1979), Provincetown (1980), and Sendai (1980).

The present book is an ideal continuation of this series, as it contains most of the invited and contributed papers of the Trieste International Conference on the Physics of Intercalation Compounds, held in Trieste, Italy during the week 6-10 July 1981.

The main emphasis is on intercalated graphite, though several interesting contributions deal with other materials, such as polyacetylene and transition-metal compounds, or with general problems, such as two-dimensional melting. The book is divided into six sections – Structure and General Properties, Electronic Porperties, Stability and Phonons, Ordering and Phase Transitions, Magnetic Resonance, and Transport Properties – reflecting the main areas of interest, and also broadly the main discussion sessions of the Conference.

We have been guided in the preparation of this Conference by the suggestions of our International Advisory Committee: B. Bergersen, J. Conard, F. F. di Salvo, M.S. Dresselhaus, J. Fischer, P. Fulde, T. Geballe, H.-J. Güntherod, C. Haas, G. Harbeke, A. Hérold, H. Kamimura, S. Solin, and A.R. Ubbelohde. Also L. Reatto, G. Scoles, and M.P. Tosi have helped us with their encouragement and advice. We are very grateful to all these people for their help.

The Conference, from which this book originates, was generously sponsored by the Italian Research Council (CNR), also through its Gruppo Nazionale di Struttura della Materia, and by the International School for Advanced Studies (SISSA) of Trieste. The hosting institution was the International Centre for Theoretical Physics of Trieste, who graciously put their facilities and staff at our disposal. On behalf of the international scientific community we wish to express our gratitude to all these institutions, as well as to those individuals—we mention here particularly Mrs Deisa Buranello, Miss Carla Carbone, and Cav. Sergio Stabile—who have contributed so much to making a successful conference, and this book as a by-product.

Baden-Dättwil · Trieste
October 1981

L. Pietronero · E. Tosatti

Contents

Part I. Structure and General Properties

Part II. Electronic Properties

Part III. Stability and Phonons

Part IV. Ordering and Phase Transitions

Part V. Magnetic Resonance

Part VI. Transport, Conductivity, Superconductivity

Part I

Structure and General Properties

Kinetic Problems with Intercalation Compounds

A.R. Ubbelohde

Department of Chemical Engineering and Chemical Technology, Imperial College
London SW7, UK

What follows refers primarily to kinetic problems in *layered* intercalation compounds. It may be assumed that mobile electrons will be found in any molecular framework when this contains conjugated valence bond sequences [1] and that crystals of such compounds are semi-conductors [2]. Graphite may be regarded as a macromolecular limit of a homologous series of planar fused ring structures on this chemical basis and behaves as a semi-conductor with zero energy gap. Already in 1950 we had established that intercalation between the carbon hexagon layers in graphite by electron acceptor molecules such as bromine or iodine monochloride, or electron donor molecules such as potassium, greatly enhances the electrical conductivity parallel to the layers. However the high conductivity of these first synthetic metals was considered two-dimensionally. Perpendicular to the layers intercalation of electron acceptor molecules generally reduces the electrical conductivity of the parent graphite even further; a recent measurement for graphite nitrate showed a reduction of conductivity perpendicular to the layers by a factor of about 40 [3].

Accordingly, at least to a first approximation, one can discuss the properties of synthetic metals based on graphite in terms of a pile of 2D macromolecules stacked in parallel. The model can subsequently be refined to allow for weak interaction between the layers whose stacking may be ordered or disordered. For any one layer, when the electrical conductivity σ_a of the parent structure can be expressed by the elementary equation $\sigma_a = n\, e\, \mu$, intercalation changes this into $\sigma_a = e(n \pm fN)\mu^*$ for the synthetic metal derived from it. To explain the marked increase of electrical conductivity, the assumption commonly made follows our original suggestion [4] that $(n \pm fN) \gg n$, as a result of fractional charge transfer f from N intercalated molecules per atom of the host structure. Such an assumption encouraged attempts to increase the conductivity of the synthetic metal by using intercalate molecules which give large values of $(n \pm fN)$. Such attempts have indeed met with some success, in the sense that partner molecules capable of forming quite strong charge transfer bonds favour the production of high electrical conductivities. However, no very direct correlation of high conductivities with strong charge transfer bonding has been established. It may be at least equally relevant to examine ways in which intercalation might enhance carrier mobility, e.g., making $\mu^* > \mu$. Such an approach stresses kinetic problems. It calls for consideration of the various mechanisms that can affect charge scattering in any synthetic metal. Unfortunately, any discussion of the principal defect structures that can influence charge scattering, and thereby affect mobility, has been far from systematic even for synthetic metals based on graphite, which have been the most fully explored to date. Accordingly, what follows is illustrative but should not be regarded as exhaustive.

Because of the versatility of bonding that can be adopted by carbon atoms, solid carbons tolerate an extraordinary variety of bond structures. Even in carbons with well-ordered crystals with, say, more than 95% of their atoms in ideal positions, the phonon spectrum could comprise a much greater breadth than, say, for a simpler inorganic crystal such as copper. Some idea of the range in carbons can be obtained from direct measurements of electrical resistivity of graphites of varying perfection parallel to the basal plane [5]. Other properties such as thermo-electric power or thermal conductivity of carbons [6] can give even more sensitive indications about the defect content of carbons than the electrical resistivity. As has often been stressed, the great versatility of thermal scattering of charge carriers in the parent graphite makes it essential to select material very critically when aiming to produce synthetic metals with optimum behaviour. According to current selection procedures [3] even the best graphites commonly used are far from the ideal limit. However, this somewhat pessimistic warning about defect structures in the parent graphite may not have correspondingly grave implications for ultimate technological superiority of the synthetic metals produced from it by intercalation. On pushing the layers apart, by the process of inserting intercalate molecules, at least some of the defects may become inoperative for charge scattering.

At the present time, various kinds of evidence that layered conductors may be 'improved' by the process of intercalation is fairly plentiful, but mostly circumstantial. It includes studies on the movement of molecules as well as charges. Direct physico-chemical measurements on the overpotential required to begin intercalation for any desired partner molecule indicate that once a threshold has been overcome [7] intercalation proceeds quite easily. There is also evidence that certain molecules which find it difficult to enter directly between the carbon hexagon sheets in graphite may be 'introduced' by prior entry of an easier species [8]. Direct measurements of kinetics of diffusion of partner molecules once they have entered show lateral movement to be surprisingly rapid for certain molecules [9] such as nitric acid.

Our present knowledge about the role of structural defects in electrical conduction of synthetic metals is mainly derived from theories about their chemical bonding, together with measurements on the temperature dependence of their electrical properties. Chemical valence theories show that the atoms in the host structure of a synthetic metal are mainly linked by covalent bonds. Defects (whose presence can hinder movement of charges, as well as diffusion of molecules through the solid) are basically different from those in close-packed crystals of many natural metals. Carbons tolerate a considerable variety of structural defects. In near-ideal graphites the most important include (the list is not exhaustive):

(a) Carbon atoms at perimeter sites at the edges of each planar macromolecule. When layered solids are annealed, the principal atomic movements lead to growth of planar macromolecules at the expense of more disordered material. This reduces the proportion of atoms in perimeter sites.

(b) Carbon atoms (such as those displaced by radiation damage) which do not form part of any regular macromolecule.

(c) Planar macromolecules may be stacked parallel in regular sequences or may tolerate various kinds of stacking disorder.

(d) When one or more covalent bonds are omitted from interior positions in a macromolecule, valence bond requirements may no longer allow neighbouring carbon hexagons to lie in the same plane identically. Instead, each omitted bond may lie at the nick of a 'claw' defect in the macromolecule, causing it to extend through neighbouring layers [10].

After intercalation of the chosen partner molecules between the layers, these may also locate themselves in distinctive ways at the different kinds of defect sites and may play more than one role in scattering of charge transport. A very obvious observation is that the temperature dependence of electrical conductivity parallel to the layers is entirely changed, in its general trend, from that of the parent graphite [11a, b]. The resistivity of various synthetic metals decreases on cooling over the range 300-50 K, showing temperature coefficients very approximately in accordance with the Bloch-Gruneisen formula. It is interesting that the characteristic temperatures θ_R calculated from such data support the view that the main source of electrical resistance parallel to the layer planes arises from thermal vibrations of the intercalate molecules.

Synthetic metal	θ_R (^oK)
C_8K	378
C_8Rb; C_8C_s	296
$C_{24}HSO_4$ $2H_2SO_4$	237
$C_{24}NO_3$ $2HNO_3$	218
$C_{24}NO_3$ $2DNO_3$	168

Lateral thermal vibrations of the molecules in any layer most probably control this kind of thermal scattering for σ_a in the temperature range from 300 K down to at least 3 K. This suggestion is supported by the observation that a steep decrease of resistivity is observed at the lambda point in graphite nitrates, when the nitric acid molecules in each intercalated layer take up an ordered pattern [12] (Fig. 1)

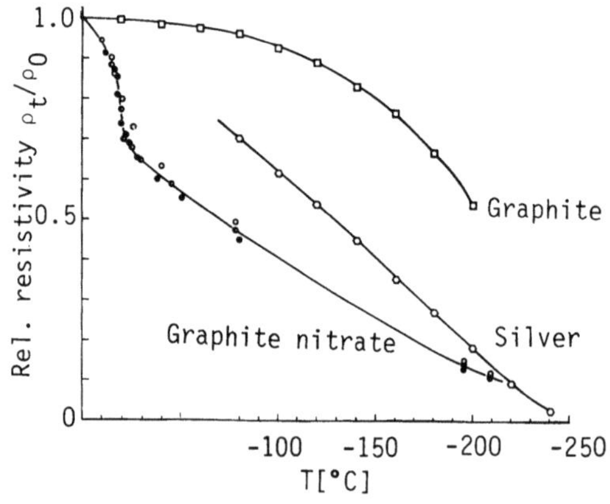

Fig.1. Relative resistivity ρ_t/ρ_0 (where ρ_0 refers to T = 0°C) as a function of temperature for graphite (□), silver (○), and graphite nitrate (•)

Somewhat surprisingly, there appears to be no clear evidence for any true 'rest resistance' in σ_a associated with non-thermal defects in near ideal graphites, but this may be because they have been specially selected to prepare the best synthetic metals. In this respect they differ from natural metals such as copper. And furthermore, as stated, the role of structural defects in charge scattering is probably smoothed out on separating the carbon-hexagon sheets by inserting the partner molecules. However, structural defects of a special kind which traverse several layers may play an important role even in the best near-ideal graphites yet used. Evidence for this suggestion has recently been obtained from accurate measurements of temperature dependence for both σ_a and σ_c, which have been made on material prepared from the same parent graphite. With graphite nitrates of the second stage, logarithmic plots permit reliable estimates of the resistance R_c of specimens perpendicular to the layers, even when the anisotropy ratio σ_a/σ_c is around 10^6. Below the lambda point a remarkable finding is that this ratio proves to be practically independent of temperature, down to at least 70 K. From various considerations [3a, 3b], this seems likely to be due to conduction perpendicular to the layers being predominantly by *extrinsic* leakage paths, which have similar dependence of conductivity upon temperature as for conduction parallel to the layers. This conclusion has so far been verified for two electron acceptor intercalates (with nitric acid and with iodine monochloride). If it can be generalised, it would strengthen the conclusion that ideal graphite free from any defects joining the layers must be truly 2D.

At least two kinds of structural defects 'of type i' have been suggested which traverse neighbouring layers and which may be the site of leakage paths whose presence is inferred from the observed temperature coefficient for the resistivity ratio σ_c/σ_a. 'Claw' structures may arise from the consequences of missing bonds in a covalent aromatic network. Alternatively, a kind of stacking disorder may arise from dislocations traversing neighbouring layers whose presence has been inferred to explain smooth intercalation by successive stages starting with a dilute synthetic metal [13,14]. Defects 'of type i' whose structure traverses more than one layer, which would otherwise be wholly independent, obviously introduce occasional correlation paths between neighbouring layers. As already stated, they are probably responsible for conduction perpendicular to the layers. They may also provide a distinctive scattering mechanism for charge carriers within any layer. The average distance $\overline{x_i}$ between neighbouring defects of this kind in any layer may be small compared with the meanfree path $\overline{\lambda_i}$ for charge scattering within any layer from all mechanisms, except for graphites of unusually high perfection. But with such graphites when $\overline{x_i} >> \lambda_i$ any trajectories with unusually high velocities parallel to the layers could persist and give rise to 'low scatter' conduction.

Reliable information about mean free paths in synthetic metals is still very scanty. Only a schematic illustration of possible novel effects need be given at this stage (cf [9]). Each conducting layer is a carbon-hexagon macro-aromatic network bounded by layers of electron acceptor molecules on either side. (Fig.2) Within each layer any charge trajectory at near grazing

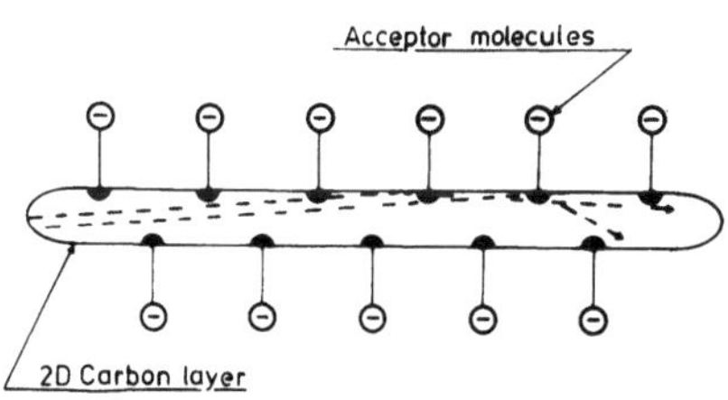

Fig.2. Schematic of carbon-hexagon macro-aromatic network bounded by layers of electron-acceptor molecules

incidence may undergo a kind of total internal reflexion and continue in much the same direction unless it is scattered somewhere within the boundary region. In near ideal conditions, depending on the absence of such scattering, unusually high mean free paths could be attained by a fraction of charge carriers, with unusual electrical consequences.

References

1. A.R. Ubbelohde (1933) *Nature*, 132, 1002

2. UK Patent No.775 465 (July 1953) A.R. Ubbelohde per NRDC

3a. I. Drummond and A.R. Ubbelohde (1980) *Proc. Roy. Soc.* A371, 309

3b. A.R. Ubbelohde and I. Drummond, 15th Carbon Conference (1981)

4. F.R. McDonnell, R.C. Pink, and A.R. Ubbelohde (1951) *J. Chem. Soc.* 151

5. A.W. Moore, A.R. Ubbelohde, and D.A. Young (1964) *Proc. Roy. Soc.* A280, 153

6. (T.E.P.) I.L. Spain, A.R. Ubbelohde, and D.A. Young (1966) *Phil. Trans. Roy. Soc.* 262, 345

   (Thermal Conductivity) C.N. Hooker, A.R. Ubbelohde, and D.A. Young (1965) *Proc. Roy. Soc.* 284A, 17

7. A.R. Ubbelohde (1969) *Carbon* 7, 523
   (1972) *ibid* 10, 201

8. M.J. Bottomley, G.S. Parry, A.R. Ubbelohde, and D.A. Young (1963) *J. Chem. Soc.* 5674

9. Examples are given in
   A.R. Ubbelohde (1981) *J. Synthetic Metals* 4, 35

10. A.R. Ubbelohde (1957) *Nature* 180, 380

11a. J.J. Murray and A.R. Ubbelohde (1969) *Proc. Roy. Soc.* A312, 371

11b. A.R. Ubbelohde (1971) *Proc. Roy. Soc.* A321, 445

12. A.R. Ubbelohde (1969) *Proc. Roy. Soc.* A309, 297

13. D.E. Nixon, G.S. Parry, and A.R. Ubbelohde (1968) *Proc. Roy. Soc.* A291, 324

14. N. Daumas and A. Herold (1969) C.R. Acad. Sci. Ser. C, 268, 373

    P. Legrange, A. Metrot, and A. Herold (1974) C.R. Acad. Sci. Ser. C, 278, 701

Comments on the Stoichiometry in the Metal-Graphite and Metal-Carbon Compounds

Albert Hérold

Laboratoire de Chimie du Solide Minéral, Associé au C.N.R.S. N° 158
Service de Chimie Minérale Appliwuée, Université de NANCY I, C.O. N° 140
F-54037 Nancy Cédex, France

Introduction

A stoichiometric compound can be defined as a chemical species which possesses an experimental composition in good agreement with its "ideal formula".

It is clear that the ideal formula of the G.I.C. cannot be forecast from chemical rules, because their formulae (such as KC_8, KC_{24} and others) cannot be related to the usual valencies of the elements but only to structural data : the G.I.C. are typical "topochemical" compounds.

The in-plane hexagonal unit cell of graphite contains two atoms and its parameter is $a_0 = 1.42 \cdot \sqrt{3} = 2.46$ Å. The surface area of 1 atom is also $(2.46)^2/2 = 2.62$ Å^2. If C_xI (I = intercalant) is the formula of a first stage compound, if S is the surface area of the in-plane unit cell of the intercalated layers, the index x will be :

$x = \frac{C}{I} = \frac{S}{2.62\ m}$ in which m is the number of atoms or molecules contained in the intercalant unit cell.

If the parameters of this latter are compatible with that of graphite, (also if the intercalated layers are epitaxial to those of the graphites) x is an integer (or a rational fraction) and the compound possesses a true ideal formula.
If the in-plane parameters of the intercalated layers are incompatible with a_0, the index x can be calculated, but it is only an integer or a rational fraction by chance ; the crystallographic formula of the compound is not a true "ideal" formula.
If there is no long-range ordering in the layers of intercalant, the index cannot be calculated from crystallographic data, only an approximate value can be obtained by using the interplanar distance and the molar volumes of graphite and the intercalant in the pure state

$$x = \frac{C}{I} = \frac{\text{molar volume of free intercalant}}{\text{molar volume of free graphite}} \cdot \frac{3.35}{d_i(\text{Å})-3.35} \qquad |1|$$

In the compounds of stage greater than one, a stoichiometric formula needs a regular distribution of the occupied and unoccupied intervals between the graphite layers, so that the stage index n is an integer.

I. Crystallographic and Experimental Formulae of the Metal-Graphite Compounds

I.1 First Stage Compounds

The stoichiometric formulae MC_8 (M = K, Rb, Cs) and MC_6 (M = Li, Sr, Ba, Eu, Yb) correspond to the well known "octal" and "hexal" dispositions of the metal atoms on the adjacent graphite sheets. In the case of the MC_8 compounds, the chemical analysis is in good agreement with the "ideal" formula corresponding to the structure (the difference is 1 or 2 percent). In the case of the MC_6 compounds, the synthesis of pure samples is difficult because of the low vapour pressure of the metals and their ability to form acetylides or interstitial carbides, and the chemical analysis is less certain : the MC_6 formulae have been established more by structural than by chemical studies |2| |3| |4|.

There are no simple and sure reasons to explain the existence of the two kinds of formula MC_8 and MC_6. One can only observe that the metallic radii of the heavy alkalis are more compatible with the octal structure and those of the other metals with the hexal one.

M-M In-plane distance : in MC_8 : 2×2.46 Å ; in MC_6 : 2×2.13 Å

	Li	Na	K	Rb	Cs	Ca	Sr	Ba	Sm	Eu	Yb
Metallic radii (Å)	1.53	1.90	2.35	2.48	2.67	1.97	2.15	2.22	1.81	1.99	1.94
	Li^+	Na^+	K^+	Rb^+	Cs^+	Ca^{2+}	Sr^{2+}	Ba^{2+}	Sm^{2+}	Eu^{2+}	Yb^{2+}
Ionic radii (Å)	0.6	0.95	1.33	1.48	1.69	0.99	1.13	1.35	1.04	1.12	1.13

However, R. SETTON |5| has shown that the interplanar distance can be calculated by addition of the graphite layers thickness to the ionic diameter of the intercalated metal. It is not very coherent to use the ionic·diameter to calculate the interplanar distance and the metallic diameter to justify the in-plane structure !

For first stage solid solutions $M_{1-x}M'_xC_8$ containing two heavy alkali metals, the theoretical value 8 for the ratio C / M+M' is also in good agreement with the analytical data |6| ; x can vary from 0 to 1.

Other stoichiometric species containing two metals are the $KHgC_4$ and $RbHgC_4$ ternary compounds, which contain triple metallic layers. The ratios C/K or C/Rb = 4 correspond to the octal epitaxy of two alkali metals layers on the graphite sheets. The ratios Hg/K and Hg/Rb = 1 can be explained by the complete occupation of the prismatic sites between the alkali metal atoms by mercury ones. One can also observe that the free alloy KHg is a defined compound but not RbHg |7|.
Solid solutions $K_{1-x}Rb_xHgC_4$ exist, in which x can vary from 0 to 1.

On the contrary, the $MTl_{1.5}C_4$ ternaries (M = K, Rb) |8| contain thick metallic layers without in-plane long-range ordering : there is not any structural explanation for the rather exact experimental value of the ratio C/K = C/Rb = 4 which have been obtained by reaction of graphite with potassium - thallium and rubidium - thallium alloys of different compositions. Nor are there any explanation for the value Tl/K = Tl/Rb = 1.5, because the free alloys $KTl_{1.5}$ and $RbTl_{1.5}$ are not defined compounds. Thus,

the formulae $MTl_{1.5}C_4$ are not true "ideal" formulae, and the indices are only approximate values ?
On the other hand, it is remarkable that the compound $RbTl_{1.5}C_4$ exists under 2 forms, which differ by their interplanar distances :
12.65 Å for the α metastable phase and 13.40 Å for the β stable phase.

I.2 Stages Higher than One

Among the metals which form intercalated layers with an hexal structure, only lithium leads to defined high stage compounds ; the stage 2 has been prepared with a stoichiometric formula LiC_{12} in the presence of a small amount of LiC_6, but its formula can vary from LiC_{12} to LiC_{25} |2|. Higher stages are not stoichiometric |9||10|.

It is well known that the heavy alkali metals do not form compounds of a second stage with an octal structure : for a long time the dodecal lacunar structure proposed by RUDORFF and SCHULZE |11| and leading to the general formulae MC_{12n} for the compounds of stage n > 2 were generally accepted. Unfortunately, recent studies of S.A. SOLIN and others, using diffuse X-ray scattering |12| |13| |14|, concluded that at room temperature the intercalated heavy alkali metals form liquid-like layers without long range ordering ; the study of these layers by M.S. DRESSELHAUS and co-workers, using the transmission electron microscope, leads to a rather different conclusion |15| : several kinds of organized domains would coexist in each layer. This observation seems to be in agreement with the previous study of PARRY |16| concerning the Cs-G system. In any case, these different kinds of organization cannot lead to stoichiometric formulae. In the case of potassium, gravimetric and analytical data are in rather good agreement with the formula KC_{24} for the second stage. However, we do not know
- if the composition depends on the conditions of synthesis
- if the composition is the same for the compounds of K, Rb, Cs.

According to M.S. DRESSELHAUS |15|, the in-plane surface area of the metal can be calculated by dividing the volume of the free metal unit cell V_m by the M-M distance, which is $a \cdot \sqrt{3} / 2$. This calculation leads to the formulae $KC_{24.1}$ $RbC_{27.76}$, $CsC_{32.26}$, which give an order of magnitude of the experimental composition, but cannot be considered as "ideal" formulae.

By lowering temperature, an important structural evolution is observed, including phase transitions at the temperature generally called T_u and T_ℓ ($T_u > T_\ell$) and leading to higher degrees of organization.

Among the ternaries of stage 2, the mercurographitides $MHgC_8$ (M = K, Rb) and the alkali metal-thallium compounds $MTl_{1.5}C_8$ (M = K, Rb) seem to contain the same type of layers as the first stage ones. So the former are stoichiometric and the latter are not true stoichiometric phases.

The sodium-barium-graphite compounds of stage 2 and 3 |17| offer another kind of structure. The triple metallic sheets (one layer of barium between two layers of sodium atoms) are biperiodic, but the parameter a = 6.36 Å of the unit cell is not compatible with the graphite one. The corresponding ratio C / Na+Ba for a second stage is 8.92 and it is only an approximate agreement with the experimental values of about 7-8.
On the other hand the ideal ratio Na/Ba is 2, but, in fact, the experimental values vary from 1.6 to 5, probably by partial substitution of one metal for the other.

This short review of the in-plane structure of the intercalated metal have shown the three types of layers announced in the introduction

- Epitaxial layers (MC_6, MC_8, $MHgC_4$, $MHgC_8$)
- Biperiodic non-epitaxial (Na-Ba-G-compounds) layers
- Layers without long-range ordering ("MC_{12n}", $MTl_{1,5}C_4$, $MTl_{1,5}C_8$)

which lead to different degrees of stoichiometry.

II. Relation Between the Stages

The mercurographitides and the alkali metal - thallium - graphite compounds contain the same type of intercalated layers for the stages 1 and 2 ; the metallic sheets are also the same in the second and third stage of the Na-Ba-G compounds. The stability of the structure in passing from one stage to another of these compounds probably corresponds to the high internal cohesion of the triple metallic layers.

The difference between the in-plane density of the first and the higher stages in the alkali metal graphite compounds and the possible coexistence of the two kinds of intercalated layers in a same sample (for instance in a graphite single crystal) is a very remarkable phenomenon.
Certainly it is a consequence of the electrostatic repulsion between the metal atoms, which increase from the first to the second stage in relation to the charge transfer. It is of interest to study the effects of pressure, temperature and nature of the parent carbon of this phenomenon.

II.1 Pressure

According to the preceding remarks on the electrostatic repulsion, one can suppose that the in-plane density in the so-called MC_{12n} compounds corresponds to an equilibrium between the electrostatic potential energy, which increases, and the spacing energy per metal-atom, which decreases with respect to the in-plane density. This equilibrium can be displaced by application of an external pressure : S.A. SOLIN and co-workers |18| have observed under about 13 kilobars the reversible transformation

$$KC_{24} \rightleftarrows KC_{24}$$

2nd stage low density layers	3rd stage compact layers

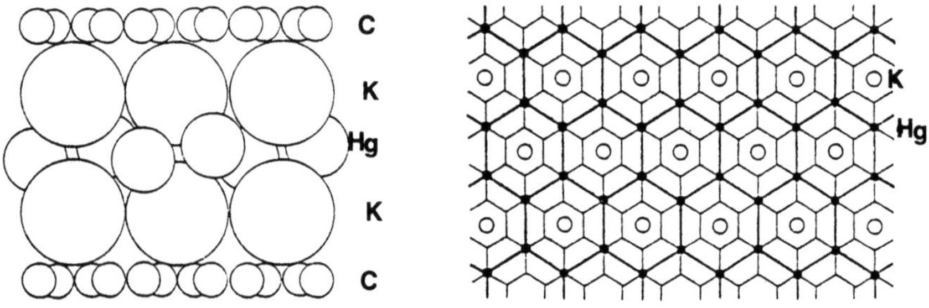

Fig.1. $KHgC_4$ structure

II.2 Temperature

The thermogravimetric study of the potassium-graphite system in a large domain of temperature |19| has shown the existence of a purple non-stoichiometric first stage phase KC_x, which has an intermediate composition between KC_8 and KC_{24}. An X-ray study concluded that this phase, which corresponds to the oblique part BC of the isobaric curves (Fig.2), is a disordered first stage.
The three phases, I_α (KC_8), I_β (KC_x) and II (KC_{24}) are in equilibrium at 370°C (Fig.3).
By cooling KC_x to room temperature it immediately decomposes into KC_8 and KC_{24}.
Also the equilibrium between these phases can be displaced in a reversible manner : over 370°C a convenient mixture of KC_8 and KC_{24} leads to a unique simple phase with non-compact layers.

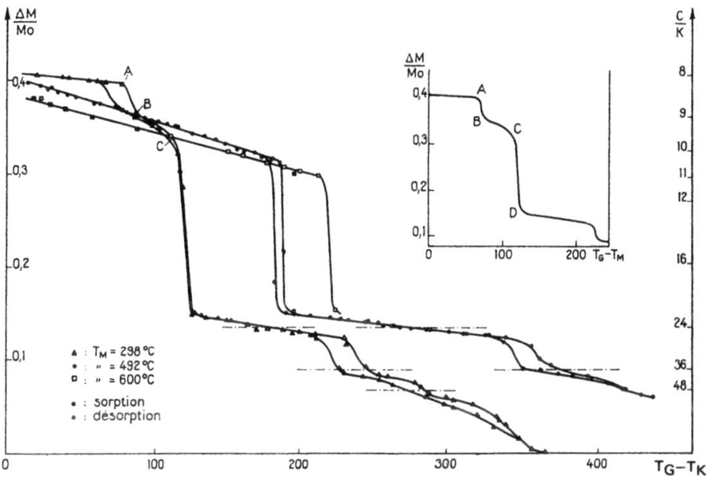

Fig.2. Isobaric curves of the potassium natural graphite system

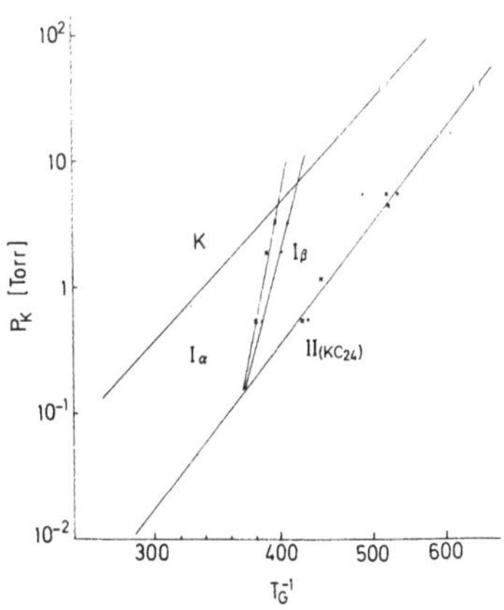

Fig.3. Phase diagram of the potassium natural graphite system

II.3 Nature of the pristine carbon

N. PLATZER studied the intercalation of potassium into the soft carbons |20| and B. CARTON extended the results to the hard carbons |21|. The study of the (hk) X-ray diffraction peaks shows that for these two kinds of "substrate" carbons, the bronze first stage compounds contain compact metallic layers, and that the in-plane K-K distance is lower than in graphite (there is no epitaxy). The diagrams of the blue second stage compounds do not show hk bands corresponding to the K atoms : so the layers probably are not compact and are disordered, as in the case of graphite.
P. LAGRANGE used the reversible absorption of hydrogen at the temperature of liquid nitrogen to study the change of the in-plane density by passing from the first to the second stage in different carbons |22|.

Of course the compact metallic layers of the first stage cannot absorb hydrogen. On the contrary the lacunar stages higher than 2 show a molecular sieve behaviour|23||24|. KC_{24} reversibly absorbs 2 moles of hydrogen per mole. If one accepts the dodecal lacunar structure of RUDORFF, each vacancy between the metal-atoms contain four moles of hydrogen under the pressure of saturation. The absorption is completely reversible.

Figure 4 shows the variation of the H_2/K ratio versus the K/C in compounds prepared from petroleum cokes.

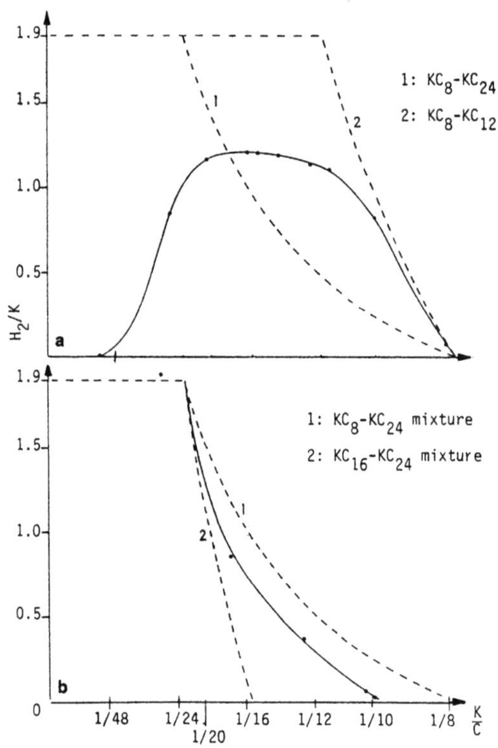

Fig.4. Hydrogen absorption in potassium petroleum cokes compounds
(a) HTT: 1250 °C; (b) HTT: 2000 °C

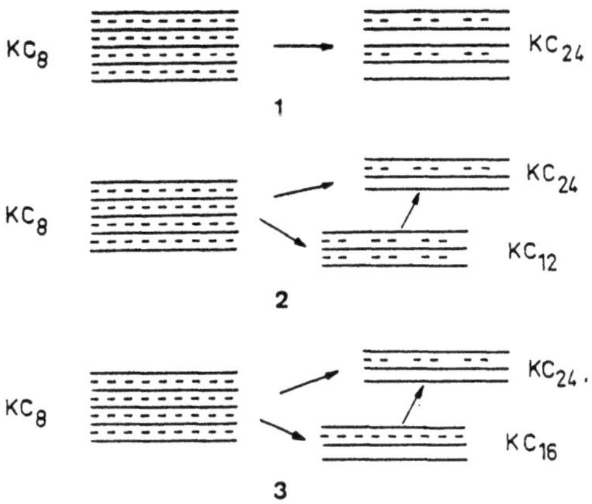

Fig.5. Evolution from KC_8-KC_{24}

- For the graphite, H_2/K is zero for K/C = 1/8 and 2 for K/C = 1/24
For intermediate x values of the K/C ratio, H_2/K exactly corresponds to that of a mixture of KC_8 and KC_{24}. (Fig. 5.1)

- For the coke of HTT = 1250°C, and for 1/12 < x < 1/8, H_2/K is higher than for graphite : the first stage is lacunar. (Fig.5.2)

- For the coke of HTT = 2000°C and for 1/16 < x < 1/8, H_2/K is lower than for graphite : in the presence of an excess of a compact first stage, the second stage also is compact. (Fig.5.3)

The change of stage and the change of in-plane density are also not simultaneous in the carbons. One can think that, in the case of graphite, the simultaneity exists by chance and not by necessity.

In conclusion, there are no sure and precise rules for forecasting the metal graphite formulae. In the best cases they are "crystallographic" formulae ; often they are only experimental ones.

References

1. A. Hérold : "Crystallochemistry of Carbon Intercalation Compounds", in F. Levy, Intercalated Materials, D. Reidel Pub. Cº, Holland,|1979|323/421
2. D. Guérard and A. Hérold : Carbon 13|1975|337
3. D. Guérard, M. Chaabouni, P. Lagrange, M. El Makrini and A. Hérold : Carbon 18|1980|203/9
4. M. El Makrini, D. Guérard, P. Lagrange and A. Hérold : Carbon 18|1980|211
5. R. Setton : Les Carbones, Editions Masson, II |1965|645
6. D. Billaud and A. Hérold : Bull. Soc. Chim. Fr. I|1971|103 ; II|1974|2042 III|1974|2407
7. M. El Makrini, P. Lagrange, D. Guérard and A. Hérold : Carbon 18|1980|203
8. M. El Makrini, P. Lagrange and A. Hérold : Carbon 18|1980|374
M. El Makrini : Thesis, Nancy|1979|
9. D. Billaud, E. Mc Rae, J.F. Marêché and A. Hérold : Synthetic Metals 3 |1981|21/26
10. P. Pfluger, V. Geiser, S. Stoltz and H.J. Güntherodt : Synthetic Metals 3|1981|27/39
11. W. Rüdorff and E. Schulze : Z. anorg. u. allg. Chem.277|1954|156

12. R. Clarke, N. Caswell, S.A. Solin and P.M. Horn : Physica B 99|1980| 457/462
13. H. Zabel, S.C. Moss,N. Caswell and S.A. Solin : Phys. Rev. Lett. 27|1979| 2018/21
14. R. Clarke, N. Caswell, S.A. Solin and P.M. Horn : Phys. Rev. Lett. 27 |1979|2022/25
15. M.S. Dresselhaus, N. Kambe, A.N. Berker and G. Dresselhaus : Synthetic Metals 3|1981|121/31
16. G.S. Parry : Mater. Sci. and Eng. 31|1977|99/106
17. D. Billaud and A. Hérold : Bull. Soc. Chim. Fr.|1978|131/34
18. N. Wada, R. Clarke and S.A. Solin : Synthetic Metals 2|1980|27/34
19. B. Carton : Thesis, Nancy|1971|
    B. Carton and A. Hérold : Bull. Soc. Chim. Fr. 4|1972|1337
20. N. Platzer : Annales de Chimie SL|1960|805
21. B. Carton : Thesis, Nancy|1971|
    Bull. Soc. Chim. 12|1970|4264 and 2|1972|521
22. P. Lagrange, M.H. Portmann and A. Hérold : Proceedings of Carbon 76, Baden-Baden, 139
23. K. Watanabe, T. Kondow, M. Soma, T. Onishi and K. Tamaru : Proc. Roy. Soc. London A|1973|51
24. P. Lagrange, A. Métrot and A. Hérold : C.R. Acad. Sc. Paris 275 C|1973| 765

Doped Polyacetylenes and Graphite Intercalation Compounds: A Comparative Study

Serge Flandrois

Centre de Recherche Paul Pascal
Université de Bordeaux I, Domaine Universitaire
F-33405 Talence, France

Abstract
Graphite intercalation compounds and doped polyacetylenes are synthetic metals which exhibit many similarities in spite of their different dimensionality (2-d for the former and 1-d for the latter, respectively). After a brief survey of the preparation, structure and properties of polyacetylenes and doped polyacetylenes, a comparison will be made with graphite intercalation compounds. A number of similarities will be shown, concerning, for example, the nature of dopant species, the mechanism of doping, the charge transfer between host and dopant, and even some features of crystal structure.

1. Introduction

Although usual metals, such as copper or aluminium, have the highest electrical conductivities, impressive progress has been made recently in the field of the so-called "synthetic metals". These materials can be divided in three classes : intercalation compounds of graphite, doped polymers and charge transfer complexes and radical ion salts. They have in common a composition of two constituents (at least) with a charge transfer from one constituent to another.

The intercalation of graphite with a donor or acceptor molecule can lead to conductivity values which still are controversial but, in any case, are higher than that of pristine graphite, close to that of copper |1|. Polymers are insulating or semiconducting, but by doping with donors or acceptors they can reach conductivities of $10^3\ \Omega^{-1}\ cm^{-1}$ or more |2|. Finally, the class of charge transfer complexes and radical ion salts has been the subject of so many studies in the last two decades that the best conductivity value have gone from about $10\ \Omega^{-1}\ cm^{-1}$ in early sixties to 10^3-10^4 at the end of the seventies, and even last year organic superconductors have appeared |3|. Figure 1 gives a few examples of room temperature conductivities and compares inorganic and organic materials.

The three families of synthetic metals exhibit some similarities. It is particularly true for graphite intercalation compounds and doped polymers, on which we will concentrate in this paper. Indeed a graphitic layer may be viewed as a kind of macromolecule. The main difference with a polymer such as polyene lies in the dimensionality : two-dimensionality for graphite and one-

Fig. 1 R.T. conductivities for some inorganic and organic materials

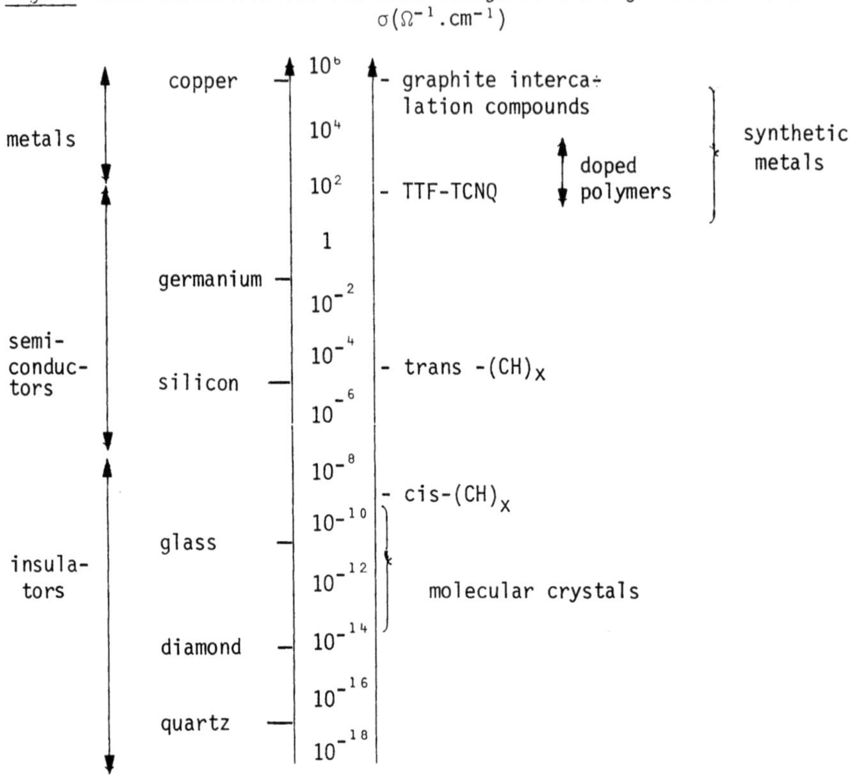

dimensionality for polyene. We will see that in spite of this inherent structural difference many similarities can be observed. As an example of conducting polymer we will choose polyacetylene, which has been (and still is) the most studied. Before comparing it to graphite intercalation compounds, we will briefly review its preparation, structure and properties.

2. Polyacetylene and doped polyacetylene

At first a remark of nomenclature must be made. The name "polyacetylene" is given to the polymer obtained by polymerization of acetylene : $x(CH{\equiv}CH) \rightarrow$ $+CH{=}CH+_x$ or, more simply, $(CH)_x$. Actually it is not a real polyacetylene $+C{\equiv}C+_x$ but a polyene, and more precisely, a polyvinylene. Obviously we will use the accepted appelation.

Its interest lies in the presence of a linear system of conjugated double bonds. It was first thought |4| that a quasi-unidimensional electron gas could be obtained. In fact, theoretical calculations |5| showed that a uniform infinite chain is unstable with respect to bond alternation. As a result a gap opens at the Fermi level, which gives a change in behaviour from metal to semiconductor. In recent years, interest in this system has been renewed with the discovery of doping and its fantastic effect on the electrical conductivity |6|.

2.1. Preparation of polyacetylene

$(CH)_x$ was first prepared as early as 1929 by JOB and CHAMPETIER |7|. It was obtained by polymerizing acetylene gas on a catalyst system consisting of iron trichloride and phenylmagnesium bromide.

In the fifties and sixties, other catalysts have been used, in particular ZIEGLER-NATTA catalysts such as $Ti(oC_4H_9)$ + $Al(C_2H_5)_3$ |8|, or other couples, for example $Co(NO_3)_2$ + $NaBH_4$ |9|, or even only by γ-irradiation of solid acetylene |10|.

A new step was taken in the early seventies with the preparation of films instead of powders by simply using more concentrated catalyst solutions |11|. Finally, the idea of doping with electron donor or acceptor molecules proved to be very fruitful, the electrical conductivity being increased by several orders of magnitude |6|.

2.2. Morphology and structure

Polyacetylene films have a low apparent density : 0.3 - 0.4g/cm³, whereas the bulk density is about 1.16g/cm³ as obtained by flotation techniques. They consist of randomly oriented fibrils with typical diameter of 200Å , as shown by electron microscopy studies.

Two isomers can be obtained (fig.2) whose ratio depends on the temperature of polymerization : at low temperature the films are rich in cis-isomer and heating cis-rich films five minutes at 180°C produces the complete cis-trans isomerization |12|. Similarly, doping may induce the cis-trans isomerization

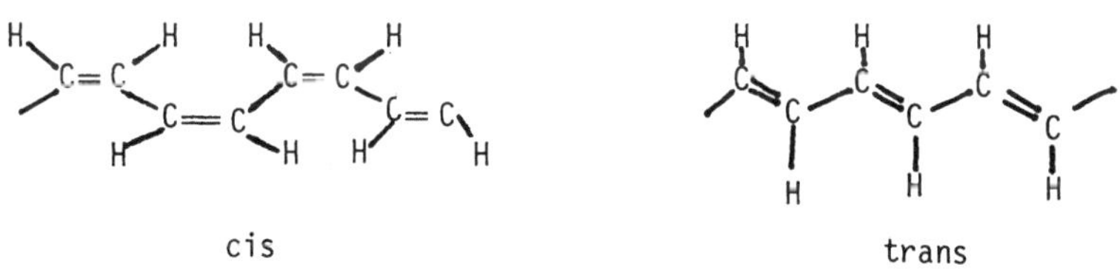

Fig.2 : Cis and trans- polyacetylene

Little information is available on the crystal structure of polyacetylene. From X-ray powder diagrams the unit-cell parameters can be determined. The values obtained for cis- and trans- isomers are shown in Table 1. As seen in this table, the agreement between various authors is not excellent. Moreover, two modifications of trans $-(CH)_x$ have been found depending on the method of preparation.

Table 1 Unit cell–parameters of cis- and trans -$(CH)_x$

Isomer	a |Å|	b |Å|	c |Å|	Ref
cis (orthorhombic)	7.61	4.39	4.47	|13|
	7.74	4.32	4.47	|14|
trans (orthorhombic)	5.62	4.92	2.59	|15|
	4.39	7.73	7.87	|16|
trans (monoclinic)	3.73	3.73	2.44 β=98°	|15|
	4.41	7.75	2.42 β=88°	|16|

2.3. Properties of pristine $(CH)_x$

From a chemical point of view, polyacetylene is hardly obtained in a pure form. Due to its insolubility, it is difficult to purify. In particular, traces of catalysts may remain in the bulk. Moreover, the films are very sensitive to oxygen:they become brittle when exposed to air.

The electrical conductivity seems to depend on the cis-trans content, ranging from 10^{-5} Ω^{-1} cm^{-1} for the trans compound to 10^{-9} Ω^{-1} cm^{-1} for the cis-isomer.

Early studies of magnetic properties |4| have shown that the susceptibility of pristine $(CH)_x$ corresponds to the core diamagnetism which can be calculated from PASCAL constants. Some Curie-like spins are visible at low temperatures, with an increase in content upon cis-trans isomerization.

2.4. Doping of polyacetylene

Polyacetylene can be doped with a variety of donors and acceptors, resulting in an increase of several orders of magnitude in the electrical conductivity. For example, AsF_5 doped $(CH)_x$ films have a room temperature conductivity of 10^3 Ω^{-1} cm^{-1}. The important point is that the maximum conductivity is obtained for very low doping levels (a few percent). At the same time, the "activation energy" of the conductivity decreases sharply but does not disappear: even at high doping levels the conductivity increases with increasing temperature.

Several models have been proposed to explain this kind of insulator-to-metal transition. The first proposal was a MOTT transition |6|, because $\ln\sigma$ vs. $T^{-1/4}$ gives more straight-line behavior than $\ln\sigma$ vs.1/T. Another model describes the insulator-to-metal transition as a percolation transition : due to the inhomogeneous distribution of dopant throughout the films, there is a mixture of metallic (doped) particles and insulating (undoped) particles ; the transition occurs when the percolation threshold is reached |17|. A third model ascribes the transition to the presence of solitons |18|: bond alternation defects form domain walls of several (7-10) formula units length, separating two regions with opposite phase. In pristine trans -$(CH)_x$ these solitons are neutral and paramagnetic. Upon doping they become charged and spinless, and with increasing doping they can overlap and thus suppress the bond alternation to give a metal. Needless to say these explanations are very controversial.

3. Comparison with graphite intercalation

3.1. Dopants

Polyacetylene can be doped like graphite with electron donors or acceptors. Table 2 gives some examples of dopants and the corresponding room temperature conductivities |19|.

Table 2 Examples of acceptors and donors able to dope $(CH)_x$

p-type (acceptors)	$\sigma \|\Omega^{-1}cm^{-1}$	n-type(donors)	$\sigma \|\Omega^{-1} cm^{-1}\|$
Br_2	4×10^{-1}	Li	200
I_2	6×10^2	Na	80
ICl	50	K	50
IBr	4×10^2		
AsF_5	10^3		
$SbCl_5$	10		
ClO_4	1000		
PF_6	30		
H_2SO_4	1.2×10^3		
$HClO_4$	10^3		

It is seen than alkali metals as well as halogens, pseudo-halogens or acids can be introduced in $(CH)_x$. Some metal halides such as $FeCl_3$ or $MoCl_5$ have been also used recently as dopants of $(CH)_x$ |20|; and it is noteworthy that, like for graphite, the stability of the products is much higher than with halogens or alkali metals.

However, contrary to graphite, iodine is a easy dopant of polyacetylene. This is probably related to the lower ionization energy of $(CH)_x$. On the other hand, chlorine, which can be intercalated into graphite (at low temperature at least), reacts with the double bonds of polyacetylene.

3.2 Doping methods

In both cases three types of methods may be used : a) doping from vapor phase, this is the most common for doping with halogens ; b) in solution, a good example is provided by nitronium salts such as NO_2PF_6 or NO_2SbF_6 which, dissolved in nitro-methane, lead to graphite or $(CH)_x$ doped with PF_6^- or SbF_6^-, and c) by electrochemical method, for example PF_6^- may be introduced into graphite or $(CH)_x$ by electrolysis of a solution of tetrabutylammonium hexafluorophosphide.

However, due to the thermal behavior of polyacetylene, it is not possible to dope it from the vapor phase of metal halides. Moreover, this method requires generally the presence of chlorine as oxidant, which is prohibited because of its reactivity with $(CH)_x$.

3.3 Kinetics and mechanism of doping

Kinetic studies are scarce for graphite intercalation as well as for $(CH)_x$ doping. Nevertheless, from studies made at Bordeaux in the case of halogens |21||16|, it appears that the kinetics is governed by the diffusion of dopant inside the host structures. As a result the products are very often inhomogeneous, with a non-uniform distribution of dopant or intercalant.

Another common feature is the existence of residue compounds : the loss of dopant is not complete. A consequence is different kinetic behavior for the first and second doping (or intercalation).

3.4 Nature and doping species

It is highly probable that some dopants are present in grahite and $(CH)_x$ as ions, such as PF_6^- or ClO_4^-. In some cases, like AsF_5 , there is a controversy. A disproportionation of AsF_5 into AsF_6^- and AsF_3 was first proposed for graphite |22| and then extended to $(CH)_x$ |23|. Other authors |24| think that the active dopant specie in polyacetylene is $HAsF_6$ or $(As_2F_{11})^-$ anions.

With halogens it is quite clear that the dopant species are different for graphite and polyacetylene. Raman studies |25| have shown that the halogens are present as X_3^- ions (and some X_5^- ions at high doping levels), whereas it is well known that bromine is in graphite as Br_2 molecules and may be Br_2^- ions.

3.5 Charge transfer

The presence of ions into doped $(CH)_x$ needs a charge transfer which can be estimated from the stoichiometry. For halogen doped films of formula $|(CH)^{+y}(X_3^-)_y|_x$ the maximum conductivity is obtained for $3y \sim 0.02$. Then a plateau is observed up to the maximum doping : $3y \sim 0.25$. This means that the average lost per C atom is between 0.01 and 0.08 electron. This is the same order of magnitude as for acceptor graphite compounds |26| : 0.01 to 0.03 electron are transferred from one C atom to the acceptor.

3.6 Crystal structure and C-C bond lengths

The structure of doped polyacetylenes is still more badly known than that of pristine $(CH)_x$. X-ray diffraction studies reveal a broadening of the Bragg reflections with the addition of a new diffraction line, which has been interpreted in the following way |13|. Polyacetylene chains form sheets comparable to the graphite layers : the dopant would be inserted between these sheets. As a result an increase of the distance between sheets is observed which depends on the size of the dopant. Table 3 shows that this assumption is corroborated by similar increases of distances in graphite and $(CH)_x$ for a same inserted molecule.

Table 3 Increase in sheet-to-sheet distances by doping

Dopant	$(CH)_x$	Graphite
Br	3.46 Å	3.70Å
AsF_5	4.98 Å	4.75Å
SO_3F	4.57 Å	4.46Å
I	4.06 Å	——

Another feature of the structure is related to the C-C bond lengths.It has been shown |26| that in graphite there is a linear relationship between the amount of charge transfer and the C-C distance. In the case of polyacetylene, no experiment has been able to measure the C-C bond lengths, but it has been shown theoretically [27] that for a charge transfer of 0.1 electron per C atom, the C-C distances should be equal. This bond change amount is of the same order of magnitude as in graphite for identical charge transfer.

Concerning the structure of the dopant or the intercalant, some differences should exist between graphite and $(CH)_x$ when the nature of the doping species is different. This should be true for bromine, for example, but probably also for metal halides. In the case of graphite a slight excess of chlorine is observed with respect to free metal halide, which has been interpreted as due to chlorine at the edges of small metal halide islands |28| |26|.

4. Conclusion

The comparison of doped polyacetylenes and graphite intercalation compounds has evidenced a number of similarities. They should allow a better understanding of one system from the knowledge gained in the study of the other.and the suggestion of new experiments and new syntheses. It is the interest of a comparative study.

REFERENCES

|1| G.M. T. FOLEY, C. ZELLER, E.R. FALARDEAU and F.L. VOGEL, Solid State Comm. 24, 371 (1977)

|2| See for example A.G. MAC DIARMID and A.J. HEEGER, Synth. Metals 1, 101 (1979)

|3| D. JEROME, A. MAZAUD, M. RIBAULT and K. BECHGAARD, J. Phys. Lett. 41 397 (1980)

|4| S. FLANDROIS, C.R. Acad. Sci. 264C, 1244 (1967)

|5| H.C. LONGUET-HIGGINS and L. SALEM, Proc. Roy. Soc. A 25, 172 (1959)

|6| C.K. CHIANG, C.R. FINCHER, Y.W. PARK, A.J. HEEGER, H. SHIRAKAWA, E.J.LOUIS, S.C. GAU and A.G. MAC DIARMID, Phys. Rev. Lett. 39, 1098 (1977)

|7| A. JOB and G. CHAMPETIER, C.R. Acad. Sci. 189, 1089 (1929)

|8| G. NATTA, G. MAZZANTI and P. CORRADINI, Atti. Acad. Naz. Lincei Rend., Cl.Sci. Fis.Mat. Nat. 25, 3 (1958)

|9| L.B. LUTTINGER, J. Org. Chem. 27, 1591 (1962)

|10| Y. TABATA, B. SAITO, H. SHIBANO, H. SOBUE and K. OSHIMA,Makromol.Chem. 76, 89 (1964)

|11| H. SHIRAKAWA and S. IKEDA, Polym. J. 2, 231 (1971)
T. ITO, H. SHIRAKAWA and S. IKEDA, J. Polym. Sci., Chem. Ed. 12, 11 (1974)
|12| B. FRANCOIS, M. BERNARD and J.J.ANDRE, J. Chem. Phys. (in press)
|13| R.H. BAUGHMAN, S.L. HSU, G.P. PEZ and A.J. SIGNORELLI, J. Chem. Phys. 68 5405 (1978)
|14| G. LIESER, G. WEGNER, W. MULLER and V. ENKELMANN, Makromol. Chem., Rapid Commun. 1, 621 (1980)
|15| G. LIESER, G. WEGNER, W. MULLER and ENKELMANN, Makromol. Chem., Rapid Commun. 1, 627 (1980)
|16| C. HAUW, B. FRANCOIS and S. FLANDROIS, to be published
|17| K. MORTENSEN, M.C.W. THEWALT, Y. TOMKIEWICZ, T.C. CLARKE and G.B. STREET, Phys. Rev. Lett. 45, 490 (1980)
A.J. EPSTEIN, H.W. GIBSON, P.M. CHAIKIN, W.G. CLARK and G. GRUNER, Phys. Rev. Lett. 45, 1730 (1980)
|18| W.P. SU, J.R. SCHRIEFFER and A.J. HEEGER, Phys. Rev. Lett. 42, 1698 (1979)
M.J. RICE, Phys. Lett. A, 71, 152 (1979)
|19| See for example A.G. MAC DIARMID and A.J. HEEGER, The Physics and Chemistry of low dimensional solids, Ed. by L. Alcacer, Reidel, Dordrecht, p.393 (1979)
|20| A. PRON, Private communication and to be published
|21| F. COURTOIS D'ARCOLLIERES, Ph. D. Thesis, Bordeaux (1971)
|22| N. BARTLETT, B. Mc QUILLAN and A.S. ROBERTSON, Mat. Res. Bull. 13, 1259 (1978)
|23| T.C. CLARKE, R.M. GEISS, W.P. GILL, P.M. GRANT, I.W. MACKLIN, H. MORAWITZ, J.F. RABOLT, D. SAYERS and G.B. STREET, J. Chem. Soc. Chem. Comm., 329 (1979)
|24| A. PRON, A.G. MAC DIARMID and A.J. HEEGER, Materials Science 7, 313 (1981)
|25| S.L. HSU, A.J. SIGNORELLI, G.P. PEZ and R.H. BAUGHMAN, J. Chem. Phys. 69 106 (1978)
|26| S. FLANDROIS, J.M. MASSON, J.C. ROUILLON, J.GAULTIER and C. HAUW, Synth. Met. 3, 1 (1981)
|27| J.L. BREDAS, R.R. CHANCE and R. SILBEY, J. Phys. Chem., in press
|28| J. GAULTIER, C. HAUW, J.M. MASSON, J.C. ROUILLON and S. FLANDROIS, C.R. Acad. Sci. 289C, 45 (1979)

Bond Length, Bond Strength and Electrical Conductivity in Carbon Based Systems

L. Pietronero and S. Strässler

Brown Boveri Research Center
CH-5405 Baden-Dättwil, Switzerland

Abstract

We discuss different properties of carbon based synthetic metals (mainly referring here to intercalated graphite) in terms of a tight binding model. In particular attention is focused on the puzzling behavior of the bondlength as a function of the charge transfer and on the shifts of phonon frequencies. Within such a model the maximum conductivity for this class of systems is also established.

1. Introduction

In this lecture we discuss the formulation of a unified picture for the description of various properties of carbon based synthetic metals. We will mainly refer to intercalation compounds of graphite, but the same analysis can be also performed for doped polyacetylene if the additional complications (i.e. instabilities) due to one dimensionality are properly treated.

The properties of the carbon-carbon (C-C) bond and of the carbon π electrons have been studied quite extensively for small molecules and (neutral) polymers. The nice thing about the π orbitals is that they are a combination of localized carbon 2p atomic orbitals and that in linear or planar systems, they are orthogonal to all the σ orbitals. It is because of these two reasons that a tight binding or Hückel description of the π states with a single p orbital per carbon atom has been very succesful in describing several properties of carbon molecules [1]. In particular the analysis of the bondlength and bondstrength on the electronic population have been considered quite extensively [2].

Before we go on to discuss how to apply these concepts to intercalation compounds we should mention that such a simplified treatment would not be appropriate for pure graphite. The reason being that in graphite the small 3-d coupling between layers has its greatest influence on the band structure just in the vicinity of the Fermi level. This results in a Fermi surface strongly dependent on the details of this 3-d coupling that has to be carefully considered [3]. Upon intercalation, as soon as the Fermi level is appreciably shifted ($|\Delta\mu| \gtrsim 0.1eV$)

from the original value (at least for the bounding layers this is always the case) the influence of the 3-d coupling becomes small and the band structure has essentially a two-dimensional structure. In this case a description in terms of 2-d π orbitals is again plausible. In addition, to generalize this model from small molecules to intercalate graphite we will describe how, within the same framework, one can describe the electron phonon scattering and define the maximum ideal conductivity for this type of systems.

The paper is organized as follows:
In Sect.2 we briefly review the derivation of the Coulson formula for C-C bondlength and discuss its failure for intercalates. In Sect.3 we generalize the method to describe intercalation compounds and show how the bondlength change can be used as a tool to determine the charge transfer. In Sect.4 we discuss the change in bond strength and phonon frequency upon intercalation. In Sect.5 we briefly review the problem of the determination of the maximum conductivity.

2. Coulson relation for C-C bondlength

In this section we discuss the dependence of bondlength on the electronic population. As we will see the simple Coulson relation does not explain the bondlength of intercalation compounds.

For a carbon molecule we write the total energy per bond between nearest neighboring atoms as [2,4]

$$E^T = E^\sigma + E^\pi \tag{2.1}$$

where E^π corresponds to the π electrons and E^σ to the σ electrons. Neglecting the terms that do not depend on bondlength, we can write

$$E^\pi = 2\,P\,J \tag{2.2}$$

where

$$J = \langle p_n | V_n | p_{n'} \rangle \tag{2.3}$$

is the tight binding matrix element between p orbitals of neighboring atoms and V is the potential of a carbon atom. The term P is the bond order for π electrons [1] defined as

$$P = 2 \sum_{i\,(\mathrm{occupied})} c_n^*(i)\; c_{n'}(i) \tag{2.4}$$

where $c_n(i)$ is the coefficient of the local orbital p_n corresponding to the i-th molecular orbital. Note that the bond order P (and therefore the total energy) is linked to the charge in the π orbitals only through the total wavefunction. Therefore systems with exactly the same charge (one electron per atom) in the π orbital but with different geometries as ethylene, benzene, polyacetylene (without bond alternation) and a neutral graphite layer have very different values of the bond order. These values are respectively 1, 2/3, 2/π and .525 [4].

Let us now start from ethane (C_2H_6) that has no π states occupied (P=0) and study the modification of the bondlength as a function of P. For small changes from the ethane equilibrium distance r_σ we can write

$$E^\sigma \simeq \frac{1}{2} K_\sigma (r-r_\sigma)^2 \tag{2.5}$$

$$E^\pi \simeq 2P \left[J + \frac{dJ}{dr} (r-r_\sigma) \right] \tag{2.6}$$

where K_σ is the ethane force constant and we have expanded J_σ around the ethane value J_σ. The equilibrium condition $dE^T/dr^\sigma = 0$ gives the Coulson relation

$$r = r_\sigma - \frac{2}{K_\sigma} \frac{dJ}{dr} P \qquad \left(\frac{2}{K_\sigma} \frac{dJ}{dr} \sim .2\text{Å}\right) \tag{2.7}$$

that reproduces rather well the experimental data of Fig.1.

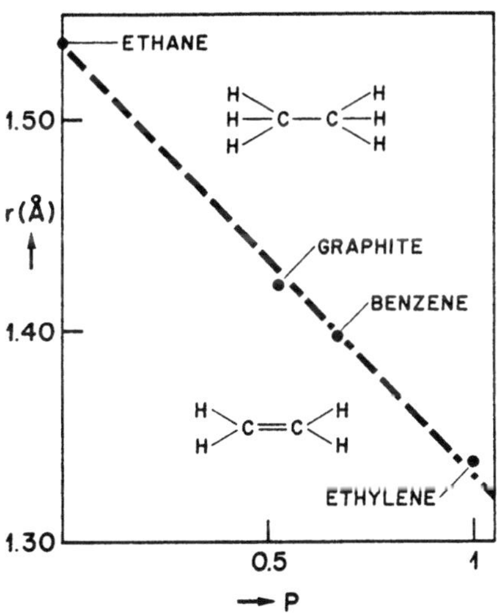

Fig.1 Dependence of the carbon carbon bondlength on the bond order P for various carbon compounds. Note that the charge per carbon atom in the π states is zero for ethane and one for graphite, benzene and ethylene.

For intercalated graphite the data of Fig.2 show an expansion for donors and a contraction for acceptors. The fact that adding electrons to a graphite layer results in an expansion might seem in contradiction with the behavior of Fig.1. This is actually not the case because the addition of electrons in antibonding states decreases the value of P [4]. A real problem exists instead for the acceptors where the depopulation of bonding states also results in a decrease of P and therefore

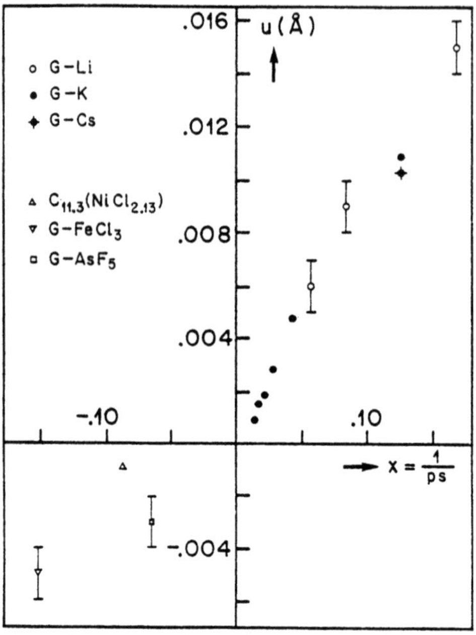

Fig.2 Experimental values for the bond length change $u=r-r_o (r_o=1.421\text{Å})$ as a function of the ratio between the number of intercalate molecules and that of carbon atoms in a given compound. The nominal formula of a compound is CA_x where $x=1/ps$, s is the stage and p=6,8,etc.. The sources of data can be found in [4].

according to Coulson relation it should produce an expansion as well. As we will see in the next section the net charge of carbon atoms in intercalation compounds is the reason of such a discrepancy.

3. Extension of the bond length theory to intercalation compounds

We generalize now the concepts of the previous section to account for the fact that carbon atoms in intercalation compounds have a net charge. As before we start from Eq.(1.1). For a general bond length r we have

$$E^{\pi}(r) = 2P\, J^{pp\pi}(r) \quad . \tag{3.1}$$

The σ bond is described in terms of bonding states between sp_2 hybrids [4]. This gives (apart from constant terms)

$$E^{\sigma}(r) = \frac{2}{3}\, J^{ss\sigma}(r) - \frac{4}{3}\, J^{pp\sigma}(r) \quad . \tag{3.2}$$

The notations $pp\pi$, $ss\sigma$ and $pp\sigma$ are as in [5]. We start by considering a single layer charged with an electronic charge f_c per carbon atom. This extra charge gives rise to the following effects:

(a) A modification of the occupancy of the π states that changes the value of the bond order: $P = P_o + \Delta P(f_c)$ where $P_o = .525$ is the value for the neutral layer.
(b) A modification of all the matrix elements due to the fact that the atomic potential is changed because of the extra charge: $J_o + \Delta J(f_c)$
(c) An additional contribution ΔE_{el} (also defined per bond) to the total energy due to the Coulomb repulsion between all the extra charges.

The total energy is then

$$E^T(f_c;r) = 2\left[P_o+\Delta P(f_c)\right]\left[J_o^{pp\pi}(r)+\Delta J^{pp\pi}(f_c;\ r)\right] + \frac{2}{3}\left[J_o^{ss\sigma}(r)+\Delta J^{ss\sigma}(f_c;r)\right]-\frac{4}{3}\left[J_o^{pp\sigma}(r)+\Delta J^{pp\sigma}(f_c;r)\right]+\Delta E_{el}(f_c;r) \quad . \tag{3.3}$$

We write now $r=r_o+u$ where $r_o=1.421\text{Å}$ is the equilibrium bond length for the neutral layer ($f_c=0$). We expand then Eq.(2.3) up to quadratic terms in u making use of the condition that r_o is the equilibrium position for $f_c=0$. The equilibrium condition for the general case gives now the displacement u as (neglecting the terms $\Delta P\Delta J$, $\Delta J\Delta J$ and $\Delta P\Delta P$)

$$u = -\frac{1}{K}\left[2\Delta P(f_c)\left.\frac{dJ_o^{pp\pi}(r)}{dr}\right|_{r_o} + 2P\left.\frac{d\Delta J^{pp\pi}(r)}{dr}\right|_{r_o} + \left.\frac{d\Delta E^{\sigma}(f_c;r)}{dr}\right|_{r_o} + \left.\frac{d\Delta E_{el}(f_c;r)}{dr}\right|_{r_o}\right] \tag{3.4}$$

where

$$\Delta E^{\sigma}(f_c;r) = \frac{2}{3}\,\Delta J^{ss\sigma}(f_c;r) - \frac{4}{3}\,\Delta J^{pp\sigma}(f_c;r) \tag{3.5}$$

and

$$K = \left.\frac{d^2E^T(f_c=0;r)}{dr^2}\right|_{r_o} \tag{3.6}$$

is the total bond stretching force constant of neutral graphite. For the evaluation of J and ΔJ we have used Slater type orbitals. The change in bond order is computed within the linear band approximation and the electrostatic term is computed from the classical selfenergy of a charged layer. For details see [4]. The final result for stage 1 is then

$$u(1)\ (\text{Å}) = .157f_c + .146|f_c|^{3/2} + .236f_c^2 \quad . \tag{3.7}$$

Note that since $f_c \leq 1/6$ the leading terms is the linear one that is due to the net charge on the carbons. It is this contribution that produces the contraction of acceptors.

In order to generalize this result to higher stages we assume (i) that the charge is localized in the bounding layers (f_c only refers to these) and (ii) that the strain is equally shared by all the layers. This gives

$$u(s) = \frac{2}{s} u(1) \qquad (s \geq 2) \quad . \tag{3.8}$$

Eqs. (3.7) and (3.8)can now be used to analyse the data of Fig.2 and determine f_c and f for the various compounds as reported in [4].

4. Phonon frequency shifts

Upon intercalation there is an interesting behavior of the 1600 cm^{-1} graphite Raman mode E_{2g_2}. For stages larger then two the mode splits into two, the relative intensities of which are roughly proportional to the ratio between bounding and internal layers. This has led to the conclusion that the two peaks belong to vibrations rather localized in either bounding or internal layers [3]. The frequency as a function of the inverse stage shows the linear behavior that we have schematically reported in Fig.3. This is deduced from measurements of a number of intercalates [3].

For acceptors the frequency increases with 1/s while for donors it decreases. This behavior suggests that the shifts of phonon frequencies are linked to the bondlength changes [see Fig.2 and Eqs. (3.7) and (3.8)] through anharmonic terms [6]. We now discuss how to compute these terms within the model described in the previous sections.

For simplicity we consider only the interior layers with large stage. What we want to compute then is the slope of the lines in Fig.3 in the vicinity of pure graphite (HOPG). We can assume that there is no charge in these layers. It is convenient to describe data of Fig.3 in terms of the quantity

$$\delta = \frac{d\left[\Delta\omega/\omega\right]}{d(1/s)} \tag{4.1}$$

where $\Delta\omega$ is the frequency shift with respect to HOPG. We have then $\delta \sim .007$ for acceptors and $\delta \sim -.007$ for donors. The cubic anharmonic contribution corresponding to (2.3) for $f_c = 0$ is

$$\Delta E^T(\mathrm{Anh.}) = \frac{1}{6}\,(2P_o\gamma_\pi + \gamma_\sigma)u^3 \tag{4.2}$$

where

$$\gamma_\pi = \left.\frac{d^3 J^{pp\pi}}{dr^3}\right|_{r_o} \tag{4.3}$$

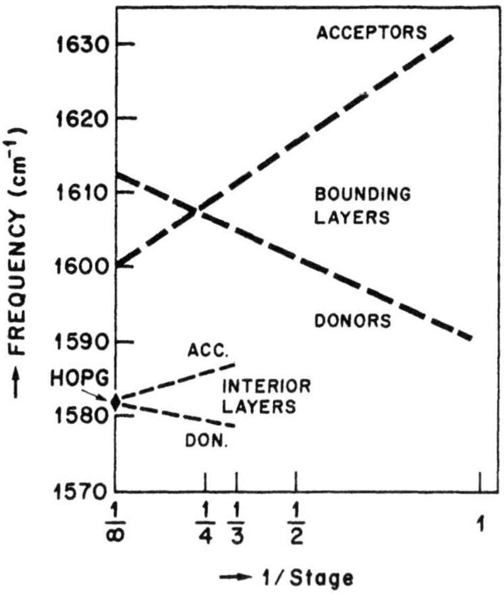

Fig.3 Sketch of the dependence of phonon frequency on the inverse stage. For more details on the experiments see [3].

$$\gamma_\sigma = \left.\frac{d^3E^\sigma}{dr^3}\right|_{r_o} = \frac{2}{3}\left.\frac{d^3J^{ss\sigma}}{dr^3}\right|_{r_o} - \frac{4}{3}\left.\frac{d^3J^{pp\sigma}}{dr^3}\right|_{r_o} \tag{4.4}$$

The corresponding shift in bond strength (force constant) is

$$\Delta K = (2P_o\gamma_\pi + \gamma_\sigma)u \tag{4.5}$$

The various terms appearing in (4.3) and (4.4) can be computed using Slater type orbitals optimized for graphite [7] as in the previous section. We obtain γ_π = 33eV Å^{-3} and γ_σ = -69eV Å^{-3}. Since

$$\delta = \frac{1}{2}\frac{d(\Delta K/K)}{d(1/s)} = \frac{1}{2K}\frac{d(\Delta K)}{du}\frac{du}{d(1/s)} \tag{4.6}$$

and from Fig.2 we have (for donors with p=8) $du/d(1/s) \simeq .012$, we obtain finally $\delta \simeq .005$ for donors (and roughly $\delta \sim -.005$ for acceptors) in reasonable agreement with observations. It is interesting to note that if one uses for γ_π and γ_σ, the values obtained by KAKITANI [2] from an empirical analysis of small molecules, the value of δ would be about ten times larger then ours. This is not surprising because the third derivative of the energy is a very sensitive quantity. Differences in the correlation energy between small molecules and graphite plus the fact that in small molecules the hydrogens (and therefore the carbon atoms) are slightly ionized (.05e [8]) can be responsible for this discrepancy.

5. Maximum Ideal Conductivity

One of the key questions concerning graphite intercalation compounds is the value of their maximum ideal conductivity. The meaning of this term is that while there are scattering mechanisms that could at least in principle be reduced or eliminated (like impurities, defects, intercalate ions etc.) other mechanisms are intrinsic and put an upper limit to the conductivity one can hope to achieve. The main intrinsic scattering mechanism is the one due to the interaction of the charge carriers with the graphite phonons. We now discuss briefly this problem along the lines of [9].

For a charged graphite layer the in-plane conductivity in a direction x within the relaxation time approximation is given by

$$\sigma(T) = e^2 N(\mu) \langle v_x^2(\mu)\rangle \tau(\mu,T) \tag{5.1}$$

where v_x is the x component of the electron band velocity, μ is the chemical potential (Fermi level), N is the density of states and τ is the "transport" relaxation time. Within the linear band description the band energy is

$$\varepsilon(\vec{k}) = \pm(\hbar v_F)|\vec{k}| = \pm \frac{3}{2} J_o r_o |\vec{k}| \quad . \tag{5.2}$$

In this section J only refers to π electrons. One has then the simple relations

$$N(\mu) = \frac{2}{\pi\sqrt{3}} \frac{\mu}{J_o^2} \quad ; \quad f_c = \frac{1}{\pi\sqrt{3}} \frac{\mu^2}{J_o^2} \quad . \tag{5.3}$$

In the high-temperature limit (not too bad for room temperature) we have [9]

$$\tau(\mu,T)^{-1} \simeq \frac{2\pi}{\hbar} k_B T \langle |g_{\vec{k},\vec{k}'}|^2 (1-\cos\theta)\rangle N(\mu) \tag{5.4}$$

where $g_{\vec{k},\vec{k}'}$ is the electron-phonon coupling and θ is the scattering angle between $\vec{k}$ and $\vec{k}'$. Note that for a graphite layer there are two Fermisurface pockets per Brillouin zone and here we consider only scattering within the same pocket due to long wavelength phonons (no umklapp). In addition, scattering can only happen between states with the same spin. Therefore the final state density of states appearing in (5.4) is effectively reduced by a factor of four. Consequently

$$\sigma(T) = \frac{9}{4\pi} \frac{e^2 (J_o r_o)^2}{d\, k_B T\, \hbar} \langle |g_{\vec{k},\vec{k}'}|^2 (1-\cos\theta)\rangle^{-1} \tag{5.5}$$

where $d=3.35\text{Å}$ is the thickness of a graphite layer and $g_{\vec{k},\vec{k}'}$ is intended as two dimensional. The problem is now reduced to the evaluation of the electron-phonon coupling.

Since the Bloch function corresponding to the graphite band are built from the p orbitals that are quite localized on the atoms, we are far from a free electron behavior (see also the linear band dispersion) and the use of the standard free ion model for the evaluation of the electron-phonon scattering is inappropriate. We start then from a tight binding picture along the lines of the previous sections. The explicit first principle calculation has been reported elsewere in detail [9] and here we only quote the results. For shifts of the Fermi level that are not too large ($|\mu| \lesssim 1\text{eV}$) we obtain for the scattering within the same pocket

$$\langle |g^{\ell}_{\vec{k},\vec{k}'}|^2 (1-\cos\theta)\rangle = \langle |g^{t}_{\vec{k},\vec{k}'}|^2 (1-\cos\theta)\rangle = \frac{9}{32\rho_c}\left(\frac{r_o\alpha}{c}\right)^2 . \tag{5.6}$$

Here $\alpha = dJ/dr$, $c = 2.1\text{x}10^6$ cm/s is the average speed of sound $\rho_c = m_c\text{x}3.82\text{x}10^{15}\ \text{cm}^{-2}$ is the two-dimensional mass density of graphite and ℓ and t refer to longitudinal and transverse modes. Note that in the rigid ion description the transverse modes would not scatter. We obtain finally

$$\sigma(T) = \frac{4e^2}{\pi\hbar}\ \frac{\rho_c c^2}{d\ k_B T}\ \frac{J_o^2}{\alpha^2} . \tag{5.7}$$

Note that (5.7) is independent on μ and therefore on the amount of charge transfer f_c. This is due to cancellations that arise because of the linear band dispersion. This expression is derived from first principles and with respect to the electronic hamiltonian it only depends on J_o and $\alpha = dJ/dr$ that also appeared in the previous sections. The values which are used for J_o vary between 2.5 and 3eV [8]. From the analysis of the optical properties of graphite within our model we obtain $J_o \simeq 2.4\text{eV}$ so we prefer to use the lower value. For the electron lattice coupling α, the situation is as follows. The analysis of various properties of small molecules gives for α values of $3.5 \sim 4.5\text{eV/Å}$. On the other hand from properties of intercalated graphite and polyacetylene one gets $6.5 \sim 7.5\text{eV/Å}$. This discrepancy is probably due to the sligth ionization of the hydrogen atoms (.05e) and different local correlation [8]. Using the value $\alpha= 6.5\text{eV/Å}$ we obtain $\sigma(T=300^oK) \simeq 13\text{x}10^5(\Omega\text{cm})^{-1}$ for a single graphite layer. The inclusion of geometry effects due to the presence of the intercalate reduces this value at least by a factor of two. We conclude therefore that the maximum values observed in intercalation compounds ($\sim 3\text{x}10^5(\Omega\text{cm})^{-1}$) are already quite close to the maximum ideal conductivity.

References

1. J.N.Murrell: The Theory of the Electronic Spectra of Organic Molecules , Chapman and Hall, London (1971).
2. T.Kaktani, Progr.Theor.Phys. 51, 656 (1974).
3. M.S.Dresselhaus and G.Dresselhaus, Adv.in Phys. 30, 139 (1981).
4. L.Pietronero and S.Strässler, Phys.Rev.Lett. 47, 593 (1981).
5. F.Bassani and G.Pastori Parravicini, Nuovo Cim. B50, 4259 (1967).
6. S.Y.Leung, G.Dresselhaus and M.S.Dresselhaus, Solid State Comm. 38, 175 (1981)
7. L.Pietronero and S.Strässler, to be published.
8. L.Pietronero, S.Strässler and P.Horsch, Proc. of the Conference on Low Dimensional Conductors, Boulder, Aug.1981, Ed. by A.Epstein and E.Conwell, to be published.
9. L.Pietronero and S.Strässler, Synth.Metals, 3, 209 (1981). L.Pietronero, S.Strässler, H.R.Zeller and M.J.Rice, Phys.Rev. B22, 904 (1980).

Intercalation Reactions of Solids by Electron/Ion Transfer

Robert Schöllhorn

Anorganisch-chemisches Institut, Universität Münster
D-4400 Münster, Fed. Rep. of Germany

Introduction and Principles

The basic concept of intercalation reactions can be described by (1) and proceeds by the topotactic uptake of mobile guest species (A) by a solid host lattice (Z) which contains accessible empty lattice sites ($\square$). This description refers only to stoichiometric and to fundamental structural aspects. With respect to their chemical reactivity, host lat-

$$x\,A + \square_x[Z] \rightleftharpoons A_x[Z] \qquad (1)$$

tices may be classified in two large groups: (i) those which are insulators (e.g. zeolites, sheet silicates [1]) are not subject to changes in the electronic nature of the host lattice matrix upon intercalation, (ii) host lattices with electronic conductivity (metals or semiconductors) are able to form intercalation compounds by topotactic redox processes [2] via reversible electron/ion transfer and concurrent changes in the electronic properties of the host lattice matrix (2,3). We shall be concerned here only with the last mentioned class of materials and their properties.

$$x\,A^+ + x\,e^- + \square_x[Z] \rightleftharpoons (A^+)_x[Z]^{x-} \qquad (2)$$

$$x\,A^- + \square_x[Z] \rightleftharpoons (A^-)_x[Z]^{x+} + x\,e^- \qquad (3)$$

The corresponding host lattices may have different structural dimensionality; those with framework structures may intercalate only metal ions and protons, while host lattices with layer or chain structure of the matrix may take up, in addition, molecular ions and cointercalate neutral polar molecules (2-5). Direct intercalation of neutral molecules

$$x\,A^+ + x\,e^- + y\,B + \square_{x+y}[Z] \rightleftharpoons (A^+)_xB_y[Z]^{x-} \qquad (4)$$

$$x\,A^- + y\,B + \square_{x+y}[Z] \rightleftharpoons (A^-)_xB_y[Z]^{x+} \qquad (5)$$

into solids with electronic conductivity has been shown to proceed by molecular redox reactions [2,3] that lead to the formation of molecular homosolvated cations or anions and electron transfer to or from the lattice matrix.

Graphite belongs to the few host lattices that are known to be redox amphoteric and may react as electron acceptor as well as electron donor (3). The reason for the strong interest in this material is in part due to the simple composition and structure, high chemical stability, particular physical properties and technological importance. In the last ten years, however, a large variety of new host lattices with appropriate electronic and structural properties of different dimensionality have been extensively studied which are able to undergo topotactic redox reactions. The great majority of these materials react by electron/cation transfer according to (2) and (4), respectively. In terms of composition these are mainly compounds M_nX_m of transition metals M with nonmetals X (oxides, chalcogenides, halides, pnictides etc.) and their structures are considerably more complex as compared to graphite. The intercalation can be performed at or close to ambient temperature either by chemical redox reactions or, in a much more elegant way, by electrochemical techniques which allow the controlled investigation of intermediate phases and yield thermodynamic and kinetic data.

On the chemistry side, central points of interest in these systems are structure and bonding, phase ranges, reaction mechanisms, kinetics and thermodynamics. Physicists are mainly concerned with electronic properties and band structure, transport phenomena (ionic and electronic conductivity), phase transitions (charge density waves, superconductivity), lattice dynamics and anisotropy phenomena in these compounds. Probably the most intriguing point is the large variability in the modification of solid host materials via intercalation by topotactic redox processes which provides the base for a systematic study of the concurrent changes in physical properties. Moreover, many of these materials are of potential technological interest, e.g. as synthetic conductors, reversible battery electrodes for energy storage, heterogeneous catalysts etc.

As a consequence of the complex nature of these systems there are, however, many open problems associated with the characterization of this class of compounds, including details of host/guest bonding, specific defect structures, ordering states, redox ranges and partial irreversibility, that will need a variety of physical methods for elucidation. More information on the aspects discussed above can be found in recent reviews on the subject [2-6]. We shall illustrate in the following a few selected examples of electron/ion transfer reactions of solids with layer and framework structures that demonstrate the influence of structure and changes in host and guest components on the reaction process as well as some problems related to our understanding of the nature of these compounds.

Layered Host Lattices

Among the layered materials, transition metal dichalcogenides MX_2 (X = S,Se,Te) have been studied rather intensively. These compounds consist of sandwich type X-M-X layers with strong ionocovalent M/X bonding; a van der Waals gap separates adjoining layers (Fig. 1). The interlayer space provides a two-dimensional manifold of empty octahedral and tetrahedral lattice positions which can be occupied by guest species. Besides metal ions, layered chalcogenides may intercalate via topotactic redox reactions a large variety of inorganic and organic molecular ions, either unsolvated or solvated by polar neutral species. Examples for simple metal ion intercalation compounds with electropositive main group and post transition metals are Li_xTaS_2, Ca_xMoS_2, Cu_xTiS_2, etc. The latter phase can be prepared, e.g. by controlled cathodic reduction

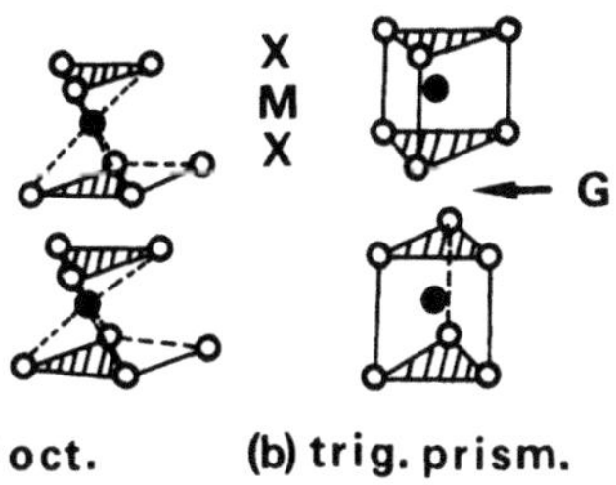

Fig. 1 Structure scheme of layered chalcogenides MX_2 (a) 1T-TiS_2, (b) 2H-NbS_2

of TiS_2 in CH_3CN/Cu^+ electrolyte (6) at ambient temperature [7]. The potential/charge transfer curve for this reaction is given in Fig. 2a.

$$x\,Cu^+ + x\,e^- + TiS_2 \rightleftharpoons (Cu^+)_x[TiS_2]^{x-} \qquad (6)$$

After a transfer of 0.9 e^-/TiS_2 (i.e. $Cu_{0.9}TiS_2$) the reaction is finished and the potential of copper metal deposition is attained. The potential steps showing up in the diagram and the changes in potential slope suggest the occurrence of two one-phase regions (I,III) and one two-phase region (II). Structural studies (Fig. 2b) confirm this conclusion; they show two nonstoichiometric regions Cu_xTiS_2 ($0 \leq x \leq 0.5$; $0.7 \leq x \leq 0.9$) and one two-phase region with the coexistence of $Cu_{0.5}TiS_2$ and $Cu_{0.7}TiS_2$. There is a significant step in the c axis (= interlayer spacing) between $Cu_{0.5}TiS_2$ (a = 3.409 Å, c = 5.83 Å) and $Cu_{0.7}TiS_2$ (a = 3.428 Å, c = 6.29 Å). $Cu_{1/2}$ and $Cu_{2/3}$ are particular compositions with respect to the distribution of Cu between the available tetrahedral sites; they are apparently also energetically favored as compared to homogeneous phases of intermediate composition. The compounds obtained thus are metastable; Cu_xTiS_2 phases prepared by thermal methods are restricted in their phase range to $0.7 \leq x \leq 1$ and have a tripled c axis [8]. If we now take the host lattice anion X as the variable parameter and look for the maximum copper content in Cu_xTiX_2, we observe substantial differences: $x_m = 0.9$ (S), $x_m = 0.5$ (Se), $x_m = 0.29$ (Te). The

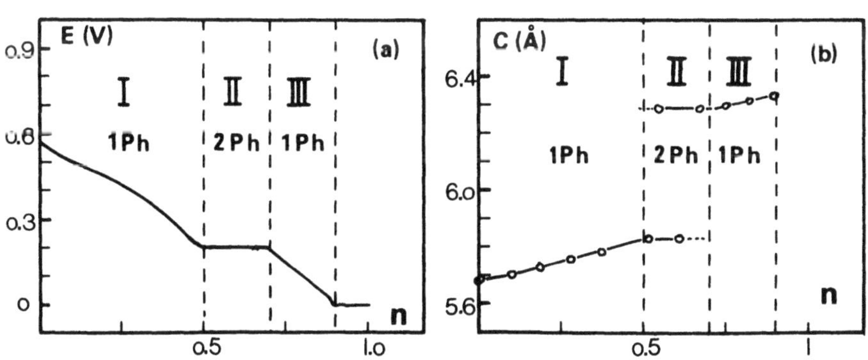

Fig. 2 (a) Cathodic reduction of TiS_2 in CH_3CN/Cu^+ at 295 K; E = potential vs Cu, charge transfer n = e^-/TiS_2, (b) change of the lattice parameter c with n

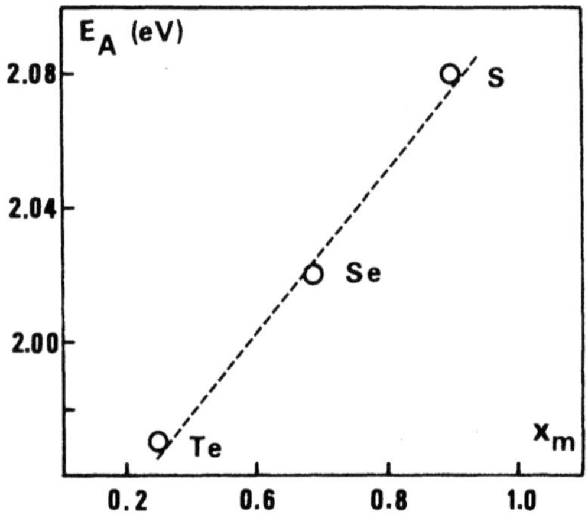

Fig. 3 Variation of the upper phase limit x_m in Cu_xTiX_2 (X = S,Se,Te) with the electron affinity E_A of the lattice anion X

starting binary phases TiX_2 exhibit a strong variation in electronic properties: TiS_2 is a semiconductor, $TiSe_2$ a semimetal and $TiTe_2$ a metal [5]. We found that a simple correlation exists between the electron affinities of the host lattice anions X and the maximum Cu content x_m (Fig. 3); obviously the increasing covalence in M-X bonding and the concurrent change in the electronic nature of the host lattice with X lead to a decreasing maximum capacity of the conduction band for the uptake of additional electrons.

In the group of solvated systems the hydrated layered chalcogenides have been investigated in more detail. The alkali metal hydrates can be obtained by cathodic reduction of the dichalcogenides in aqueous alkali ion electrolytes according to the overall reaction given in (7). Depending upon the cation charge radius ratio Z^2/r, they form either monolayer

$$x\,A^+ + x\,e^- + y\,H_2O + MX_2 \rightleftharpoons A_x^+(H_2O)_y[MX_2]^{x-} \qquad (7)$$

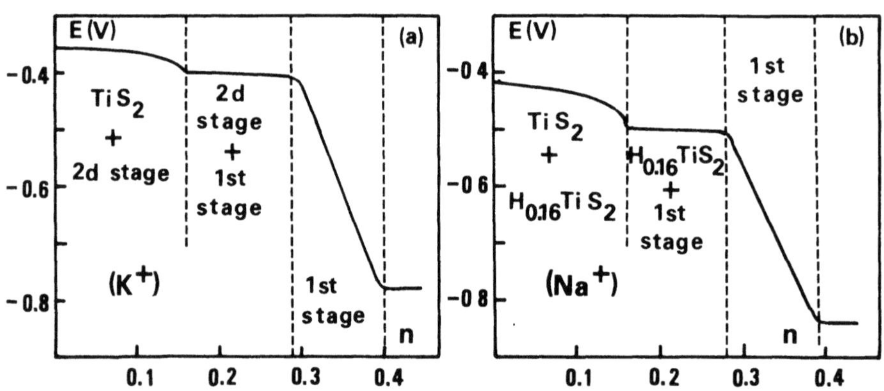

Fig. 4 Cathodic reduction of TiS_2 (a) in K^+/H_2O, (b) in Na^+/H_2O. E = potential vs Ag/AgCl, charge transfer n = e^-/TiS_2

(K^+, Rb^+, Cs^+) or bilayer hydrates (Li^+, Na^+). We shall consider here the influence of the alkali cation A and of the host lattice metal M on the formation of these phases.

In Fig. 4a the potential/charge transfer curve is given for the cathodic reduction of TiS_2 in K^+/H_2O electrolyte. After a transfer of 0.4 e^- the potential of H_2 discharge is reached; the final product is $K^+_{0.4}(H_2O)_y[TiS_2]^{0.4-}$ with a tripled c axis (3s form). The changes in potential and the structural data reveal three intermediate regions:
(i) two phase region: TiS_2 + 6s-$K^+_{0.16}(H_2O)_y[TiS_2]^{0.16-}$ (2d stage intermediate with K^+/H_2O intercalated in every second van der Waals gap, 6s stacking)

(ii) two phase region: 2d stage phase + 1st stage phase 3s-$K^+_{0.28}(H_2O)_y[TiS_2]^{0.28-}$

(iii) nonstoichiometric region: 3s-$K^+_x(H_2O)_y[TiS_2]^{x-}$, $0.28 \leqslant x \leqslant 0.4$

In Fig. 4b the corresponding potential/charge transfer curve is given for the reduction of TiS_2 in Na^+/H_2O. Although both curves have apparently close similarity, the intermediate stages are not identical. The three regions correspond to:
(i) two phase region TiS_2 + $H_{0.16}TiS_2$

(ii) $H_{0.16}TiS_2$ + 1st stage 1s-$Na^+_{0.27}(H_2O)_y[TiS_2]^{0.27-}$

(iii) nonstoichiometric region 1s-$Na^+_x(H_2O)_y[TiS_2]^{x-}$; $0.27 \leqslant x \leqslant 0.4$

The differences in layer stacking change are due to the preference of hydrated K^+ ions for trigonal-prismatic coordination. The fact that in the Na^+ case the first product is a hydrogen bronze can be explained in terms of kinetic reasons: the elastic deformation energy for second stage phases must be much higher for bilayer hydrates (Na^+) as compared to monolayer hydrates (K^+) and the system avoids this stage by intercalating protons from the electrolyte, which induce very small changes only in interlayer spacing. The intermediate stages become even more complicated if hydrated bivalent or trivalent cations are intercalated or if the hydrogen ion concentration is different from pH = 7. Changes in physical properties of hydrated phases have been studied so far only with respect to superconducting behavior [5].

We shall now consider an example of the influence of a change in the host metal cation M on the reaction process given in (7), while the intercalated cation will remain the same (K^+). The reduction of 1T-TaS_2 is illustrated in Fig. 5; only a first stage product (3s type) is observed. The reduction of 2H-TaS_2 and 2H-NbS_2 in K^+/H_2O has been studied earlier by neutron diffraction [9], while the reaction of TiS_2 has been described above. A comparison of the intermediate stages presents the following sequence:

1T-TaS_2 (oct.)			→ 1st stage (3s)
1T-TiS_2 (oct.)		→ 2d stage	→ 1st stage (3s)
2H-NbS_2 (trig.-pr.)	→ 3d stage	→ 2d stage	→ 1st stage (2s)
2H-TaS_2 (trig.-pr.)	→ 3d stage	→ 2d stage	→ 1st stage (2s)

1T-TaS_2 and TiS_2 have identical structures (CdI_2 type, octahedral coordination of M); they differ, however, in their electronic properties and show a different reaction mechnism. 2H-NbS_2 and 2H-TaS_2 both are metals with identical structures (trigonal-prismatic coordination of M)

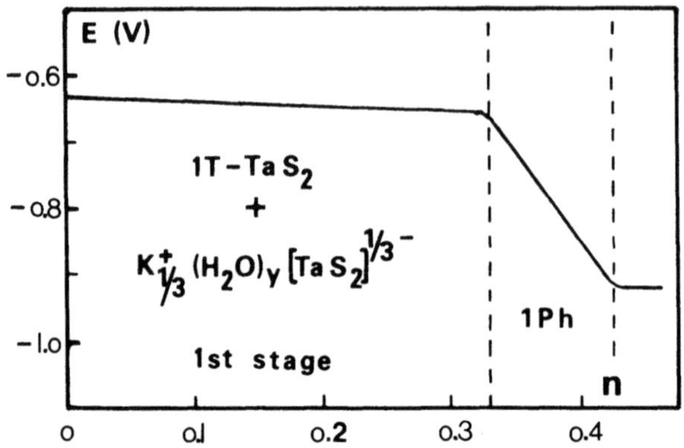

Fig. 5 Cathodic reduction of $1T\text{-}TaS_2$ in K^+/H_2O electrolyte. Change in potential E (vs Ag/AgCl) with the charge transfer $n = e^-/TaS_2$

and identical mechnisms. It can be concluded thus that electronic aspects play a role in the mechanism of formation of these phases, although the correlations are not yet understood. Contrary to graphite, no stages higher than third stage have been observed in these compounds by X-ray and neutron diffraction techniques.

Host Lattices with Three-Dimensional Structures

A series of compounds with framework structures that are able to undergo reversible topotactic electron/ion transfer reactions has been studied in recent years. These systems are characterized by interconnected or isolated "channels" of empty neighboring lattice sites which may accommodate metal guest cations up to a critical maximum size that is determined by the channel radius. The effective maximum electron/ion transfer to the host lattice and the appearance of intermediate stages are governed by the structure and electronic nature of the host lattice as well as by the type of guest cation. A most interesting example are the Mo_6 cluster chalcogenides Mo_6X_8 (X = S,Se) and the corresponding ternary phases $A_xMo_6X_8$. The strong regard given to these compounds originates from their unusual structural, electronic, magnetic and superconducting properties [10]. The binary compounds contain intersecting channels of empty lattice positions along the three rhombohedral axes (Fig. 6). They are able to take up small main group and transition metal cations (critical upper radius ca. 1.2 Å) according to the formal eq. (8) to give ternary phases [2]. Compounds with larger cations can

$$x\,A^+ + x\,e^- + Mo_6X_8 \rightleftharpoons A_x^+[Mo_6X_8]^{x-} \qquad (8)$$

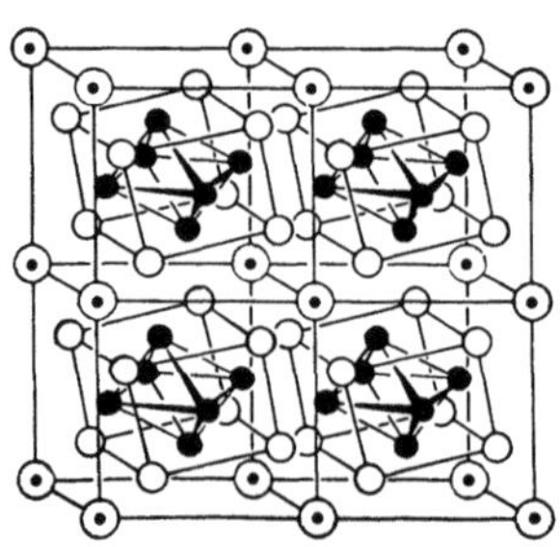

Fig. 6 Scheme of Mo_6S_8 structure; ● = Mo, ○ = S, dotted circles refer to unoccupied channel positions at the origin of the unit cell

only be prepared by thermal reactions. YVON [10] has proposed an ionic model which predicts that the maximum number of electrons that can be transferred to the host lattice matrix is equivalent to 4, i.e. $[Mo_6X_8]^{4-}$. This model is supported by band structure calculations. All experimental data obtained so far including ternary phases with two types of guest ions are in agreement with this model on the assumption that the intercalated guest ions are present in their usual oxidation states, e.g. $Cu_4^+[Mo_6Se_8]^{4-}$, $(Zn^{2+})_2[Mo_6S_8]^{4-}$, $(Zn^{2+})_{1.65}(Ag^+)_{0.7}[Mo_6S_8]^{4-}$. The following example will demonstrate, however, that there are problems in some cases as concerns the oxidation state of the guest ion (i.e. the effective electron transfer to the host lattice).

We observed recently in a study on the electrochemical intercalation of Ni^{2+} from acid aqueous electrolyte (pH = 3.5) into Mo_6S_8 a total formal electron transfer of 6 e^-/Mo_6S_8 (i.e. 2 electrons more than expected) [11]. Fig. 7 shows the potential/charge transfer curve for this reaction. Region I is a two phase range $Mo_6S_8/Ni_{1.0}Mo_6S_8$ (9); an earlier study on the corresponding thermal phase suggests that Ni is present as Ni^{+1} [10]. On further reduction a nonstoichiometric region is found with continuous changes in lattice parameters. Analytical, ^{1}H-NMR and mass spectrometric investigations revealed that in addition to one further Ni atom, two hydrogen atoms are taken up into the lattice which is consistent with the consumption of 4 additional electrons (10).

$$Ni^{2+} + 2\,e^- + Mo_6S_8 \rightleftharpoons Ni_1^+[Mo_6S_8]^- \qquad (9)$$

$$Ni^{2+} + 2\,H^+ + 4\,e^- + NiMo_6S_8 \rightleftharpoons (Ni_2H_2)^{n+}[Mo_6S_8]^{n-} \qquad (10)$$

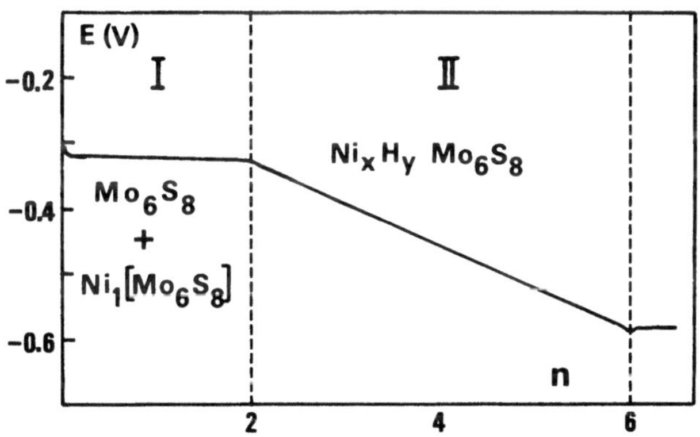

Fig. 7 Cathodic reduction of Mo_6S_8 in Ni^{2+}/H_2O. E = potential vs Ag/AgCl, charge transfer $n = e^-/Mo_6S_8$

Since ternary hydrogen compounds $H_xMo_6S_8$ are not existent even under high hydrogen pressure [12], it is concluded that the hydrogen in the new quaternary phase is stabilized by the Ni guest atoms, presumably as hydrogen bridged $(Ni_2H_2)^{n+}$ clusters. Clusters with this type of structure have been found in molecular hydride complexes of nickel [13].

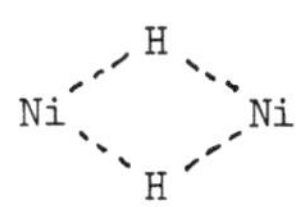

Since transition metal hydride complexes require a low oxidation state of the metal ion, it is likely that the total charge of the complex is n = +2, i.e. Ni^{+1} and 3 center-2 electron Ni/H bonding. As concerns the effective electron transfer to the Mo_6S_8 matrix, this scheme would be consistent with the model of YVON. The formation of guest ion clusters in Mo_6X_8 phases seems not to be restricted to the case discussed here; we found evidence lately that metal-metal bonding between guest ions also occurs, e.g. in the zinc and cadmium phases $(Zn_2)^{2+}[Mo_6S_8]^{2-}$, $(Cd_2)^{2+}[Mo_6S_8]^{2-}$.

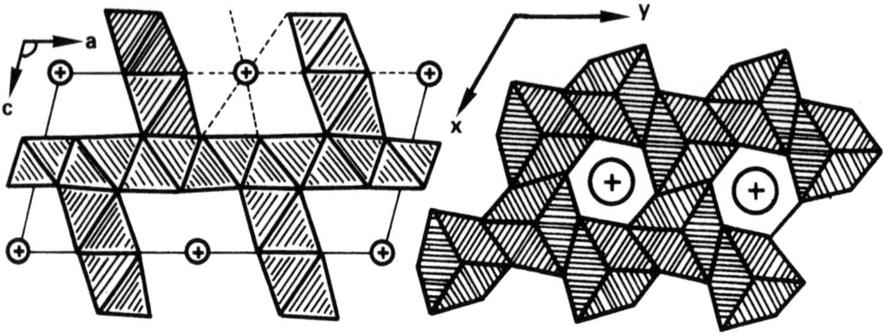

Fig. 8 Structure scheme of $Tl_xV_5S_8$ (left) and $A_xNb_3S_4$; cross section normal to channel axis

Some of the host lattices with three-dimensional structures are characterized by the presence of isolated parallel channels, e.g. Nb_3S_4, Nb_3Se_4 (Fig. 8), which can accept guest metal atoms via electron/ion transfer. In addition, a series of ternary phases are known, prepared by thermal methods, that contain mobile cations in channels and which can be totally or partially (reversibly) oxidized, e.g. $K_xTi_3S_4$, $Tl_xV_6S_8$ and TlV_5S_8 [2]. The guest ions are confined in these lattices to one-dimensional diffusion. We shall compare here the oxidation of the two metallic Tl/V/S phases which has been studied by Tl and V NMR [11]. The structure of $Tl_{0.6}V_6S_8$ [14] is rather similar to that of Nb_3S_4 (Fig. 7); anodic oxidation proceeds in a nonstoichiometric phase range up to the end phase which corresponds to $Tl_{0.2}V_6S_8$. The ^{205}Tl resonance shows that thallium is clearly present as Tl^+ independent of the stoichiometry; with decreasing Tl content a resonance shift is observed (see scheme) which is thought to be due to the decreasing concentration of conduction electrons in the V_6S_8 matrix and to small changes in the position of Tl^+ towards the octahedron centre. The ^{51}V resonance is dominated by second-order quadrupole coupling and shows as expected a clear resonance shift with decreasing Tl content.

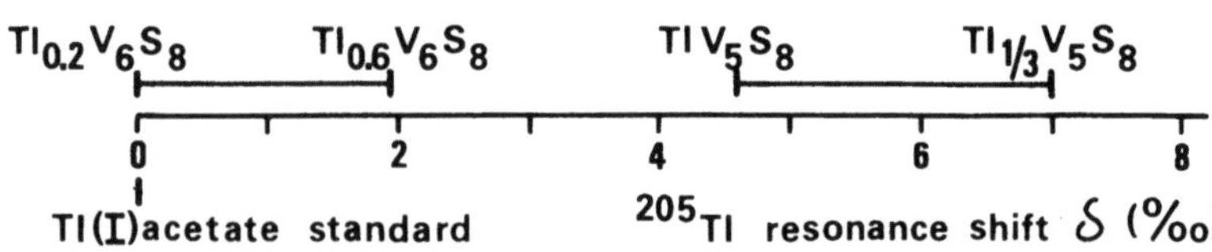

The situation is quite different for $Tl_1V_5S_8$, which has a channel structure of monoclinic symmetry [15] (Fig. 7). We find here the interesting case of a stoichiometry dependent change in the oxidation state of the guest ion at constant host matrix charge. Anodic oxidation of $Tl_1V_5S_8$ leads again to nonstoichiometric phases with the lower phase limit $Tl_{0.33}V_5S_8$. The NMR data, however, show here an unexpected, almost linear, strong increase in the ^{205}Tl resonance shift with decreasing Tl content (see scheme above and Fig. 9). Inversely the ^{51}V NMR data exhibit no resonance shift with decreasing Tl content, and the temperature dependent paramagnetic susceptibilities of TlV_5S_8 and $Tl_{0.33}V_5S_8$ are identical. We conclude from these experimental observations that on the oxidation of TlV_5S_8 the host lattice charge remains constant, while the Tl guest ions change their oxidation state continuously from Tl^{1+} in $Tl^{1+}[V_5S_8]^{1-}$ to Tl^{3+} in $(Tl^{3+})_{0.33}[V_5S_8]^{1-}$. It is clear from these observations that the unlike behavior of the two thallium vanadium sulfides originates from differences in the electronic nature of the host matrix which in turn must result from the structural differences.

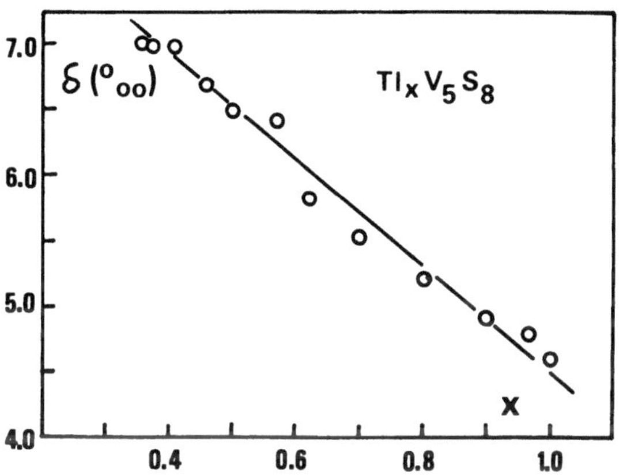

Fig. 9 Correlation of the resonance shift δ of the ^{205}Tl signal with the Tl content x in $Tl_xV_5S_8$ [11]

Conclusions

The examples given above have been concerned with only a few aspects typical for systems with two- and three-dimensional structures. Materials with quasi one-dimensional lattice elements have also been investigated, e.g. transition metal trichalcogenides MX_3 (M = Ti,Zr,Hf, Nb,Ta; X = S,Se) which consist of chains of trigonal MX_6 prisms sharing basal planes and may intercalate metal ions [4,6]. Polysulfur nitride $(SN)_x$ may take up Br_2 between the sulfur/nitrogen chains to give $(SN)_x^{n+}(Br_y)^{n-}$ [16]. A new challenge is created by the observation that amorphous highly conducting organic polymers, e.g. polyacetylenes undergo reversible electron/cation, as well as electron/anion, transfer in nonprotolytic electrolytes [17], their use as secondary battery electrodes has been proposed.

An exceptionally interesting guest cation is the proton H^+, which may react by electron/proton transfer with host lattices of different dimensionality to hydrogen bronzes (12), e.g. H_xMoO_3, H_xWO_3, H_xTaS_2,

$$x\,H^+ + x\,e^- + Z \rightleftharpoons H_x[Z] \qquad (12)$$

$H_{0.5}ZrCl$ [2]. These are of considerable interest with respect to hydrogen storage, heterogeneous and electro-catalysis, photoelectrochemical H_2O decomposition and corrosion processes. Recently electron/proton transfer reactions have also been observed for molecular solids (e.g. for the one-dimensional metal $[Pt(NH_3)_4PtCl_4]^{0.5+}[H_{1.5}SO_4]^{0.5-}(H_2O)_y$ [18].

The foregoing brief account has attempted to demonstrate that topotactic electron/ion transfer processes provide a large variety of attractive reactions that allow the systematic modification of solids by reactions at ambient temperature under retention of the structural elements of the host matrix. As was pointed out above the detailed characterization of the nature of these complex materials presents a most interesting challenge to chemists as well as to physicists.

References

1. R.M. Barrer in L. Mandelcorn, Nonstoichiometric Compounds, Academic Press, New York (1964) 309
2. R. Schöllhorn, Angew. Chem. 92, 1015 (1980); Angew. Chem. Int. Ed. Engl. 19, 983 (1980
3. A.R. Ubbelohde in F. Lévy, Intercalated Layered Materials, D. Reidel Co, Dordrecht (1964) 1
4. J. Rouxel, Reviews in Inorganic Chemistry 1, 245 (1979)
5. G.V. Subba Rao, M.W. Shafer in F. Lévy, Intercalated Layered Materials , D. Reidel Co, Dordrecht (1979) 99
6. M.S. Whittingham, Progr. Solid State Chem. 12, 41 (1978)
7. P. Buller, F. Pötzing, R. Schöllhorn, in preparation
8. N. le Nagard, O. Gorochov, G. Collin, Mat. Res. Bull. 10,1287 (1975)
9. C. Riekel, H.G. Reznik, R. Schöllhorn, J. Solid State Chem. 34, 253 (1980)
10. K. Yvon, Curr. Top. Mat. Sci. 3, 54 (1978)
11. W. Schramm, H. Eckert, P. Dolscheid, R. Schöllhorn, in preparation
12. B. Stritzker, H. Wühl, F.J. Culetto, F. Pobell, U. Poppe, Verh. DPG (VI), 13, 343 (1978); K. Yvon, personal communication
13. B.L. Barnet, C. Krüger, Yi-Hung Tsay, Chem. Ber. 110, 3900 (1977)
14. M. Vlasse, L. Fournès, Mat. Res. Bull. 11, 1527 (1976)
15. L. Fournès, M. Vlasse, M. Saux, Mat. Res. Bull. 12, 1 (1971)
16. J.W. Macklin, G.B. Street, W.D. Gill, J. Chem. Phys. 70, 2425 (1979); H. Temkin, G.B. Street, Solid State Comm. 26, 455 (1978)
17. D. Mac Innes, M.A. Drury, P.J. Nigrey, D.P. Nairus, A.G. Diarmid, A.J. Heeger, Chem. Comm. (1981) in press
18. R. Schöllhorn, K. Wagner, H. Jonke, Angew. Chem. 93, 122 (1981); Angew. Chem. Int. Ed. Engl. 20, 109 (1981)

Domain Structures in Graphite Intercalation Compounds

S.A. Safran

Corporate Research-Science Laboratories, Exxon Research and Eng., P.O. Box 45
Linden, N.J. 07036, USA

Abstract

Domain structures have been proposed for both c-axis ordering (staging) of the intercalant layers, as well as for the in-plane superlattices that exist at low temperatures in alkali metal intercalation compounds. In this paper, the kinetics of domain growth in intercalation compounds is examined theoretically. The simultaneous growth of staged islands of intercalant is calculated for an initially homogeneous, dilute, stage-one compound. Upon quenching, the intercalant atoms cluster within a plane to form islands whose positions in different planes are correlated as a result of the interlayer interaction. The kinetics of island growth for in-plane superlattice structures are also analyzed. It is suggested that in a two-dimensional approximation, kinetically induced disorder may occur in rapidly quenched samples, leading to an amorphous or glassy structure.

Introduction

Recent structural studies [1-5] of graphite intercalation compounds have suggested that simple models which assume infinite, well-ordered planes of intercalant layers should be re-examined in light of evidence for the existence of domains of intercalant. A domain model for staging was proposed by DAUMAS and HEROLD [1] in order to rationalize the kinetics of the staging process. Their suggestion of finite (but large) domains of staged material, with the domains arranged so that every pair of carbon layers has the same density of intercalant (averaged over the whole sample) has received recent attention because of the pressure experiments of CLARK and co-workers [2,4]. The staging transition (e.g. from stage 2 to stage 3) observed by these workers as a function of hydrostatic pressure, with the constraint of a constant number of intercalant atoms, could only be understood in the context of the domain model [4]. In addition to proposals for domains in connection with staging, KAMBE et al. [3], have interpreted their electron diffraction experiments in terms of islands of intercalant atoms with coexisting mixtures of in-plane ordering [5].

This focus on domain structures in intercalated graphite coincides with a general interest in domains in the condensed matter physics community. Domains have been invoked to understand the nature of one-dimensional magnets [6], doped polyacetelyne [7], two-dimensional structures of adsorbed layers on surfaces [8], and the kinetics of phase transitions [9-11]. In this paper, domain structures in intercalated graphite will be discussed with an emphasis on the kinetics of domain growth. Although the intercalation process itself involves the time evolution of islands of intercalant atoms, it is probably quite sensitive to the nature of the surfaces of the graphite host [12]. A microscopic treatment of the realistic case of the intercalation of an empty graphite crystal which treats both transport across the surfaces and interactions in the bulk is beyond the scope of this work. Instead, the existence of an intercalated crystal is assumed, and the time evolution of the structure is examined for quenching from high to low temperatures.

It is important to note the distinction between domains for equilibrium and non-equilibrium (time dependent) phenomena. In equilibrium, the domains arise because of a competition between two types of forces. For example, for the well-studied case of adsorbed atoms on surfaces [8,13], the competition between the adatom-adatom interactions (which favor a particular value of the lattice constant) and the sustrate-adsorbate interaction (which favors values of the adatom lattice constant which are commensurate with the substrate lattice) results in phase transitions from incommensurate to commensurate adatom lattices. The incommensurate phase is characterized by islands of adatoms which are almost commensurate with the substrate. The island size is related to the incommensurability of the two competing lattice constants [8,13] and is time independent. Although such structures are obviously relevant to the in-plane, equilibrium, properties of intercalated graphite, their properties can be understood in the context of the work on adsorbed layers with an allowance for interlayer coupling. On the other hand, even systems with non-competing interactions (e.g. simple Ising antiferromagnets) [11] which are quenched from high (disordered state) to low temperature (ordered state) approach the equilibrium ordered state through the growth of domains of ordered regions. The domain sizes are time dependent and approach macroscopic lengths as the time goes to infinity.

In this paper, the time evolution of domains in quenched systems is examined for both interlayer (staging) and in-plane ordering in intercalated graphite. Section I analyzes the time evolution of stage ordering in a dilute, stage-one compound quenched from high to low temperature. The evolution of the c-axis staging correlations are shown to be related to the changes of density within the plane. In Section II, the domain coarsening kinetics of a single layer of intercalant (or adatoms) is examined for a quench from a high-temperature disordered phase to a low-temperature commensurately ordered phase. It is suggested that in a two-dimensional approximation, a disordered arrangement of domains may be effectively frozen into the lattice, leading to a glassy structure.

I. Staging Transitions and Island Formation

A recent model for phase transitions in staged intercalation compounds [14, 15] has predicted the instability of the staged phases to a dilute, stage-one phase at high temperatures. Experimental observations of temperature-induced staging transitions (e.g. stage-two at low temperatures and stage-one at high temperatures) at constant intercalant concentration have also been reported [2,16]. The kinetics of such phase transitions involve the growth of staged islands of intercalant. As an example, consider the kinetics of domain growth for a dilute stage-one compound which is quenched to low temperatures where pure stage two ordering is the equilibrium state. Working at constant intercalant concentration (e.g. 50% of stage-one saturation concentration of intercalant), and assuming that the intercalant atoms diffuse only between the carbon layers, one can use a modified theory of spinodal decomposition [15,17] to calculate the structure factor as measured by X-ray or neutron scattering. The structure factor, $S(\vec{q},k,t)$, is the spatial Fourier transform of the non-equilibrium, density-density correlation function, describing the instantaneous distribution of intercalant atoms in the graphite host. The wavevector $\vec{q}$ is in the hexagonal basal plane and $\vec{k}$ is perpendicular to this plane in reciprocal space, while t is the time elapsed from the initial quench.

The details of the calculation of the structure factor are described in [15]. Fig. 1 shows the initial and final intercalant distributions in the graphite host. Upon quenching from a uniform, stage-one distribution, the intercalant atoms tend to cluster to form high-density islands within each plane. Due to the interplanar repulsion of the intercalants, the positions of islands in different planes are correlated, and equilibrium corresponds to two macroscopic regions of stage-two material. The reduced structure factors $S_{\parallel}(\vec{q},t)$ and $S_{\perp}(k,t)$, defined by the integrals of $S(\vec{q},k,t)$ with $\int dk$ and $\int dq$, are shown in Fig. 1 for some intermediate time after the quench. It can be shown that $S_{\perp}(k,t)$ is a measure of the c-axis correlations of atoms at the same in-plane positions, while $S_{\parallel}(\vec{q},t)$ is a measure of the correlations within a single plane, averaged over all planes. In Fig. 1, q_o and k_o correspond to the in-plane and c-axis wavevector cutoffs (of the order of the inverse of the in-plane and c-axis lattice constants), respectively.

With increasing time after the quench, the peak in $S_{\parallel}(\vec{q},t)$ moves to smaller wavevectors. This signifies the growth of two-dimensional islands of inter-

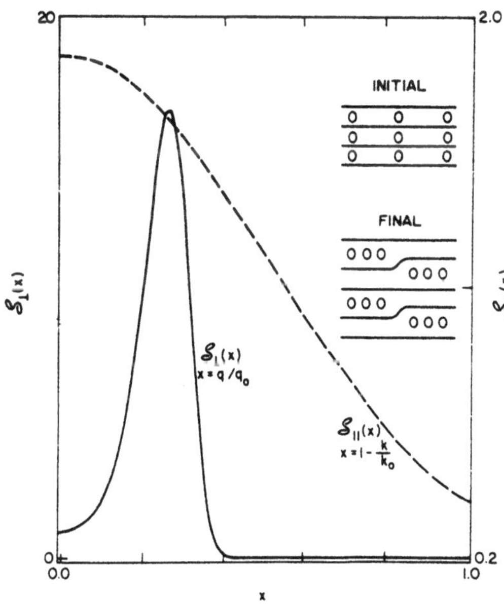

Fig.1. Reduced structure factors $S_{\parallel}$ and $S_{\perp}$ describing the in-plane and c-axis correlations defined in the text plotted as function of reduced wave vectors at some intermediate time (τ=600 see [15]) after the quench. Also shown are the initial and final (equilibrium) configurations of intercalant atoms for the quenching process discussed in the text.

calant with in-plane densities approximately equal to the equilibrium densities which are double their initial values. The inverse of the wavevector (q_{max}^{-1}) at which $S_{||}(\vec{q},t)$ is a maximum can be interpreted as being proportional to the average island size. For long times, the islands grow since [15] $q_{max}^{-1} \sim t^{1/5}$. In addition, the islands are correlated in their interlayer positions due to the interplanar repulsive interactions. The width of $S_{\perp}(k,t)$ decreases as a function of time and is a measure of the development of long-range stage-two ordering. This phenomena is thus an example of island formation in a quenched system, which involves the redistribution of intercalant atoms - i.e. changes in the spatial distribution of the conserved intercalant density.

II. In-Plane Ordering and Domain Growth

In contrast to the previous example, which required the redistribution of the intercalant atoms over large distances to go from the homogeneous stage-one structure to the equilibrium stage-two configuration, domain growth can also occur in quenched systems where no large-scale motion of the atoms is involved [11]. For example, an Ising spin system, disordered at high temperatures, proceeds to equilibrium after quenching to low temperatures by the formation of small domains of spin-up and spin-down material. As time evolves, small islands of spin-up in mostly spin-down regions disappear (and visa versa). The islands of single-domain ordered material grow larger, until at long times after the quench, there are regions of spin-up and spin-down that are macroscopic in size. The same physical description applies to the time evolution of quenched systems with superlattice structural ordering. Depending on the spatial dimensionality and the degeneracy of the superlattice structure, the kinetics of domain growth can be fast (power-law in time) or slow (logarithmic in time) [9].

Considering only the structure of a single layer (spatial dimensionality d=2), an example of a three-fold degenerate structure (degeneracy p=3) in graphite intercalation compounds is the commensurately ordered low-temperature structure of LiC_6 [18], where the degenerate structures are denoted A, B, and C (see Fig. 2). Since these structures all have the same free energy, the energy to form domain walls between finite regions of A and B (for example) is small. Thus, a short time after a quench from a high-temperature disordered state (equal occupation of the A, B, and C sublattices) small domains of A, B, and C ordered material should co-exist.

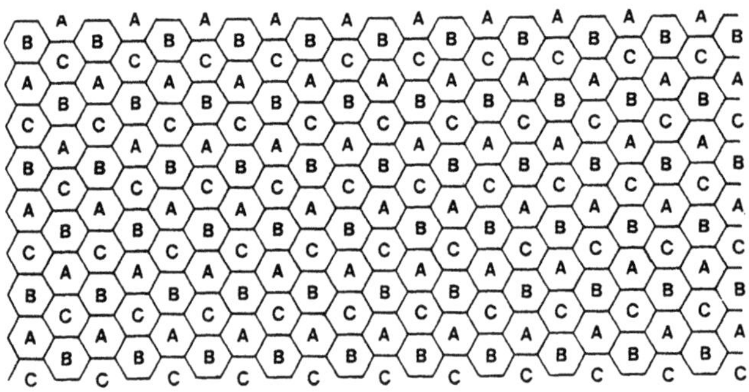

Fig. 2. Single layer of graphite showing a superposition of the intercalant atoms in the A, B, and C positions. In equilibrium, a macroscopic region of the crystal with only one type of superlattice exists.

In the limit of highly localized domain walls, the free energy of a degenerate system is essentially given by the "surface tension" of the walls. For a single domain of A phase material surrounded by an infinite domain of B type ordered material in two spatial dimensions (see Fig. 3), a small change in the domain size L,($L \to L_0 + \Delta L$) results in a change of the free energy by an amount proportional to L_0, which provides a driving force for the shrinking of the minority domain. On the other hand, for the three-fold degenerate system with the domain configuration shown in Fig. 4, the shrinking of the A domain results in a reduction of the perimeter of the A domain by $6\Delta L$, while at the same time, the walls separating domains B and C increase in length by $6\Delta L$. There is thus no net change in the free energy for small displacements, and the system is not strongly driven to either shrink or grow the minority phase and proceed towards equilibrium (macroscopic regions of a single phase). Thus, in two spatial dimensions, it is expected that the kinetics of the two-fold degenerate system are "fast" and those of the three-fold degenerate system are "slow", corresponding to the, respectively, large and small driving forces [9,19].

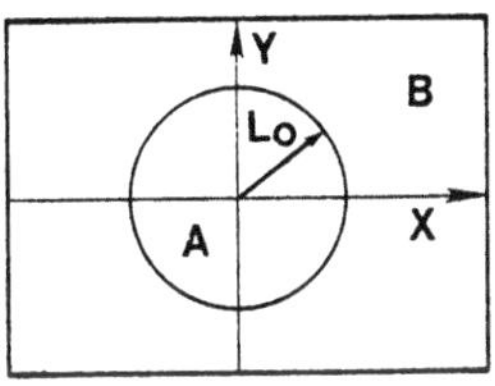

Fig.3. A circular domain in two dimensions (d=2) of A phase material, surrounded by an infinite matrix of B phase superlattice.

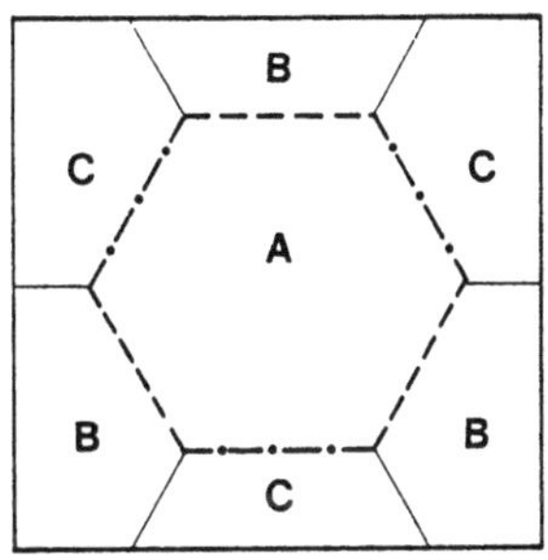

Fig.4. A domain for p=3 and d=2. The solid line separates the B and C domains while the dashed line separates the A domain from the B domain and the dot-dashed line separates the A and C regions.

In [9], a more quantitative treatment of the domain growth of degenerate phases is given (see also [19]) and it is shown that the fast kinetics are characterized by a power-law dependence of the average domain size on time, while the slow processes are characterized by a logarithmic dependence on time of the domain size. For a system with an initial distribution of domain sizes and arrangements, the slow equilibration of highly degenerate systems ($p \geq d+1$ [9,19]) results in an effective freezing-in of the late time domain structure. These systems are thus simple examples of the formation of an amorphous or glassy structure due to the slow growth of domains, which are frustrated from equilibrating by the other degrees of freedom of the system, in qualitative agreement with recent Monte Carlo computer simulations of quenched systems [20-22].

The effects of thermal fluctuations on the kinetics of these systems is under current investigation. Of particular interest is the study of the [23] existence of a "roughening temperature" above which the domain walls are no longer sharp and the kinetics of the highly degenerate systems ($p \geq d+1$) change from slow (effectively glass-like) to fast (equilibrium, ordered superlattice). While Monte Carlo computer simulations are a test of the validity of the calculations, the ultimate relevance of these concepts depends on their applicability to experiment. Structural studies of intercalated graphite - e.g. LiC_6, quenched from high to low temperatures - could measure the kinetics of the ordering process and determine the applicability of the simple model described above to the development of amorphous or glassy structures. In addition, such experiments might help resolve some of the controversies regarding materials preparation, since different kinetics for the intercalation and subsequent quenching processes (slow or fast) can result in qualitatively different stable or metastable structures.

The author wishes to acknowledge P.C. Hohenberg and J.S. Langer for useful discussions.

References

1. N. Daumas & A. Herold, C. R. Acad. Sci. 268, 373 (1969).
2. R. Clarke, N. Caswell, S. A. Solin, Phys. Rev. Lett. 42, 61 (1979).
3. N. Kambe, H. Mazurek, M. S. Dresselhaus, G. Dresselhaus, Physica (to be published).
4. R. Clarke, N. Wada, S. A. Solin, Phys. Rev. Lett. 44, 1616 (1980).
5. A. N. Berker, N. Kambe, G. Dresselhaus, M. S. Dresselhaus, Phys. Rev. Let. 45, 1452 (1980).
6. J. Villain, in Ordering in Strongly Fluctuating Condensed Matter Systems, ed. T. Riste (Plenum, N.Y., 1980). P. 91.
7. W. P. Su, J. R. Schrieffer & A. J. Heeger, Phys. Rev. Lett. 42, 1698 (1979).
8. J. Villain in [6], p. 222.
9. S. A. Safran, Phys. Rev. Lett. 46, 1581 (1981)
10. J. S. Langer, Annals of Physics 65, 53 (1971).
11. M. K. Phani, J. L. Lebowitz, G. H. Kalos, O. Penrose, Phys. Rev. Lett. 45, 366 (1980).
12. J. Hooley, Mat. Sci. and Eng. 31, 17 (1977).
13. J. P. Mctague, M. Nielsen and L. Passel in [6], p. 195.
14. S. A. Safran, Phys. Rev. Lett. 44, 937 (1980).
15. S. A. Safran, J. Synth. Metals 2, 1 (1980).
16. J. Rossat-Mingnod, D. Fruchart, M. J. Moran, J. Milliken, and J. E. Fischer, J. Synth. Metals 2, 143 (1980)
17. T. Tomita, Prog. Theor. Phys. 59, 1116 (1978).
18. N. Kambe, M. S. Dresselhaus, G. Dresselhaus, S. Basu, A. R. McGhie, and J. E. Fischer, Mat. Sci. Eng. 40, 1 (1979).
19. I. M. Lifshitz, Sov. Phys. JETP 15, 939 (1962).
20. J. Chalupa (unpublished).
21. P. Sahni & J. Gunton (unpublished).
22. J. R. Banavar, G. S. Grest, D. Jasnow, Phys. Rev. Let. 45, 1424 (1980).
23. S. A. Safran (unpublished).

Surface Study of Vacuum Synthesized Cesium Graphite Intercalation Compounds CsC_n

M. Lagues, D. Marchand, A.P. Legrand

Laboratoire de Physique Quantique ERA CNRS 676
E.S.P.C.I. - 10, rue Vauquelin
F-75231 Paris Cedex 05, France

1. Introduction

This work is mainly directed by the two following objects.

Numerous physical measurements on graphite intercalation compounds (GIC's) require the transfer under neutral atmosphere of the sample into an ultrahigh vacuum vessel. Due to the extremely high reactivity of main GIC's, the surface must then be prepared by cleaving the sample : it is thus very difficult to obtain well defined surfaces. We show in this paper that "in situ" synthesis of the GIC can prevent all transfer and preparation troubles.

The equilibrium between adsorbed and intercalated cesium is studied in different cases of thermal and Cs deposition processes. The Auger transitions of the cesium determine the cesium content, while the 285 eV satellite of the carbon Auger peak is a measurement of the insertion stage in the first layers of the sample.

In part 2, we present the method of synthesis show that in Highly Oriented Pyrolytic Graphite (HOPG), the diffusion of cesium takes place mainly along grain boundaries. In part 3, we compare the deposition kinetics of cesium on graphite, HOPG and natural single crystals, at 150 K and 300 K. We show that Cs intercalation takes place only at room temperature. In part 4, we follow the equilibrium between adsorbed cesium Cs^a and intercalated cesium Cs^i. The variation of Cs^i versus Cs^a presents sharp steps, assigned to different stages of intercalation. The lowest identified stage corresponds to the stoichiometry CsC_6 which could be stabilized in the vicinity of the surface.

Finally, we describe in part 5 the segregation of intercalated cesium towards the surface, upon cooling of the sample. We propose a simple explanation for this phenomenon.

2. Grain Boundary Diffusion in HOPG

Cesium is first deposited on a HOPG sample at room temperature : no intercalation is observed (see part 3). After a one hour heating at 750 K, a second deposit involves intercalation of cesium.

Figure 1 presents a micrography of the sample after a total amount of deposited cesium of about one hundred Langmuir.

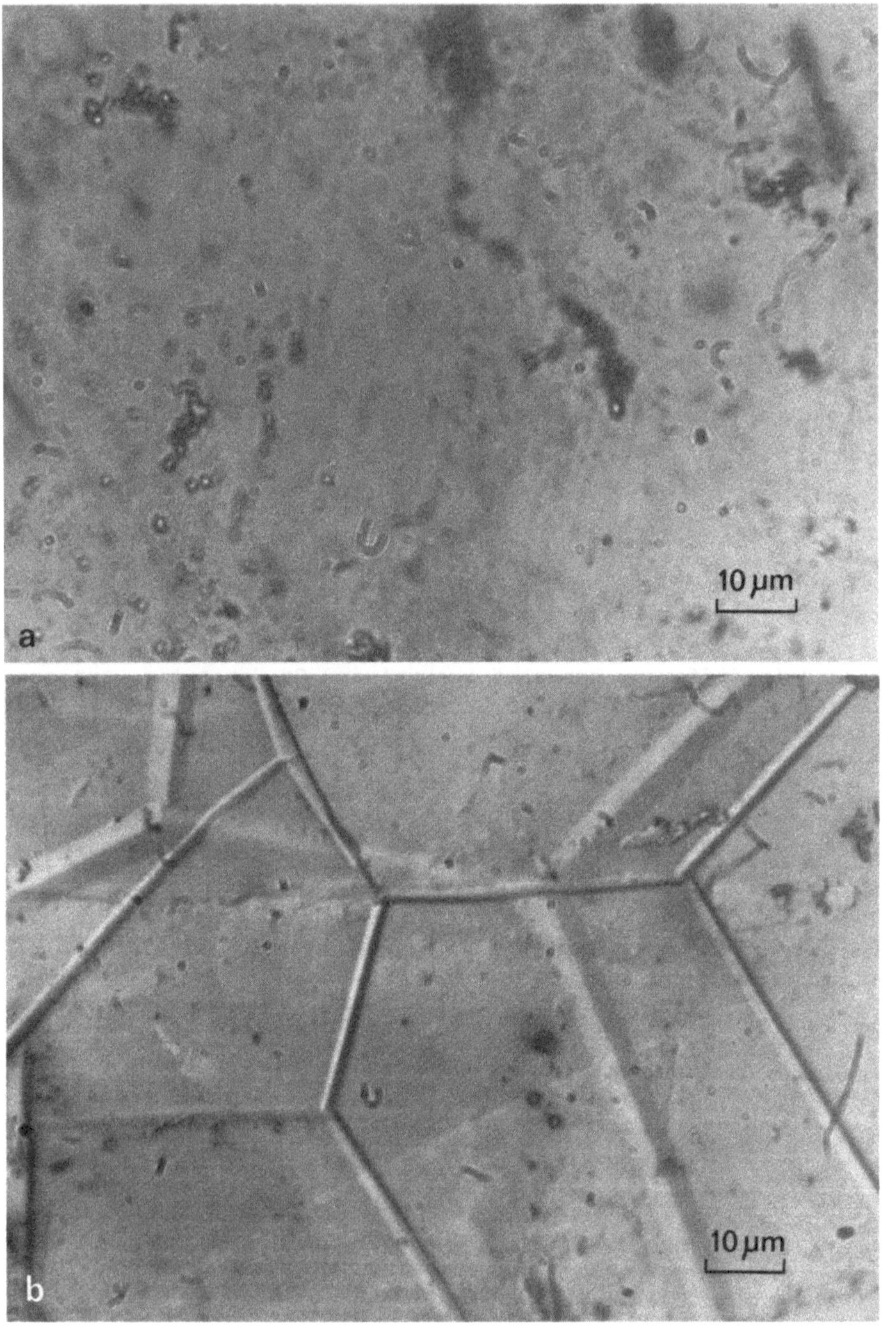

Fig.1 a) Micrography of the surface of a HOPG sample
b) Micrography of the surface of a HOPG sample after deposition of 100 Langmuir of cesium

It is quite obvious from this micrograph that cesium diffuses preferably along grain boundaries which appear clearly, both at the surface of the sample and at some microns deep in the bulk. This explains the difference between the kinetics of cesium deposition on HOPG and on natural single crystal, as it is discussed in the next part.

3. Cesium Deposition Kinetics on Graphite

The figure 2 presents a typical Auger spectrum of a) pure graphite, b) after a cesium deposit.

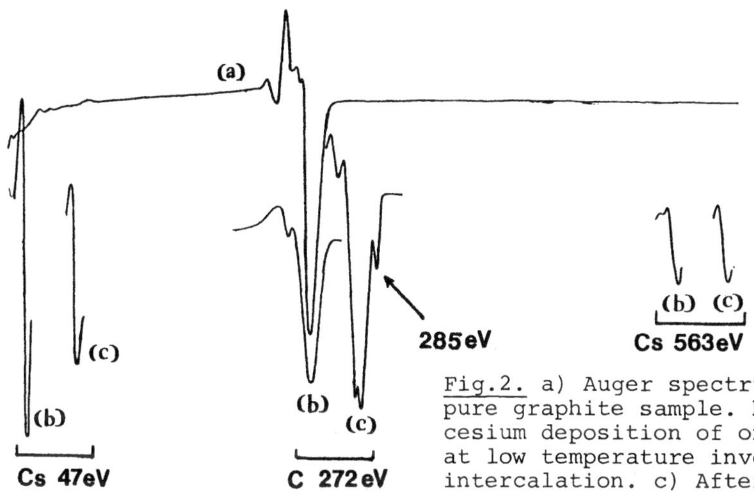

Fig.2. a) Auger spectrum of a pure graphite sample. b) After a cesium deposition of one Langmuir at low temperature involving no intercalation. c) After a cesium deposition at room temperature

The surface coverage of cesium is characterized by the cesium peak height divided by a constant linear combination of the C peak height and the Cs peak height. In practice the 563 eV peak varies much more linearly that the 47 eV peak with cesium content. This is explained by the variation of the Auger cross section with cesium content, the 47 eV transition involving "O" levels which are affected by the bond state of the cesium atom. Thus the ratio of the 47 eV and the 563 eV peaks is affected first by the cesium content and secondly by the cesium intercalated in the first twenty angströms, the escape depthes being respectively of the order of 5 angströms and 10 angströms. Fig.2c presents the case of a cesium deposit involving intercalation : the carbon peak is then distorted and a satellite peak appears at 285 eV. This feature has been both experimentally observed on Auger and UPS spectrums of GIC's [1] and theoretically predicted in the case of donor GIC's [2].

Our equipment allows the measurement of the Auger peak heights during the deposition of cesium. Figure 3 presents three typical cases of registered kinetics.

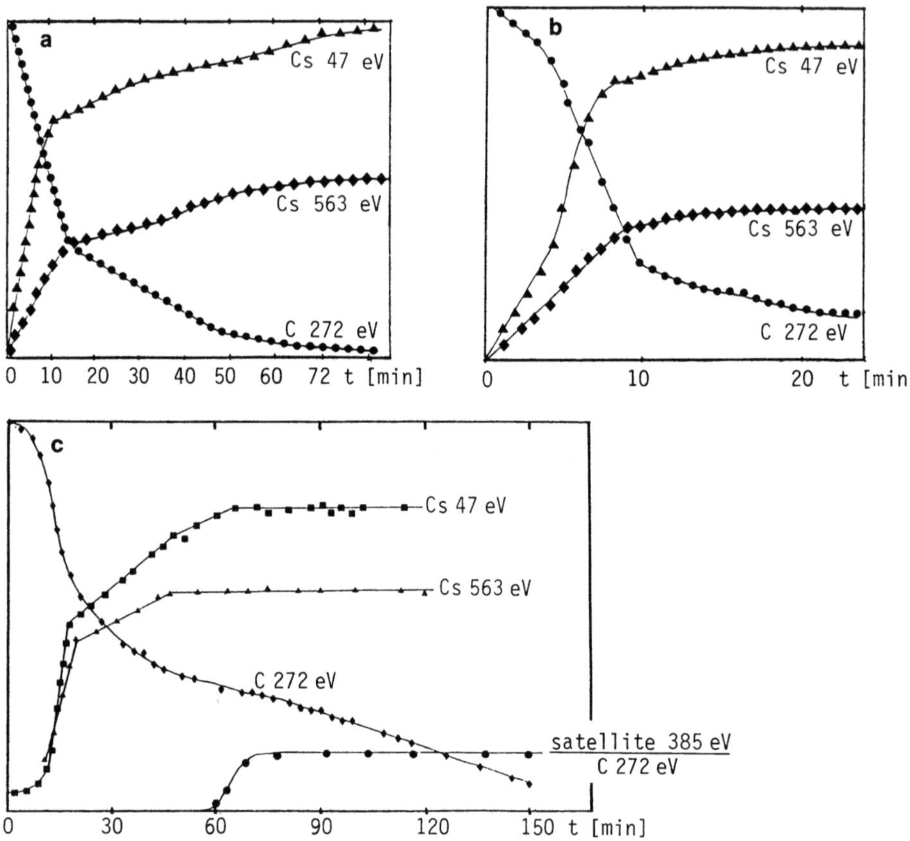

Fig.3 a) Low temperature deposition (150 K) of cesium on a natural graphite single crystal
b) Low temperature deposition (150 K) of cesium on a HOPG sample
c) Room temperature deposition of cesium on a single crystal of graphite

The first situation a) is a low-temperature deposition (150 K) of cesium on a natural single crystal of graphite. The 285 eV peak keeps at a zero value during the whole process, meaning that no intercalation arises at all. The other peaks vary linearly, showing clearly a Franck-Van der Merwe growth process, each break in the curves showing the formation of a new adsorbed layer. In the second case b) the cesium is deposited at low temperature (150 K) but on a HOPG sample. No intercalation is observed but the kinetics are not linear, due to the diffusion of cesium along the grain boundaries. The last situation c) is a deposition on a single crystal of graphite at room temperature. The process is here much slower, the curves are not linear, and the satellite peak at 285 eV appears when the second adsorbed layer is completed. If the deposition goes on for some hours, the 285 eV satellite (divided by the graphite peak) reaches a value of about .5. The sample has then a golden colour typical of first intercalation stage. This means that the sample is a stage one GIC over a minimum thickness of about 1 000 angströms.

4. Adsorption versus Intercalation Equilibrium

A GIC synthetised as described in part 3 is now heated up step by step and the equilibrium intercalation stage is recorded versus the adsorption coverage of cesium. Fig.4 presents this curve compared with the variation of the ratio of the two Auger transitions of the cesium.

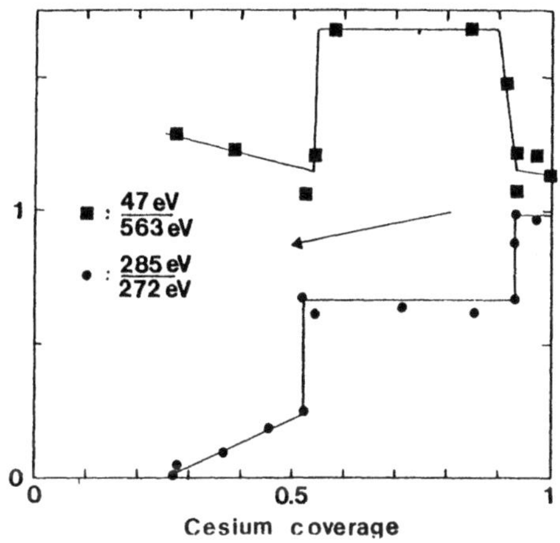

Fig.4 Heating of a stage one GIC : the lowest curve presents the variation of the Cs intercalation Auger peak (285 eV) (divided by the C Auger peak) versus the surface content of cesium (563 eV peak of the cesium divided by a constant linear combination of this peak with the carbon peak). The upper curve is the variation of the ratio of the 47 eV to the 563 eV Auger peak of the cesium

A first point is to be noticed : the intercalation stage presents two main steps. The widest one is assigned to the first stage CsC_8 because of its typical golden aspect. The second step, reached when the cesium content is about one, corresponds precisely to the value which would be expected for the intercalation stage CsC_6. Although this stoichiometry is in principle allowed by the ionic radius of Cs+, CsC_6 has never been observed in bulk GIC's. We suggest that the vicinity of the surface could stabilize some layers of CsC_6, for instance because the adsorbed structure has this stoichiometry.

Why are not more steps, corresponding to higher stages, observed ? The reason is that the escape depth of Auger electrons is 5 angströms, which is about the distance between cesium layers in CsC_8.

The Auger measurements are thus not sufficient to separate clearly higher intercalation stages.

The aspect of this curve agrees with the theoretical models [3]. When the material is a pure stage, the chemical potential varies with the surface concentration up to the appearance of a new stage. The material is then a mixture of two stages. The chemical potential keeps constant with the surface coverage, while the ratio of the two stages concentration varies. This explains the step-like shape of this curve.

The ratio of the two Auger peaks of the cesium shows clearly the variation of the cross section with the surface concentration. The low energy peak (47 eV) being less sensitive to the bulk than the high energy peak (563 eV), their ratio should increase when bulk concentration decreases. This is indeed observed for the transition from CsC_6 to CsC_8, but an opposite variation (decrease) is observed for the transition from CsC_8 to CsC_{24}. The chemical environment of the adsorbed cesium atoms changes very much at this last transition and this explains the variation of the cross section.

A last thing to notice is that the transition from CsC_8 to CsC_{24} is observed at a temperature of about 350°C, which is the fusion temperature of the intercalated cesium in CsC_8 [4]. This is another confirmation of the stage corresponding to the main step of this curve.

5. Surface Segregation of Intercalated Cesium

A stage-one GIC is heated up to 700 K for some hours : at this temperature, the concentration of adsorbed cesium is not measurable, neither is the concentration of intercalated atoms.

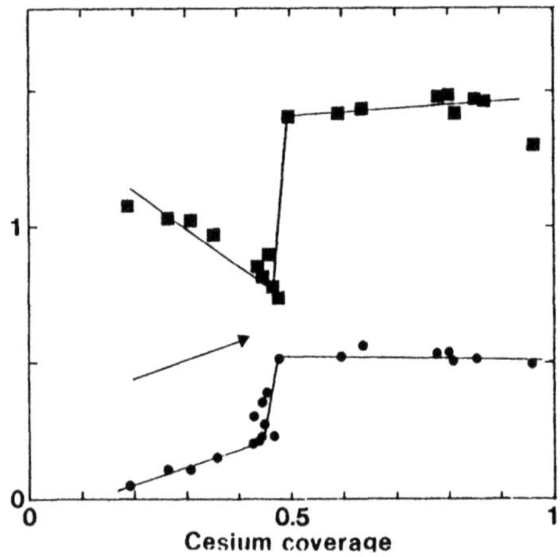

Fig.5 Cooling down of a stage one GIC after heating some hours at 700 K : the lowest curve is the intercalation concentration and the upper curve the ratio of the two Auger peaks of the cesium (see the caption of Fig.4)

When the sample is cooled down to room temperature, both surface segregation and intercalation are observed (Fig.5).

The variation of the intercalation stage versus the cesium surface content is the same as in the case of the heating of a low stage GIC presented in part 4. The lowest stage CsC_6 is not reached in this segregation process, probably because the chemical potential of the cesium is not sufficient.

This observation is quite surprising because it could mean that near the surface of a high stage GIC, due to surface segregation, the material would always be a stage-one GIC in the first atomic layers. This is not true in practice : as we showed in part 3, a high-stage GIC synthetised from a graphite sample is stable and presents no surface segregation. The segregation is observed only if the sample is a stage-one GIC before heating. This means that heating at 700 K is not sufficient to reconstruct the initial graphite structure of the sample ; in the remaining spaces between the graphite layers, the surface segregation is thus possible. This is indeed confirmed by the difference of shape between the carbon Auger peaks of a new graphite sample and of a GIC purified from the intercalated Cs by heating.

REFERENCES

1 - P. OELHAFEN, P. PFLUGER, E. HAUSER and H.J. GUNTHERODT Phys. Rev. Lett 44 (1980) 197

2 - G. VOLPILHAC, L. DUCASSE, F. ACHARD and J. HOARAU Phys. Rev. B 23 (1981) 4236

3 - S.A. SAFRAN and D.R. HAMANN Phys. Rev. B 23 (1981) 565

4 - R. CLARKE, N. CASWELL and S.A. SOLIN Phys. Rev. Lett 42 (1979) 61

Irradiation Induced Defects in Layered Dichalcogenides

H. Mutka, L. Zuppiroli

Section d'Etude des Solides Irradiés
Centre d'Etudes Nucléaires de Fontenay-aux-Roses, F-92260, France

P. Molinié

Laboratoire de Chimie des Solides, Université de Nantes, France

1. Introduction

1.1 Why Study Defects in Layered Dichalcogenides

The layered dichalcogenides of group Vb transition metals are well known not only as suitable intercalation materials but also for the charge density waves (CDW). This instability of the electron-phonon system commonly distorts their crystal structure and changes their electronic properties |1,2|. Ranging from hardly visible inflexions on conductivity curves to metal-insulator transitions, the manifestations of the CDWs have been for years a fruitful subject of research |3|.

It is easy to imagine that the response to a defect potential is strong in a material whose structure is unstable with respect to the charge perturbation associated with phonons. In fact it has been shown theoretically that in the CDW materials the screening of a charged defect is not very effective : the Friedel oscillations known from the normal metals have a long range and far from the defects they are indistinguishable from the inherent CDW |2,4,5|.

This intimate coupling between the defects and the CDW leads to difficulties in separating the intrinsic properties of the materials from those due to defects that are always present. One can even ask if a point defect in the classical sense really exists as a local perturbation in a CDW metal. The charge oscillations characteristic of the pure material and those pinned to defects are forced to search for a compromise in finding the energetically most favourable configuration. The ideal CDW distortion becomes disordered in a large volume, the sharp phase transitions give place to gradual temperature dependence of the amplitude of the CDW clouds pinned to defects. The image of a defect as a punctual scatterer does not necessarily work.

Our results give examples of the extent to which defects modify the physical properties thus helping to understand the intrinsic character of these materials. They provide also phenomenology for a clearer picture of defects in the CDW materials. For these studies irradiation is a unique tool.

1.2 Production of Defects by Irradiation

Irradiation with energetic particles is the most convenient method for producing point defects. The fast electrons (~ 1MeV) from a Van de Graaff accelerator that we have mainly used in this work lose their energy by electronic excitations and elastic collisions with atoms. The latter mechanism is responsible for the production of permanent defects by displacing atoms |6|.

The greatest advantages of irradiation are that it permits studies in situ and that at low irradiation doses the number of defects is always proportional to irradiation dose. Thus it has been possible for us to follow, e.g. the resistivity and the Hall effect as a function of defect concentration up to 10^{-2} atomic fraction and to compare accurately different concentrations at the level of 10^{-5} atomic fraction. This can be done at low temperatures (liquid He or H_2) to assure that the defects are frozen in random positions. Subsequently, it is possible to study the defect reactions induced by thermal excitation.

To know the absolute value of the defect concentration it is necessary to study the energies needed for displacing atoms. We have observed that in 1T-TaS_2 |7| and in a related material, monoclinic TaS_3 |8|, the damage production begins when more than the threshold energy of about 16 eV can be transferred in a electron-Ta collision. It was also found that the damage produced in the S-sublattice recovered completely around 10 K in 1T-TaS_2. This leads to the conclusion that the defects stable around room temperature are associated with the displacements of Ta atoms. The above-introduced threshold energy describes the strength of the atomic interactions of the displacement process. As the electron-atom collision follows well the MOTT-RUTHERFORD formalism |9|, the threshold energy is the only parameter needed for the calculation of the displacement cross-section |10|. When the displaced atoms have enough energy to induce secondary displacements they can be taken into account by using calculations of displacement cascades. These have been recently extended by LESUEUR |11| to polyatomic targets that we study.

We assume that value of the threshold energy of Ta measured in 1T-TaS_2 and TaS_3 is representative for the dichalcogenides. We use it for the calculation of the fraction of displaced metal atoms (e.g. dp Ta, displacements per tantalum atom) which we believe accurate within a factor of two. The number of dpa serves us as an estimate of defect concentration.

1.3 Structural Properties of Defects

At the moment we hold the point of view that the most effective defects are those created by displacing metal atoms. Next thing to ask is what is the crystallography of these defects. High-resolution electron microscopy on 1T-TaS_2 did not reveal any large (> 10 Å) clusters in the irradiated samples, so many of the displaced atoms can even remain as single intenstitials and the metal layers can have isolated vacancies.

Crystal parameters measurements on 1T-TaS_2 and 1T-VSe_2 give a closer view on the defect crystallography. Fig. 1 shows that in both of these materials the interlayer c-parameter decreases due to irradiation, even strongly enough to induce a lattice contraction in spite of the increase of the a-parameter. These results are compared to the non-stoichiometric 1T-$V_{1+x}Se_2$ |12| in Fig. 1. The excess vanadium believed to occupy the interlayer sites does not contract the unit cell even though the c-parameter decreases.

This difference might be attributed to the fact that the irradiated material contains both interlayer interstitials and vacancies in the layers. If this is true it is possible to calculate the relaxation volumes for both types of defects. We get $\frac{\Delta V}{V} = 2\frac{\Delta a}{a} + \frac{\Delta c}{c} = -0.5$ for a vacancy and $\frac{\Delta V}{V} = 0.15$ for an interstitial. Both defects decrease the c-parameter thus making the material more isotropic, like in the case of 1T-TaS_2 |13|. The 'intrinsic intercalation'

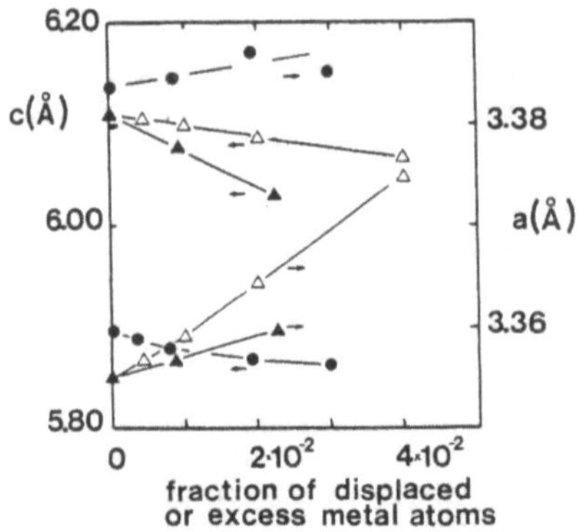

Fig.1 - The crystal parameters of irradiated 1T-TaS_2 (●), 1T-VSe_2 (▲) and non stoichiometric 1T-$V_{1+x}Se_2$ (Δ) |12| as a function of the fraction of displaced or excess metal atoms.

of metal atoms is a plausible hypothesis for the defect crystallography. It has been also called upon to explain magnetic properties to which we will return in Section 2.3.

2. The Effects of the Defects

2.1 The CDW in the Disordered Material

We have used electron diffraction observation on irradiated 1T-TaS_2 to follow the lattice distortion at different defect concentrations for temperatures ranging from 100 K to 400 K. These studies have been completed by 'in situ' observation in a 1 MV electron microscope at the temperature of 7 K. We will give here some main points of the results which will be reported in detail elsewhere |14|.

An example of the pinning effect of the defects on the CDW is given in Fig. 2 where the low temperature (100 K) distortion of 1T-TaS_2 is shown at

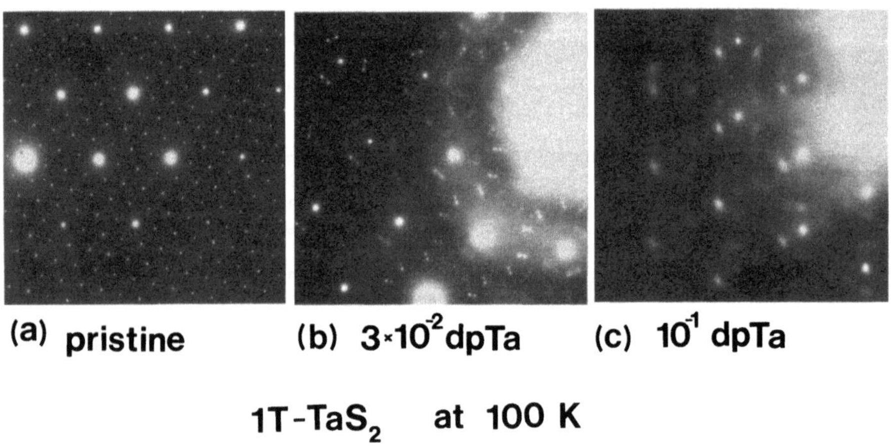

Fig. 2 - Electron diffraction patterns of 1T-TaS_2 at different defect concentrations. The commensurate superstructure of (a) is suppressed already at 10^{-3} dpTa.

different defect concentrations. The commensurate superstructure of the pure material, Fig. 2(a), is suppressed already at 10^{-3} dp Ta. At 3.10^{-2} dp Ta we have observed that three different distortions are superposed at low temperatures Fig. 2(b). This is due to the CDW breaking into domains where the defects impose either the incommensurate or one of the two possible nearly commensurate variants |15| of the distortion. Finally at the high dose of 10^{-1} dpTa a very incoherent distortion is observed, Fig. 2(c). At this dose the diffraction pattern is similar between 100 K and 400 K, the configuration of the distortion pinned to the defects has become static.

2.2 Electronic Transport

Conductivity curves of Fig. 3 give an impression of the capability of defects to smear out the CDW phase transitions. Up to now our most detailed study has been made on 1T-TaS_2 in which the conductivity, the Hall effect and the CDW distortion were followed to get a clearer image of the intrinsic properties of the material. In 1T-TaS_2 the defects drive an insulation metal-insulation transition sequence which can be interpreted as a change from MOTT to ANDERSON localization |16|.

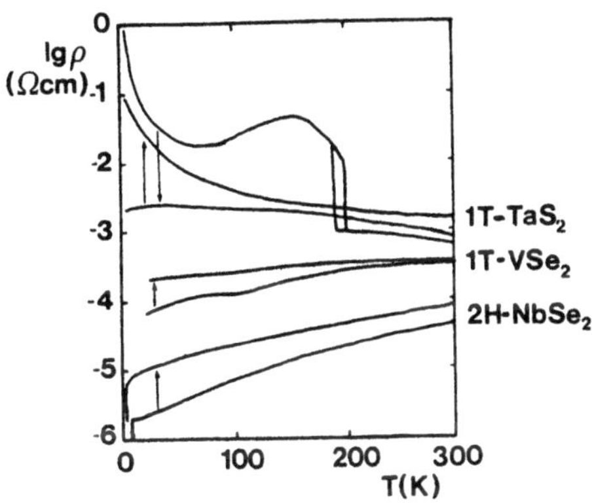

Fig.3 - The resistivity of pure and irradiated 1T-TaS_2, 1T-VSe_2 and 2H-$NbSe_2$. The arrows show the variations induced by a few percent of displaced metal atoms.

Less striking effects are observed in the materials with weaker CDW distortions. For example in 2H-$NbSe_2$, which is a good metal with a high conductivity and a superconducting transition at 7 K, the defects obey the Matthiessen's rule. A temperature independent contribution due to defects adds to the intrinsic resistivity. This is true also for 2H-TaS_2 |17|. Between 1T-TaS_2 and 2H-$NbSe_2$ the 1T-VSe_2 shows an intermediate behaviour : the defects add three times more to the resistivity at 20 K than at 300 K.

2.3 Magnetic Susceptibility

The susceptibility curves of Fig. 4 show the effects of disorder in 1T-VSe_2 and 2H-TaS_2. Here, too it is easy to observe the effect of the defects on the CDW anomalies. What is more interesting is that in irradiated 1T-VSe_2 the low temperature rise of the susceptibility is proportional to the defect concentration. This confirms the role of the interlayer V atoms as carriers of localized moments |12, 18, 19, 20|. In 1T-TaS_2, analogous defects have been invoked to explain a sample dependent susceptibility rise at low temperatures |21|. A strongly enhanced paramagnetism was also observed in

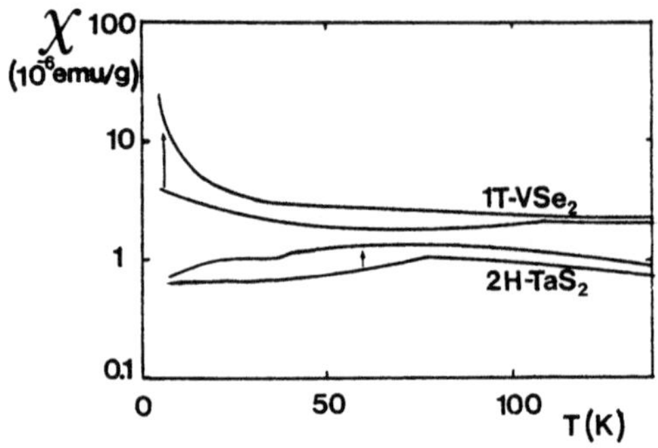

Fig. 4
Magnetic susceptibility of pure and irradiated 1T-VSe_2 and 2H-TaS_2.

irradiated 1T-TaS_2, but the effect was not stable and further experiments are carried out to clarify the role of defects in the magnetic properties of 1T-TaS_2.

3. Conclusion

The irradiation is shown to be a practical tool for studying the structural perfection of the layered dichalcogenides. The strong effect of the defects on the coherence of the charge density waves and the anomalous transport properties of the defects with respect to those typical of normal metals or semiconductors are well demonstrated in our study.

REFERENCES

1 J.A. Wilson, F.J. Di Salvo, S. Mahajan: Adv. Phys. *18*, 193 (1969)
2 R.H. Friend, D. Jerome: J. Phys. C *12*, 1441 (1979)
3 See e.g. Proceedings of the International Conf. on Layered Materials. Physica *99B* (1980)
4 W.L. Mc Millan: Phys. Rev. B *12*, 1187 (1975)
5 T. Tsuzuki, K. Sasaki: Solid State Commun. *33*, 1063 (1980), *34*, 219 (1980)
6 J.W. Corbett, J.C. Bourgoin in: Point Defects in Solids eds. J.H. Crawford, L.M. Slifkin (Plenum N.Y., 1975)
7 H. Mutka: Thesis, University of Paris, Orsay (1980)
8 G. Mihaly, N. Housseau, H. Mutka, L. Zuppiroli, J. Pelissier, P. Gressier, A. Meerschaut, J. Rouxel: C.R. Acad. Sc. Paris, *292*, 875 (1981)
9 N.F. Mott, H.S.W. Massey: The Theory of Atomic Collisions. (Oxford University Press, 1949)
10 L. Zuppiroli in: Defauts Ponctuels dans les Solides (Editions de Physique, 1978)
11 D. Lesueur: To be published in Phil. Mag. (1981)
12 L.F. Schneemeyer: Thesis, Cornell University, 1978
13 H. Mutka, P. Molinie: Solid State Commun. *33*, 1083 (1980)
14 H. Mutka, N. Housseau, L. Zuppiroli, J. Pelissier: To be published
15 J. Van Landuyt, G. Van Tendeloo, S. Amelinckx: Phys. Stat. Sol. (a) *36*, 757 (1976)
16 H. Mutka, L. Zuppiroli, P. Molinie, J.C. Bourgoin: Phys. Rev. B *23*, 5030 (1981)
17 H. Mutka, D. Lesueur, L. Zuppiroli: Rad. Effects *45*, 219 (1980)
18 C.F. Van Bruggen, C. Haas: Solid State Commun. *20*, 251 (1976)
19 A.H. Thompson, B.G. Silbernagel: Phys. Rev. B *19*, 3420 (1979)
20 F.J. Di Salvo, J.V. Waszczak: Phys. Rev. B *23*, 457 (1981)
21 F.J. Di Salvo, J.V. Waszczak: Phys. Rev. B *22*, 4241 (1980)

The Role of Kinetic Effects in Voltammetry Studies of Intercalation Systems

J.R. Dahn and R.R. Haering

Department of Physics, University of British Columbia
Vancouver, B.C., Canada

1. Introduction

The techniques of electrochemical potential spectroscopy [1,2] and linear sweep voltammetry [3,4] have recently been shown to be of value in studying electrochemical cells based on the intercalation of lithium in transition metal dichalcogenides. Fine structure in the voltage (V) versus charge (Q) relation is best seen when the dependence of -dQ/dV on V or Q is examined. It has been shown [4,5] that such structure can be caused by phase transitions, including those arising from lithium ordering. Thus linear sweep voltammetry can in principle be used to yield the phase diagram of an intercalation system such as Li_xMX_2. Earlier studies of such phase systems [6,7,8] usually employed the techniques of X-ray or neutron diffraction.

In practice, the quantitative interpretation of linear sweep voltammograms in terms of phase transitions is complicated by kinetic effects which disguise the features to be expected from the thermodynamic aspects of the Li_xMX_2 system. In this paper we examine these kinetic effects in detail and we show that a quantitative analysis is possible in spite of their presence. We then apply our results to the Li_xVS_2 system and show that the phase diagram of this system, as determined by voltammetry, is in good agreement with results obtained by X-ray diffraction.

In linear sweep voltammetry the cell voltage is swept linearly with time and the cell current is recorded. It is easily seen that in the absence of any internal impedances or overvoltages the current, I, measured under conditions of constant sweep rate, $\dot{V}$, is directly proportional to dQ/dV, the thermodynamic quantity of interest, since

$$I = \dot{Q} = \dot{V}\,(dQ/dV). \qquad (1)$$

Consider now an idealized V(Q) behaviour, $V(Q)=V_o$, $Q<Q_o$, which represents a system which exists as a two phase mixture between Q=0 and $Q=Q_o$. Equation (1) predicts that for this case the voltammogram (i.e. I as a function of V) consists of a delta function current spike at $V=V_o$,

$$I(V) = -Q_o\,\dot{V}\,\delta(V-V_o) \qquad (2)$$

where δ denotes the Dirac delta function. This prediction is drastically modified when kinetic effects are included.

2. Kinetic Effects

When an electrochemical cell is charged or discharged, the measured cell voltage, V_c, is not equal to the equilibrium voltage V(Q). This is due to the presence of various overvoltages associated with the finite ionic conductivity of the electrolyte, the finite electronic conductivity of the electrodes, ion transfer across the electrode-electrolyte interfaces and - in the case of intercalation electrodes - solid state diffusion of the intercalate in the host [4,9 10]. In order to discuss the role which kinetic effects play in voltammetry studies of intercalation systems, we shall model a typical intercalation cell (terminal voltage V_c) by means of an ideal cell V(Q) and a series impedance R which represents all internal losses other than those associated with the diffusion of the intercalating atoms in the host electrode material. The latter cannot be modelled even approximately by means of a series impedance and are taken into account explicitly in our calculation.

We confine our attention to the case of systems which display one or more phase transitions over the allowable range of variation of the charge, Q, because it is in this case that voltammetry can be of greatest value in clarifying the nature of the intercalation system. The simpler case of a single phase system has been considered previously [4,9].

In the case of a two phase system, the intercalation is characterized by a phase boundary which separates two regions of different concentration of intercalant and which moves through the host lattice as intercalation proceeds. The motion of this boundary is governed by solid state diffusion. For the moment we assume that diffusion is very rapid and that diffusion overvoltages are negligible. This limit is seldom of practical interest even at very low sweep rates. However it is instructive to examine this simple limiting case first.

When all internal losses are modelled by a series impedance R and the cell voltage $V_c(t)$ is swept linearly with time, we have

$$V_c(t) = V_c(0) \pm \alpha t \tag{3}$$

and

$$V(t) - (dQ/dV)\, R\, dV/dt = V_c(t). \tag{4}$$

In (3) α is the constant sweep rate and the + and - signs apply to charge and discharge, respectively. dQ/dV is the quantity to be determined. For the simple model two phase system discussed above, we clearly have

$$dQ/dV = -Q_o\, \delta(V-V_o). \tag{5}$$

Using (3) and (5), (4) is easily solved graphically. The solution for the current as a function of $V_c(t)$ and Q(t) is shown in fig.1. It can be seen

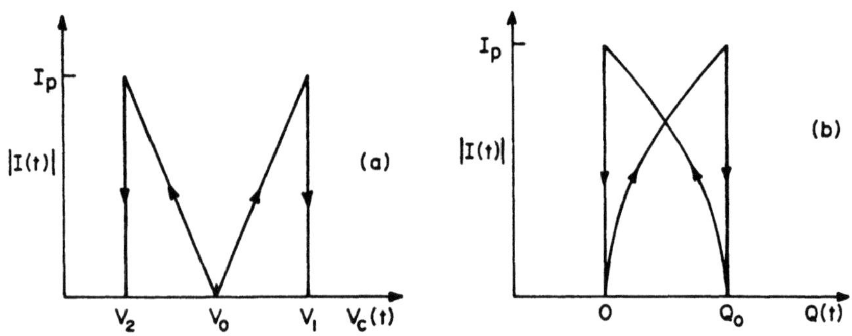

Fig.1. Cyclic voltammograms based on (3),(4) and (5); (a) current versus voltage, (b) current versus charge (see text).

that the current spikes expected in the absence of kinetic effects (2) become saw-toothed-shaped peaks when a series impedance R is included. The dependence on sweep rate of the positions, heights and widths of these peaks is of interest. Referring to fig.1 it is easily shown that

$$V_1 - V_o = V_o - V_2 = (2\alpha Q_o R)^{\frac{1}{2}} \tag{6}$$

and

$$I_p = (2\alpha Q_o/R)^{\frac{1}{2}} . \tag{7}$$

We note that the current peak height, I_p, and the peak width (in voltage) are both proportional to the square root of the sweep rate, $\alpha^{\frac{1}{2}}$. This fact is of considerable help in interpreting complex voltammograms since features in the current associated with single phase intercalation normally are proportional to the sweep rate [4]. We note further that the equilibrium voltage V_o which characterizes the two coexisting phases, coincides with the position of the "notch" between the two peaks in the voltammogram shown in fig.1.

When solid state diffusion in the intercalation electrode is considered, the simple results given in fig.1 are modified. As the intercalation proceeds the boundary which separates the two coexisting phases moves through the host lattice. We denote the concentrations of the intercalant in phases 1 and 2 by ρ_1 and ρ_2,respectively ($\rho_1<\rho_2$). The boundary is assumed to be sharp and its position at time t is denoted by $\tilde{r}(t)$. We consider three geometrical cases for the shape of the average cathode particle, identified by a "dimensionality" parameter γ: $\gamma = 1$, the infinite slab of thickness 2L; $\gamma = 2$, the infinite cylinder of radius L; and $\gamma = 3$, the sphere of radius L. We define $r = 0$ at the middle of the slab, the axis of the cylinder and the center of the sphere, so that in all cases $r = L$ at the surface. We assume that the material is initially in equilibrium in phase 1 at $t = 0$. As intercalate is added, the material at the surface undergoes a first-order phase transition to phase 2 and then the phase 1 - 2 boundary moves inward. A concentration gradient is set up across the phase 2 region in response to the current flow. We assume that the current is small enough to use the steady state approximation [11]. As the boundary moves past any point the concentration jumps by $\Delta\rho = \rho_2 - \rho_1$. The rate of motion of the phase boundary can be determined by the steady state concentration profiles and the currents needed to sustain them. The steady state concentration profiles are [12]

$$\begin{aligned} \rho(r,t) - \rho_2 &= J_s/D\ (r - \tilde{r}) && \gamma = 1 \\ &= J_s L/D\ \ln(r/\tilde{r}) && \gamma = 2 \\ &= J_s L^2/D\ (1/\tilde{r} - 1/r) && \gamma = 3 . \end{aligned} \tag{8}$$

In these expressions, J_s is the surface number current density and D is the diffusion constant, assumed to be concentration independent, of the intercalate in the host in phase 2. The rate of motion of the boundary, $d\tilde{r}/dt$, is determined by the number of intercalated atoms needed to increase the concentration from ρ_1 to ρ_2 in the region $\tilde{r}$ to $\tilde{r} + d\tilde{r}$ (note $d\tilde{r} < 0$). We find

$$(\tilde{r}/L)^{\gamma-1}\ d\tilde{r}/dt = -J_s/\Delta\rho . \tag{9}$$

The open circuit voltage is determined by the concentration of intercalate at the surface of the cathode particles, ρ_s. We assume that $d\rho/dV$ is a constant in phase 2 near the phase transition. Using (4) we then obtain

$$\rho_s(t) - \rho_2 = d\rho/dV\ (V(t) - V_o) = at - bJ_s(t) \tag{10}$$

where

$$a = |d\rho/dV|\ \alpha \qquad \text{and} \qquad b = |d\rho/dV|\ AeR . \tag{11}$$

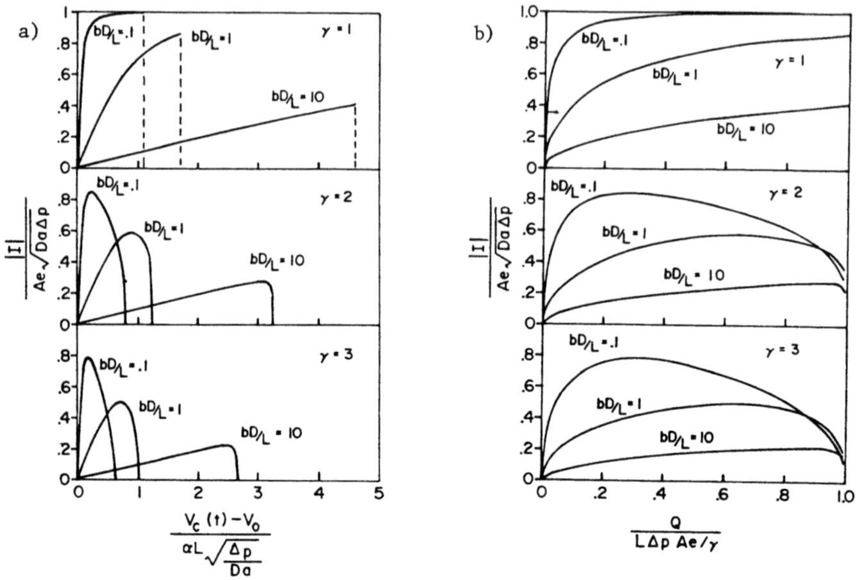

Fig. 2. Voltammograms for different values of the parameter bD/L (a) I versus V_c, (b) I versus Q.

In (11) A is the total area of the cathode particles and e is the electronic charge. We have assumed that the equilibrium cell voltage is equal to V_o when t = 0 and that a singly ionized species is involved in the intercalation. Equations (8), (9) and (10) may now be used to obtain J_s, $\int_0^t J_s\,dt$ and t as a function of $u = \tilde{r}/L$. The solutions are given in [13] and may now be used to obtain voltammograms of I versus voltage, V_c or charge, Q for different values of the dimensionless parameter bD/L. Typical curves are shown in fig.2 and should be compared with the simpler results shown in fig.1. The inclusion of diffusion has resulted in a rounding of the saw toothed peaks in fig.1. It is easily seen that the parameter bD/L can be expressed as the ratio of resistive to diffusive time constants. That is, $bD/L = \gamma RC/\tau$, where $C = dQ/dV = eAL|d\rho/dV|$, and $\tau = L^2/D$.

As long as R is finite, the current, I, will always initially increase linearly from zero, regardless of the value of γ or D. Thus the rise of the current peaks is controlled by the series resistance, R. This result provides one with a method for determining R. We note furthermore that the height and width of the current peaks in fig.2 still scale proportional to $\alpha^{\frac{1}{2}}$, as was the case in fig.1. The effect of diffusion is primarily to alter the fall of the current.

3. Application to Li_xVS_2

In order to demonstrate the usefulness of the previous analysis, we shall apply the results to Li_xVS_2 and show that a quantitative determination of the complex phase diagram of this system is obtained. The phase diagram of Li_xVS_2 has been reported previously by MURPHY et al. [7,14].

A linear sweep voltammogram of a Li/1M $LiClO_4$,PC/Li_xVS_2 cell taken at 23°C is shown in fig.3a. The presence of "notches" corresponding to first-order

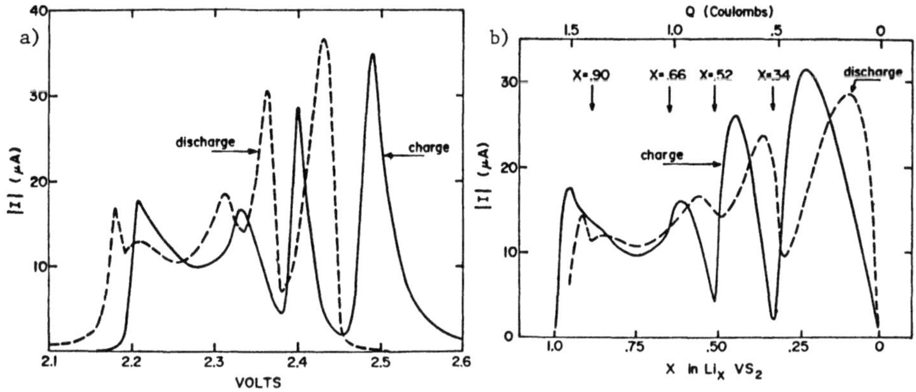

Fig.3. Linear sweep voltammogram of Li/$Li_x VS_2$ cell taken at 23°C. The sweep rate is 2.82μV/s. (a) I versus V, (b) I versus Q.

phase transitions is clearly seen at 2.380 ± .005 and 2.460 ± .005 volts. Because these transitions are close together in voltage, the linear ramp of the 2.38 volt transition lies on top of the diffusion portion of the 2.46 volt transition on discharge. Less obvious first-order transitions at 2.190 ± .005 and 2.325 ± .010 volts are superimposed on a background of single phase capacity. Another sweep taken on this cell, plotted against Q, is shown in fig.4b. The data shows that two phase regions exist between 0.00 ± .01 < x < .33 ± .01, .33 ± .01 < x < .50 ± .02 and .90 ± .02 < x < 1.00 ± .01. The range of the additional transition at 2.325 V is less well known and is given by .66 ± .02 < x < .55 ± .05.

It is of interest to compare our results (see fig.4b) with the X-ray determined phase diagram given by MURPHY et al.[7] (see fig.4a). Our data suggest that the widths of the α and β phases near x = 1/3 and x = 1/2 are somewhat narrower than observed by MURPHY et al.. A two phase region between x ≃ .55 and x ≃ .66 corresponding to a mixture of β and 1T phases is observed by us. We believe that the 3S phase exists only for x very near 1 and that the two phase region corresponding to 1T and 3S phases extends from x ≃ .90 to x ≃ 1.0 at 23°C. Apart from these differences, agreement with the phase diagram of MURPHY et al. is good.

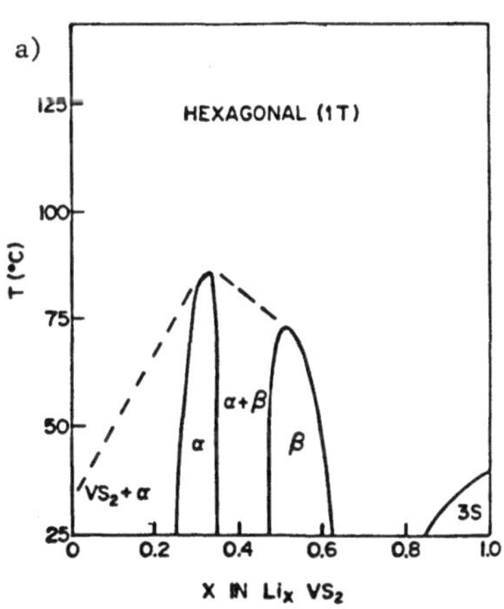

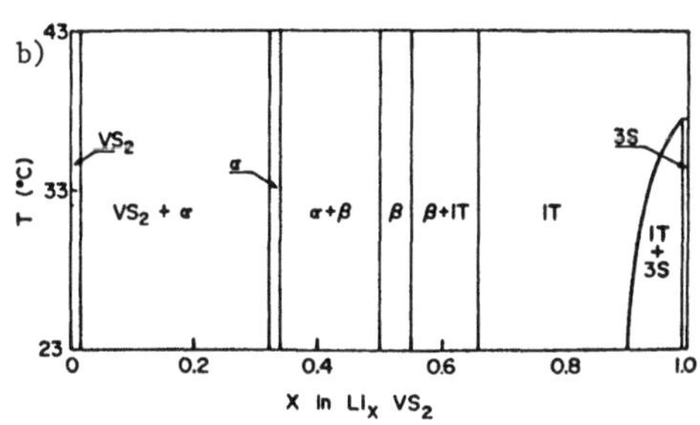

Fig.4. Phase diagram of $Li_x VS_2$ (a) after MURPHY et al. [7], (b) as determined by voltammetry using the analysis presented here.

4. Conclusions

We have presented an analysis of linear sweep voltammograms for intercalation systems. We have shown that kinetic effects play an important role in such an analysis and that these effects result in apparent hysteresis in cyclic voltammograms which is easily mistaken for thermodynamic hysteresis since it is present even at very low sweep rates. In spite of the complications associated with series impedance and diffusion, it is possible to interpret such voltammograms quantitatively. A specific application to Li_xVS_2 establishes four ranges of x in which this system is characterized by two coexisting phases. One of these phase mixtures is absent above 40^oC.

References

1. A.H. Thompson, Phys. Rev. Letters 40,1511,(1978)
2. A.H. Thompson, J. Electrochem. Soc. 126,608,(1979)
3. T. Jacobsen, K. West and S. Atlung, J. Electrochem. Soc. 126,2169,(1979)
4. W.R. McKinnon, Ph.D. Thesis, University of British Columbia (1980)
5. A.J. Berlinsky, W.G. Unruh, W.R. McKinnon and R.R. Haering, Solid State Comm. 31,135,(1979)
6. J. Rouxel, M. Danot and J. Bichon, Bull. Soc. Chim. France 11,3930,(1971)
7. D.W. Murphy, C. Cros, F.J. DiSalvo and J.V. Waszczak, Inorg. Chem. 16, 3027,(1977)
8. J.R. Dahn, W.R. McKinnon, R.R. Haering, W.J.L. Buyers and B.M. Powell, Can. J. Phys. 58,207,(1980)
9. S. Atlung, K. West and T. Jacobsen, J. Electrochem. Soc. 126,1311,(1979)
10. E. Peled, J. Electrochem. Soc. 126,2047,(1979)
11. J. Crank, Mathematics of Diffusion, Clarendon, Oxford, 1975
12. H.S. Carslaw and J.C. Jaeger, Conduction of Heat in Solids, Clarendon, Oxford, 1959
13. J.R. Dahn and R.R. Haering, Solid State Ionics 2,19,(1981)
14. D.W. Murphy, J.N. Carides, F.J. DiSalvo, C. Cros and J.V. Waszczak, Mat. Res. Bull. 12,825,(1977)

Part II

Electronic Properties

Experiments on the Electronic Structure of Graphite Intercalation Compounds

U. Gubler, J. Krieg, P. Oelhafen*, P. Pfluger** and H.-J. Güntherodt

Institut für Physik, Universität Basel
CH-4056 Basel, Switzerland

E. Cartier and F. Heinrich

Laboratorium für Festkörperphysik, Eidgen. Techn. Hochschule Zürich
CH-8093 Zürich, Switzerland

Abstract

The experimental information on the electronic structure of Graphite Intercalation Compounds (GIC's) both in reciprocal and real space is reviewed. The focus is on electron spectroscopy (photoemission and secondary electron spectroscopy) and positron annihilation. A combination of these techniques provides information about the changes of the electronic structure of graphite on intercalation. The main emphasis is on the differences found between the electronic structures of high- and low-stage donor compounds, and between stage 1 compounds of the C_8M (M = K,Rb,Cs) and the C_6Li type.

1. Introduction

This paper deals with experiments to reveal the electronic structure of Graphite Intercalation Compounds (GIC's) in terms of Density of States (DOS) and the character of electrons in the region of Fermi energy E_F. The main question to be explored is to what extent does the electronic band structure of graphite change on intercalation.

Graphite itself has 4 valence elctrons per C-atom, three of which (2s, $2p_x$, $2p_y$) form the covalent in-plane σ-bonds. The $2p_z$ electrons give rise to the well-known π-states dominating the neighbourhood of E_F. In two-dimensional graphite, the π-valence band states are just completely occupied, while the antibonding π*-conduction band states are completely unfilled. In a three-dimensional picture, graphite is a semimetal with nonzero DOS at E_F. The graphite planes produce a small overlap of valence and conduction bands.

In a simple picture donor intercalants raise E_F, while acceptor species lower it. However, more details than just the shift of E_F on intercalation are required in order to establish the electronic structure of GIC's. These details are: the total and local DOS of valence and conduction band, the DOS at E_F, the character of electrons near E_F, contributions from backfolded states due to Brillouin zone changes and from the intercalant. In addition to this problems in reciprocal space, the distribution of charges in real space are of interest.

present addresses: *IBM, Thomas J.Watson Research Center, Yorktown Heights 10598, USA
**IBM, Research Laboratory, San José, USA

It turns out that several of these questions can be answered quite successfully by studying electron spectroscopy in terms of photoemission (LEPS: Low Energy Photoelectron Spectroscopy; UPS: Ultraviolet Photoelectron Spectroscopy; XPS: X-ray Photoelectron Spectroscopy), Secondary Electron Spectroscopy (SES) and Positron annihilation experiments. The results of these experimental studies are compared with recent theoretical predictions or calculations. The main emphasis is on the Alkali GIC's (AGIC's).

2. Electronic Structure of AGIC's studied by UPS and XPS

Energy band structures of these GIC's were first published by Inoshita et al. [1] from a tight binding calculation of C_8K and by Holzwarth et al. [2] from a KKR Muffin-tin calculation for C_6Li. The energy bands of these GIC's in the region of E_F could essentially be described by a superposition of graphite π- and alkali-like s-bands. For C_6Li, the bottom of the 2s band was predicted to be well above E_F, while the lower part of the K 4s band already comes closer to E_F in C_8K. However, several experiments [3] indicate a partial charge transfer from the alkali to graphite in the GIC's of the heavier alkali metals, which could shift the bottom of the alkali-like conduction band below E_F. In C_6Li and C_8M (M: K, Rb, Cs), formation of different superlatices on the graphite matrix in real space leads to correspondingly reduced Brillouin zones into which the graphitic bands are backfolded. In C_6Li, the crossed π-bands from the zone edge of pristine graphite are folded into the center of the new Brillouin zone. This is not the case in C_8K, because its unit cell is different from that of C_6Li.

Experimentally, the electronic structure of AGIC's was directly investigated in a series of electron spectroscopy experiments (UPS, XPS, AES: Auger Electron Spectroscopy) by Oelhafen et al [4] on C_8M and higher stage $C_{12n}M$ compounds, while Eberhardt et al. studied the C_6Li compound by angle-resolved photoemission [5].

Figure 1 shows conduction and valence band spectra by UPS in annealed C_8K compounds and in pure graphite. The occurence of a conduction band with a high $N(E_F)$ value, the shift of E_F with respect to structures in the graphitic valence bands, and the small changes in the mainly rigid graphitic DOS structures can be seen clearly. The data obtained from C_8M compounds and from the concentration dependence of photoemission spectra in C_xCs (n= 1-7) GIC's have led to the following interpretation of the conduction band in low-stage heavy alkali metal GIC's: alkali-like s-type states are superimposed on the graphitic π^* antibonding states, giving appreciable s character to the Fermi level states in the C_8M compounds. These considerations imply that the transfer of electrons from the K 4s, Rb 5s or Cs 6s into the graphitic π^* bands is by no means complete.

In contrast to the C_8M AGIC's, photoemission by synchrotron [5] and X-ray [6] radiation have indicated a transfer of charge near unity into the graphite π^* bands for the case of C_6Li. The backfolded graphite bands were found by probing the dispersion of the energy bands [5]. However, in C_6Li an enhanced DOS near

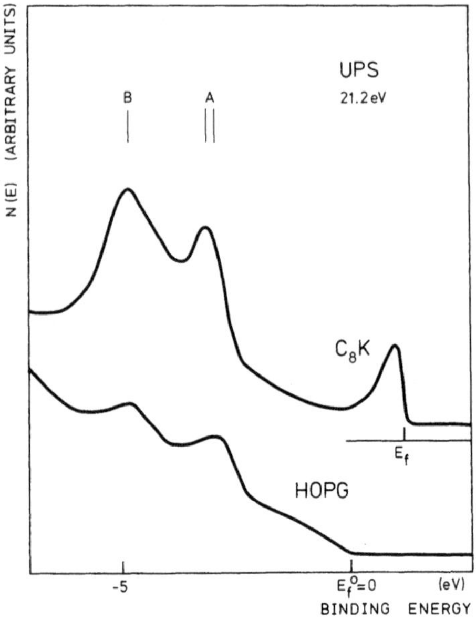

Fig.1 Valence band spectra of pure graphite and C_8K, obtained from UPS ($h\nu$=21.2 eV) measurements [18]. Structure A is due to flat π bands at the Q point and structure B is mainly due to flat σ bands at the Γ point in the energy band structure of graphite.

E_F with respect to the pure graphite π^* conduction band, expected from backfolding effects and also from experimental determinations of $N(E_F)$ [7], was not measured in the synchrotron and X-ray photoemission experiments [5,6], which is probably due to the unfavourable photoelectric cross-sections at the used excitation energies. On the other hand, the strong DOS peak belonging to the conduction band of C_6Li was found by 21.2 eV UPS (see Fig. 2), but unlike in the case of heavy alkali GIC's, an unambiguous assessment concerning the character of the Fermi level states could not be made from a series of electron spectroscopy measurements.

A conclusive picture regarding the character of the states near E_F in C_6Li and the difference in electronic structure of C_8M and C_6Li can be obtained by LEPS and positron annihilation experiments.

3. Electronic Structure of AGIC's studied by low-energy photoelectron spectroscopy

Photoemission experiments with varying incident low photon energy are able to reveal distinct differences in the stage 1 GIC's of heavy alkali metals and Li.

We have studied the conduction bands of various stages of different AGIC's by low-energy photoelectron spectroscopy (LEPS) with excitation energies $h\nu$ < 6.6 eV. A pronounced peak near E_F is observed in the energy distribution curves (EDC's) of the photoelectrons which is related to the conduction bands of these

Fig.2

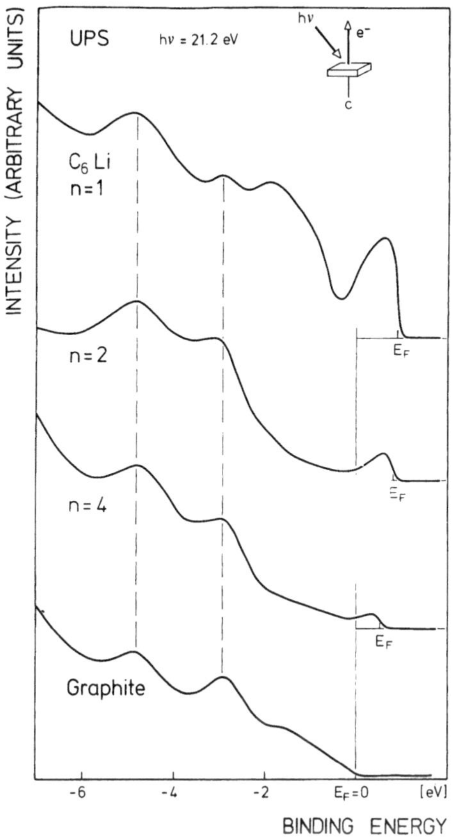

Fig.3

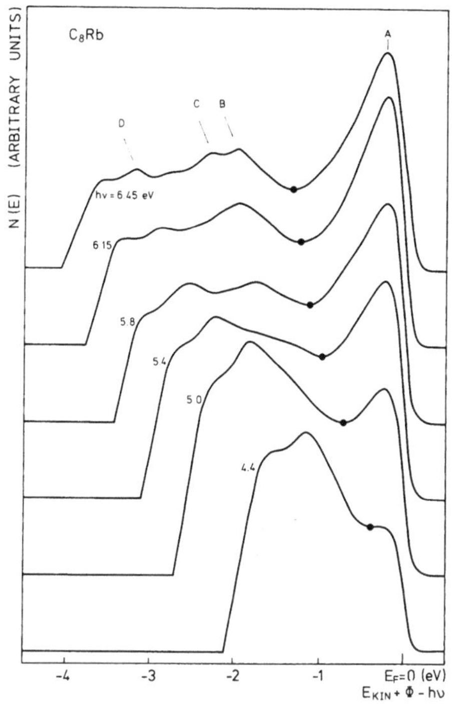

Fig.2. UPS valence band spectra of pure graphite and Li-GIC's of stages 1,2 and 4.

Fig.3. Energy distribution curves of C_8Rb measured with different photon energies hν.

compounds. The width of this peak depends characteristically on the excitation energy hν and the stage number of the AGIC's. In particularly, LEPS shows drastic differences for C_8M and C_6Li.

In Fig. 3 EDC's of C_8Rb are shown which were measured with different incident photon energies. The conduction bandwidth β is determined by the energy difference of E_F and the position of the minimum in the spectrum below E_F (marked by a dot). For stage 1 and 2 compounds with K, Rb and Cs one finds a parabolic dependence of the bandwidth β with respect to the photon energy hν. A comparison with pure alkali metals [8] shows a very similar dependence of the conduction bandwidth. This behavior has been explained by direct ($\vec{k}$-conserving) optical transitions in connection with the nearly free electron model. In this model the bandwidth β becomes a quadratic function of the excitation energy, indicating that the conduction band peak is mainly determined by s-like bands. In contrast, the bandwidth β of C_6Li (see Fig. 4) is independent of the excitation energy and behaves quite different from C_8M. Therefore, we conclude that the conduction band of C_6Li consists of non-s-like bands.

Fig. 5 shows the bandwidth β as a function of photon energy for higher stages of C_xK. A similar parabolic dependence of β as in stage 1 compounds has been found for stage 2 AGIC with K and Cs. In contrast, the stage 4 and 5 compounds $C_{48}K$ and $C_{60}K$ have a weak and linear energy dependence of β, determined by non-s-like bands. This indicates a change in character of the conduction bands taking place between stage 2 and 4. Since a rigid band behavior is expected in dilute compounds [9], we conclude from our results that the change from a partical to complete charge transfer occurs between stage 2 and 4 compounds.

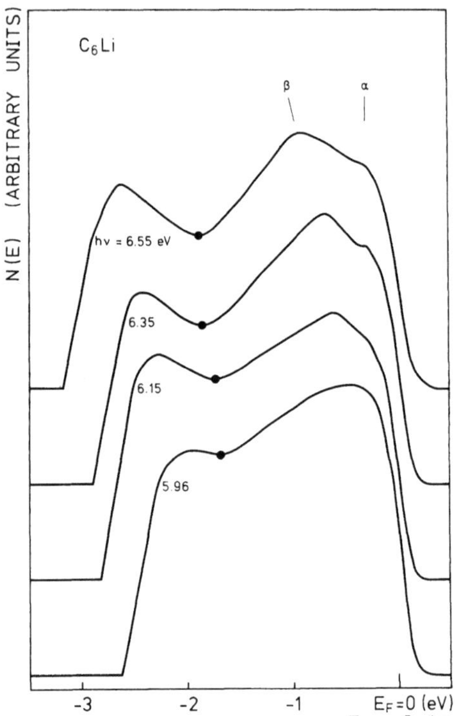

Fig. 4 Energy distribution curves of C_6Li measured with different photon energies hν

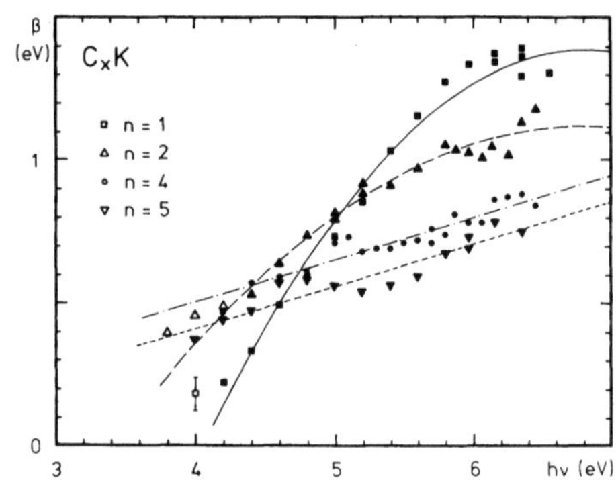

Fig. 5 Conduction bandwidth β as a function of incident photon energy hν and stage number n of C_xK

These results are in good agreement with recently performed self-consistent recalculations of the C_8K [10] and C_6Li [11] band structure. In particularly, in the calculations for C_8K the bottom of the K 4s band was found to be below E_F.

4. Electronic Structure of AGIC's studied by secondary electron spectroscopy (SES)

Secondary electron spectroscopy (SES) is a less common technique among the methods of electron spectroscopy. Nevertheless, from secondary electron emission (SEE) measurements one gets interesting information about the electronic band structure of graphite and graphite intercalation compounds, GIC's. There are fine structures superimposed on the typical secondary electron (SE) "cascade slope" of graphite, which were observed first by Willis et al. [12]. SE's are preferentially excited into final states with high density of states. As a consequence, maxima in the density of states above the Fermi level (E_F) give rise to structure in the SE-energy distribution. Thus, SES is a method to study the band structure above the vacuum level. In combination with photoemission experiments one obtains information about the changes of the electronic structure of graphite upon intercalation.

The excitation of the SE was made by an ultraviolet resonance lamp ($h\nu$ = 21.2 eV), X-rays ($h\nu$ = 1253.6 eV) or by electrons of different primary energies. The energy positions of the different structures do not depend on the excitation source.

In Fig. 6 photoelectron and SE spectra of HOPG are compared with the band structure calculated by Painter and Ellis [13]. An intense, narrow peak is observed at E = 7.6 eV in the SE energy distribution. This peak is related to a maximum in the density of states due to σ-bands with weak dispersion at the Γ^+_{3u}-point. In a derivative spectrum of the SE energy distribution, it is possible to resolve some more structures, which can also be interpreted in terms of the density of states maxima [14]. Three of the four valence electrons of graphite, the s-, p_x- and p_y-orbitals are responsible for binding in the C-layers, and they form the σ-bands.

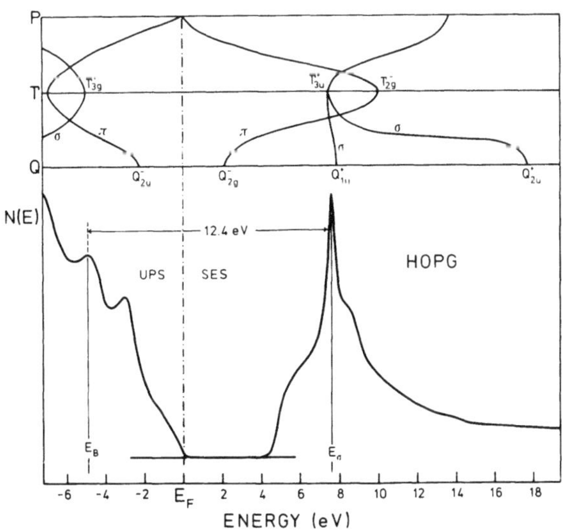

Fig. 6 UPS ($h\nu$=21.2 eV) and SES (primary energy E_p=80 eV) spectra of pure graphite. In the upper part of the Fig., a section of the graphite band structure, calculated by Painter and Ellis [13] is shown

The p_z-orbitals are related to the π-electrons, which determine the binding between the C-layers. The binding characteristics in the C-layers are changed little on intercalation. Therefore, one expects essentially an unchanged σ-band gap. As in pure graphite, the excitation of electrons into the unfilled σ-band (Γ^{+}_{3u}) causes structures in the SE spectra of alkali-metal GIC's.

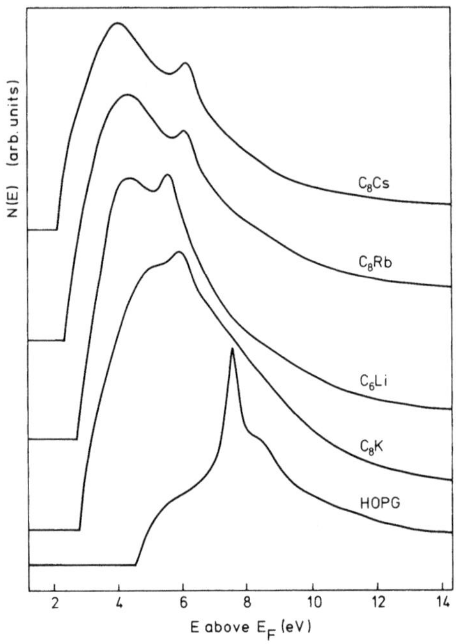

Fig. 7 Secondary electron distribution of stage 1 alkali graphite intercalation compounds

Fig. 7 shows the SE energy distributions of various stage 1 alkali-metal GIC's. When the energies of the unfilled σ-bands measured by SES are compared with the energy levels of filled σ-bands, measured by ultraviolet photoemission (UPS), it is seen that their relative position is not rigid on intercalation. The σ-band gap $E_\sigma - E_B$ in the Γ-point (Fig. 6) decreases by about 0.3-1.0 eV. Therefore, the shift in the Fermi level ΔE_F becomes a relative quantity depending on the reference point. The energies of the unoccupied σ-band and the binding energy of the $\sigma_{2,3}$-band for the various alkali-metal GIC's are listed in column one and two of table I. The σ-band gap energies ($E_\sigma E_B$) are shown in the third column. In the fourth column, n is the electron density. The changes of the σ-band gap energies are small in the K, Rb and Cs GIC's, but a shift of 1 eV is observed in the case of C_6Li. Non-rigid-band effects in the occupied bands upon intercalation were already observed by W. Eberhardt et al. [5] in a C_6Li intercalation compound.

In addition, a study of the SEE dependence on the stage number has been done on C_xK (stage 1-5,7,9). As one would expect,

Table 1. Fermi energies and non-rigid band effects in stage 1 AGIC's. The symbols are explained in Fig.6

GEV	E_F (eV)	E_D (eV)	$E_F - E_D$ (eV)	n 10^{21} El/cm^3
C_6Li	5.6 ± 0.1	-5.8	11.4	17.1
C_8K	6.0 ± 0.1	-6.0	12.0	8.8
C_8Rb	6.1 ± 0.1	-6.0	12.1	8.4
C_8Cs	6.2 ± 0.2	-5.7	11.9	8.0
HOPG	7.6 ± 0.1	-4.8	12.4	-

the Γ^+_{3u} energy is shifted to lower values with increasing alkali electron density. The potassium GIC's of higher stages show a similar behavior like the stage one alkali-metal GIC's, i.e. the σ-band gap is not constant. In stage 4 and 5 a splitting of the σ-structure into two peaks has been observed in the second derivative of the SE-energy distribution curve. (Fig.8). It appears that changes in the electronic structure occur at electron concentrations corresponding to stage 4 or 5. Also in positron annihilation experiments a drastic change of the 2γ angular correlation distribution is measured in GIC's of C_xK between n=4 and n=5 [16]. These data might be related to the existence of inequivalent C-layers.

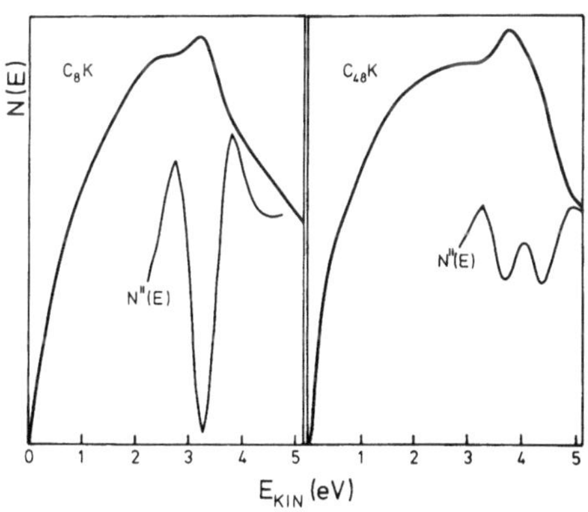

Fig. 8 Secondary electron energy distribution in the K-GIC'ь of stages 1 and 4. The characteristic feature on top of the SE energy diьtiibution is a two-peak structure in thc stage 4 compound, as can be seen from the second derivative of the SES curve

5. Positron Annihilation Studies of AGIC's

The contribution from this studies is twofold:

1. Angular correlation distribution (ACD) measurements have shown a pronounced difference between the two GIC's C_6Li and C_8K and allow a classification of GIC's into s(+π)-band metals like C_8K and π-band metals like C_6Li [16].

2. A highly inhomogeneous charge distribution has been detected by Coulomb repulsion of the positron from the intercalant [15].

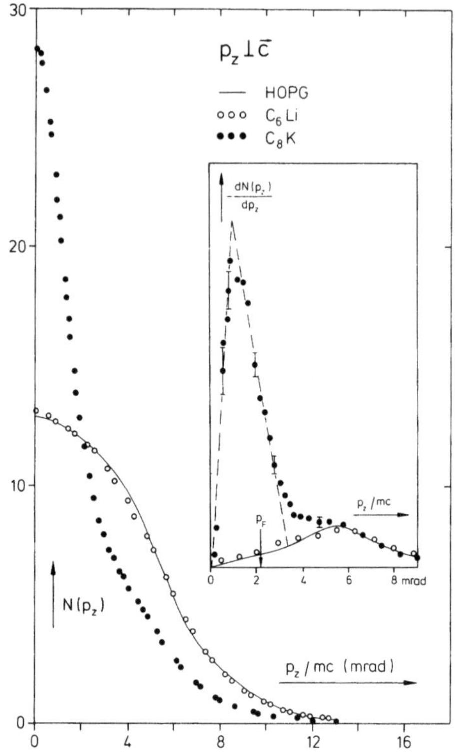

Fig.9 2γ-angular correlation distribution (ACD) of positron annihilation radiation in C_6Li, C_8K and HOPG for p_z perpendicular to the crystallographic c-axis

We present the result of 2γ-angular correlation measurements on C_8K and C_6Li compounds in two crystallographic directions (c-axis parallel and perpendicular to the pair momentum component p_z). Fig. 9 shows the ACD's of C_6Li, C_8K and HOPG for $p_z \perp c$. All spectra are normalized to equal area. In the case of C_8K, an extremely intense narrow component is superimposed on a broad graphite-like background. This distribution as well as its first derivative are very similar to those observed in pure heavy alkali metals [17]. This observation strongly supports the existence of an alkali-like free electron gas in the compound C_8K. The broad background in the ACD is due to annihilations with graphite-like π- and σ-electrons. The quantitative determination of k_F yields a value of 0.57 $Å^{-1}$ with a corresponding Fermi energy of 1.24 eV, in good agreement with photoemission data [4]. A hypothetical free electron gas of K would have a Fermi energy of 1.56 eV. The difference between this value and the experimentally determined value of 1.25 eV corresponds to a difference in charge density of 0.3 electrons per K atom.

Obviously, this amount of charge is transferred to the graphitic π-bands.

Fig. 10 shows the ACD's of C_6Li, C_8K and HOPG for $p_z||c$. Comparing the spectra for $p_z \perp c$ and $p_z||c$ in Fig. 9 and 10, the ACD's of C_8K show a significant anisotropy. Although the broad part of the ACD's can be well approximated by that of HOPG, we were unable to separate an identical free electron parabola in both orientations, as could be expected for an isotropic free electron gas. The observed anisotropy might reflect a non-isotropic character of this electron gas. In sharp contrast, the ACD's of C_6Li give no indication of a comparable free electron contribution. The distributions in both directions are similar to those of HOPG. Thus, we conclude that there are no contributions from alkali-like states in C_6Li.

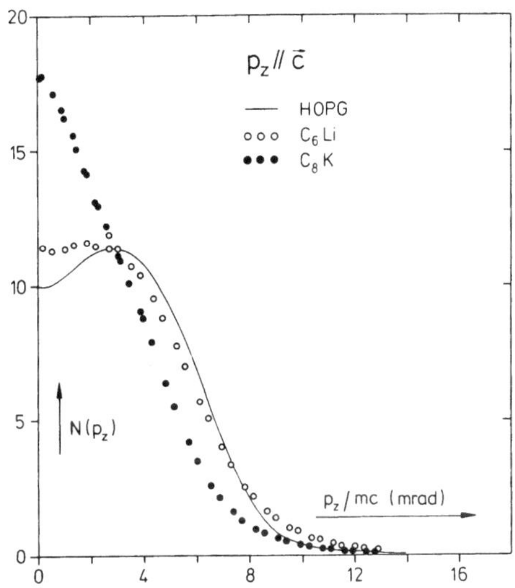

Fig. 10 ACD's of C_6Li, C_8K and HOPG for $p_z||c$

In addition, these experiments can probe the c-axis charge distributions. The first experiments [15] clearly demonstrated the long-range nature of the intercalant-layer screening. In these studies, positrons (e^+) were annihilated in K-GIC's of various stages up to n=10 (see Fig. 11). The positively charged e^+ particles are Coulomb repelled from the likewise positively charged K-layers and thus sample the center of the n graphite layers. For stages > 5, the positron finds pure graphite-like layers in the compound, as evidenced by the ACD curves identical to those of pristine graphite. While the ACD of the stage 5 compound $C_{60}K$ is still almost graphite-like, a dramatic change to a completely different behavior occurs on going to the stage 4 compound $C_{48}K$. The foregoing observations imply that next to an intercalant layer, the two C-layers on each side are strongly influenced by the intercalant layers, whereas the third C-layer behaves almost as a pure graphite layer. Adding even more C-layers by

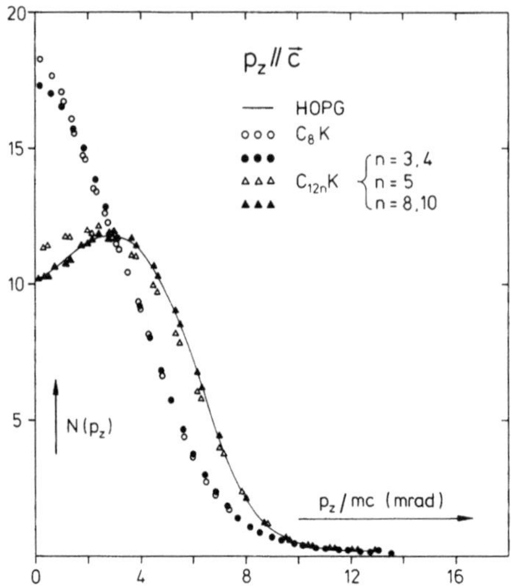

Fig. 11 ACD's in pure HOPG and K-GIC's of various stages for $p_z || c$

increasing n, a purely graphitic portion is built up in the center of the graphite-subsystem. Obviously, the K-intercalant is fully screened within a distance of ~8.5 Å from the intercalant/ graphite interface.

6. Conclusions

The experimental results on the electronic structure of AGIC's are very consistent with recently performed self-consistent recalculations of the C_8K and C_6Li band structure. There is a tendency towards considering C_6Li as a π-band metal and C_8K as a (s+π)-band metal.

Several experiments reveal a discontinuity in the electronic structure of $C_{12n}M$ GIC's as a function of stage for n = 4-5. Below that limit (s+π) character is found, whereas above stage 5, AGIC's can probably be described in a proper way by a system of charged graphite layers.

A highly inhomogeneous charge distribution is detected by Coulomb repulsion of the positrons from the intercalant.

Non rigid band effects are effective in all AGIC's of stage 1, the most important departure occuring in C_6Li.

Acknowledgement

We would like to thank V. Geiser for her help with sample preparation work and Dr. A. Moore of Union Carbide Corporation for the HOPG. Financial support of the Swiss National Science Foundation is gratefully acknowledged.

References

1 T. Inoshita, K. Nakao and H. Kamimura, J. Phys. Soc. Jap. 43, 2137 (1977)

2 N.A.W.Holzwarth, S. Rabii and L.A. Girifalco, Phys. Rev., B18, 5190 and 5208 (1978).

3 See e.g. L.E. Campbell, G.L. Moutet and G.J. Perlow, Phys. Rev. B15, 3318 (1977).

4 P. Oelhafen, P. Pfluger, E. Hauser and H.-J. Güntherodt, Phys. Rev. Lett., 44 197 (1980).
P. Oelhafen, P. Pfluger and H.-J.Güntherodt, Sol.St.Comm., 32, 885 (1979)

5 W. Eberhardt, I.T. McGovern, E.W. Plummer and J.E. Fischer, Phys. Rev. Lett., 44 200 (1980).

6 G.K. Wertheim, P.M.Th.M. van Attekum and S. Basu, Sol.St. Comm., 33 1127 (1980).

7 P. Delhaes, J.C. Rouillon, J.P. Manceau, D. Guérard and A. Hérold, J. Physique (Letters), 37 L127 (1976).
P. Delhaes, J.P. Manceau and D. Guérard, Proceedings Second Conference on Intercalation Compounds of Graphite, Provincetown, Mass. USA. Synth. Met. 2, p. 277 (1980).

8 N.V. Smith, W.E. Spicer, Phys. Rev. 188, 593 (1969).
N.V. Smith, G.B. Fisher, Phys. Rev. B3, 3662 (1971).

9 M.S. Dresselhaus, G. Dresselhaus, J.E. Fischer, Phys. Rev. B15, 3180 (1977).

10 T. Ohno, K. Nakao, H. Kamimura, J. Phys. Soc. Jap. 47, 1125 (1979).

11 T. Ohno, Physica B, to be published.

12 R.F. Willis, B. Feurbacher, B. Fitton, Phys. Rev. B4, 2441 (1971)

13 G.S. Painter, D.E. Ellis, Phys. Rev. B1, 4747 (1970).

14 J. Krieg, Ph. D. thesis, University of Basel, Switzerland, 1981
J. Krieg, to be published.

15 E. Cartier, F. Heinrich, P. Pfluger, H.-J. Güntherodt, Phys. Rev. Lett. 46, 272 (1981).

16 E. Cartier et al., Solid State Commun., in press.

17 A.T. Stewart in Positron Annihilation , Ed. A.T. Stewart and L.O. Roellig, p. 29, Academic Press, New York 1967.

18 U.M. Gubler, Ph.D. Thesis, University of Basel, Switzerland, 1981
U.M. Gubler et al., to be published.

Electronic Properties of Graphite Intercalation Compounds

Hiroshi Kamimura

Department of Physics, University of Tokyo, Bunkyo-ku
Tokyo 113, Japan

Abstract

During the past several years the graphite intercalation compounds (GIC) have aroused a great deal of interest in the field of solid state physics. In the present paper a review is given on two topics on the electronic properties of GIC about which intensive studies have been recently made. These are the electronic structures of higher stage GIC and the mechanism of superconductivity in the first stage alkali-metal GIC. In the first part of the present paper (Part I) the first principles non-empirical band structure calculation of higher stage GIC recently done by OHNO, SHIMA and KAMIMURA [1] is presented, with a particular emphasis being laid on the determination of the non-homogeneous charge distribution along the c-axis. In the second part of the present paper (Part II) the theory of superconductivity of the first stage alkali-metal GIC, C_8M, recently developed by TAKADA [2] is presented. We clarify the long-standing question on why C_8M become superconductors, although neither graphite nor the alkali metals are superconducting.

Part I. Electronic Structures of Higher Stage GIC

1.1 Introduction

Recently several band structure calculations of higher stage GIC have been reported, with the purpose of exploring the underlying mechanisms for the existence of stage order in higher GIC, that is, the c-axis superlattice of n contiguous graphite layers and one intercalant layer in the n stage compounds [3,4,5]. In these calculations, however, only the π band electrons are taken into account and the effects of the σ band electrons are ignored. In the present paper it is shown that the effects of σ bands are essentially important in determining the non-homogeneous charge distribution along the c-axis. For this purpose we present a self-consistent method to calculate the band structures of higher stage GIC within the framework of the density functional formalism [6]. Then the method is applied to the 3rd and 4th stage GIC, and the calculated results of band structures, density of states and the c-axis charge distribution are presented. Finally the effects of σ bands on the c-axis charge distribution are clarified.

1.2 Methodology

In this subsection we describe the method of calculating the energy bands of higher stage GIC self-consistently and non-empirically. The system under consideration is the n graphite layers bounded by two intercalant layers, just like a thin film. Let us denote f the charge transfer per intercalant atom. We assume that the in-plane spatial variation of the electrostatic potential due to the charged intercalant layers is neglected. Thus the intercalant layers are treated as charged sheets.

We solve the one-electron equations for higher stage GIC within the local density functional (LDF) formalism. Such one-electron equation is written in the following form in the atomic unit:

$$\{-\frac{\nabla^2}{2} - \sum_{j\alpha} \frac{Z_\alpha}{|\vec{r}-\vec{R}_{j\alpha}|} + \int \frac{\rho(\vec{r}\,')}{|\vec{r}-\vec{r}\,'|} d\vec{r}\,' + V_{xc}[\rho(\vec{r})]\}\psi_n(\vec{k},\vec{r}) = \varepsilon_n(\vec{k})\psi_n(\vec{k},\vec{r}). \quad (1)$$

Here $\vec{k}$ is a two-dimensional wave vector defined in the two-dimensional hexagonal Brillouin zone, and Z_α denotes the nuclear charge of a carbon atom situated at site $R_{j\alpha}$, where j and α represent the unit cell and the position of an atom in the unit cell, respectively. The second, third and fourth terms in (1) represent the potential between nuclei and electrons, and between electrons and the exchange-correlation potential, respectively.

The eigenfunctions $\psi_n(\vec{k},\vec{r})$ are related to the total charge density by

$$\rho(\vec{r}) = \sum_{\text{occupied } n,k}\sum \psi_n^*(\vec{k},\vec{r})\psi_n(\vec{k},\vec{r}), \quad (2)$$

which, in turn, determines self-consistently the LDF potential in (1). As regards the exchange-correlation potential, we adopt the local approximation and it is expressed as

$$V_{xc}[\rho(\vec{r})] = -3\alpha[(3/8\pi)\rho(\vec{r})]^{1/3}, \quad (3)$$

where α is taken to be 0.7 in the present paper.

In the present paper we adopt the LCAO approximation and choose the 1s, 2s and 2p orbitals of carbon atoms as the bases. We begin with the construction of the crystal potential in each step of the iterative self-consistent calculation. On doing so we express the total charge density by the superposition of a charge density at each site. Namely,

$$\rho(\vec{r}) - \sum_{j\alpha} \rho(\vec{r}-\vec{R}_{j\alpha}) = \sum_{j\alpha} [\rho_\alpha^s(|\vec{r}-\vec{R}_{j\alpha}|) + \rho_\alpha^d(\vec{r}-\vec{R}_{j\alpha})], \quad (4)$$

where the charge density has been expanded in terms of spherical harmonics and the second-order term has been neglected. The first and second terms in (4) correspond to the spherical and dipole charge densities, respectively. The crystal potential, which is refined in each step of the iterative calculation, is obtained by solving Poisson's equation for $\rho(\vec{r})$ and by using the exchange and correlation functionals with this total charge density.

The crystal wave functions $\psi_n(\vec{k},\vec{r})$ are expanded in terms of Bloch functions $\phi_{\ell\alpha}(\vec{k},\vec{r})$ in a standard form:

$$\psi_n(\vec{k},\vec{r}) = \sum_{\ell\alpha}\sum C_{n\ell\alpha}(\vec{k})\phi_{\ell\alpha}(\vec{k},\vec{r}), \quad (5)$$

where $\phi_{\ell\alpha}(\vec{k},\vec{r})$ is defined in terms of the ℓ-th basis orbital $\chi_{\ell\alpha}(r)$ situated at site $R_{j\alpha}$,

$$\phi_{\ell\alpha}(\vec{k},\vec{r}) = (1/\sqrt{N})\sum_j \exp[i\vec{k}\cdot\vec{R}_{j\alpha}]\chi_{\ell\alpha}(\vec{r}-\vec{R}_{j\alpha}). \tag{6}$$

As the LCAO basis orbitals we use the numerical atomic-like LDF orbitals, and in the present paper we take 1s, 2s and 2p orbitals as this numerical set. These are obtained by the following atomic LDF equation:

$$\{-\frac{\nabla^2}{2} - \frac{z_\alpha}{r} + \int\frac{\tilde{\rho}_\alpha(r')}{|\vec{r}-\vec{r}'|}\,d\vec{r}' + V_{xc}[\tilde{\rho}_\alpha(r)]\}\chi_{\ell\alpha}(\vec{r}) = \varepsilon_{\ell\alpha}\chi_{\ell\alpha}(\vec{r}), \tag{7}$$

where $\tilde{\rho}_\alpha(r)$ is the single site charge density obtained by averaging spherically the total charge density $\rho(\vec{r})$ in (4) around the site α.

Once the crystal potential and the basis set are obtained, the usual linear secular equation

$$\sum_{\ell'}\sum_{\alpha'}[H_{\ell\alpha,\ell'\alpha'}(\vec{k}) - S_{\ell\alpha,\ell'\alpha'}(\vec{k})\varepsilon_n(\vec{k})]C_{n\ell'\alpha'}(\vec{k}) = 0 \tag{8}$$

is solved, where $H_{\ell\alpha,\ell'\alpha'}(\vec{k})$ is the Hamiltonian matrix element defined by

$$H_{\ell\alpha,\ell'\alpha'}(\vec{k}) = \langle\phi_{\ell\alpha}(\vec{k},\vec{r})|H(\vec{r})|\phi_{\ell'\alpha'}(\vec{k},\vec{r})\rangle \tag{9}$$

and $S_{\ell\alpha,\ell'\alpha'}(\vec{k})$ the overlap matrix element defined by

$$S_{\ell\alpha,\ell'\alpha'}(\vec{k}) = \langle\phi_{\ell\alpha}(\vec{k},\vec{r})|\phi_{\ell'\alpha'}(\vec{k},\vec{r})\rangle. \tag{10}$$

In calculating the matrix elements (9) and (10), we have to carry out a large number of multi-centre integrals. For this purpose we adopt a three-dimensional Diphantine integration scheme with the numerical form of the potentials and Bloch functions, in which a set of pseudorandom integration sampling points $\{\vec{r}_i\}$ within a unit cell and an associated set of weight function $\{w(\vec{r}_i)\}$ are chosen to evaluate the values of multi-centre integrals [7,8]. For example, for the 3rd and 4th stage GIC to which OHNO et al [1] have applied the present method, they have needed about 5000 Diophantine integration points.

The solutions of the secular equation (8) provide us with the energy eigenvalues $\varepsilon_n(\vec{k})$ and the eigenfunctions $\psi_n(\vec{k},\vec{r})$, which lead to the crystal charge density in (2). For the 3rd and 4th stage GIC, the secular equation (8) has been solved at 66 points in the irreducible section of the Brillouin zone (BZ) which is obtained by dividing the hexagonal BZ into the 12 parts. With use of the Mulliken gross population analysis the crystal charge density which consists of the net atomic charge densities and the overlapping contributions is expressed as the sum of the gross atomic charge densities. By neglecting the term of quadrupole charge density, the charge density is expressed in the form (4). This iterative procedure is repeated until self-consistency is accomplished. For 3rd and 4th stage GIC in which the amount of charge transfer f is 1.0, sufficient convergence is obtained in 5 cycles of iteration for the charge density and the band structure.

1.3 Band Structures of 3rd and 4th Stage GIC with f = 1

In this subsection the theoretical results by OHNO et al.[1] are presented for the band structures, the densities of states and the distributions of charge density in each carbon layer in 3rd and 4th stage alkali-metal GIC

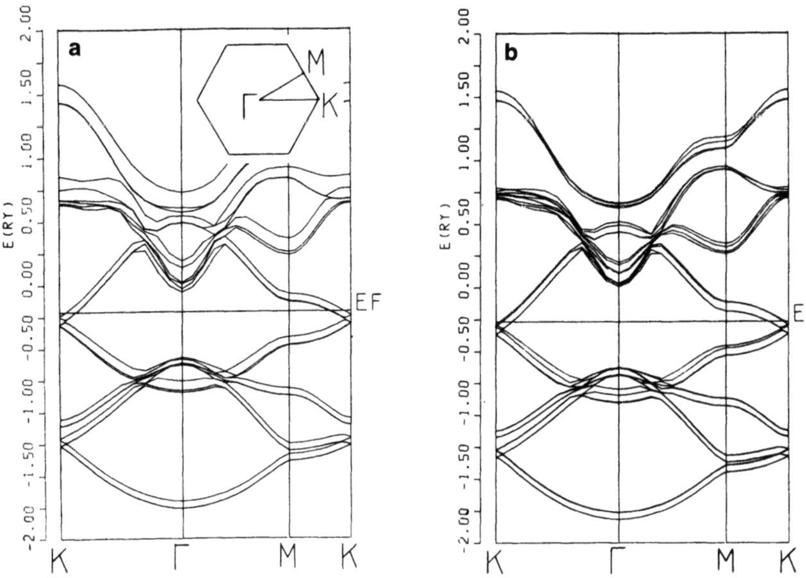

Fig.1 Band structures of 3rd and 4th stage GIC with f= 1 [1]: (a) the Brillouin zone and 3rd stage, (b) 4th stage

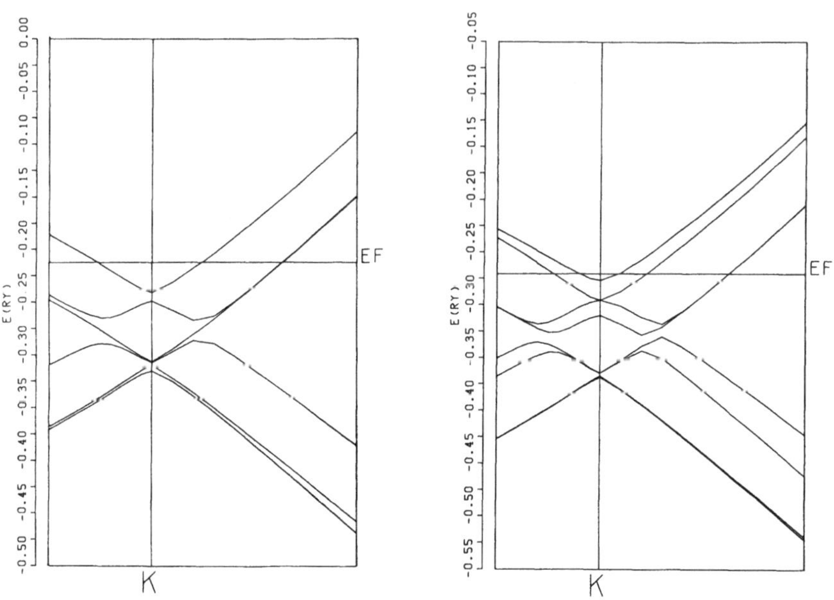

Fig.2 Expanded section of the π band structures: (a) 3rd stage GIC (left) and (b) 4th stage GIC (right) [1]

for which the charge transfer from the alkali-metal layers to the carbon layers is 100 percent, that is, $f = 1.0$.

The self-consistent non-empirical band structures of 3rd and 4th stage GIC are shown in Figs. 1(a) and (b), respectively, for various symmetry directions in the BZ. In the inset of Fig. 1(a) the two-dimensional BZ is also shown. The expanded sections of the π band structures are also shown in Figs. 2(a) and (b). The overall features of the π band structures are similar to those calculated by BLINOWSKI and RIGAUX [3], SAFRAN [9] and SHIMAMURA [5], although their band structures include phenomenological parameters. The calculated densities of states for 3rd and 4th stage GIC are shown in Figs. 3(a) and (b), respectively. The calculated density of states at the Fermi energy E_F, $D(E_F)$, are 0.062 and 0.049 states/eV·atom for 3rd and 4th stage GIC. These values are very close to the observed values of $D(E_F)$, 0.081 and 0.061, obtained from the specific heat measurement for 3rd and 4th stage alkali-metal GIC for which $f = 1.0$ [10], and by comparing the calculated $D(E_F)$ with the observed values, the dimensionless electron-phonon coupling constant λ are estimated to be 0.3 and 0.24, respectively. The former is in good agreement with the calculated value of 0.25 by KAMIMURA et al.[11].

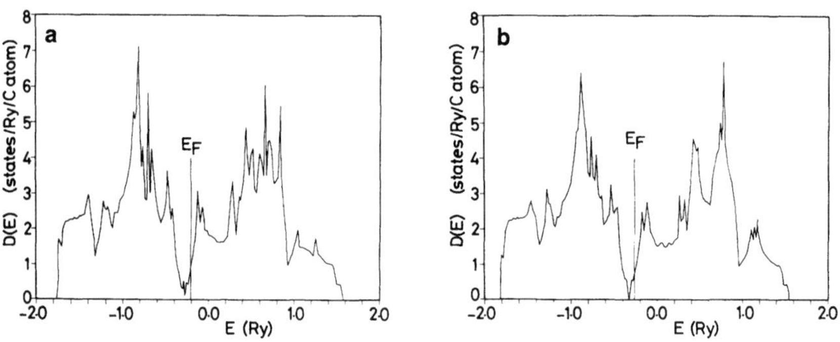

Fig.3 Density of states of 3rd and 4-th stage GIC with $f = 1$: (a) 3rd stage, (b) 4th stage. The Fermi energy E_F is determined for $f = 1$ [1]

1.4 The c-axis Charge Distributions of 3rd and 4-th Stage GIC

Now we present the calculated results for the c-axis charge distribution for 3rd and 4th stage GIC with $f = 1$ by OHNO et al. in Table 1. This table lists the fractional charge distribution for each layer, f_i for the ith layer, where $\sum f_i = f = 1.0$ and layer 1 is the graphite layer adjacent to the intercalant layer (bounding layer). The c-axis charge distribution was first calculated by PIETRONERO et al.[12] and later extended by SAFRAN and HAMANN [13] by using the semiclassical Thomas-Fermi approximation which neglects the discrete nature of the graphite layers and the hopping of electrons among the graphite layers. They pointed out the important role of screening effects.

Since then, a number of quantum-mechanical calculations for the π band electrons have been made. In Table 1 we also list some of these results as well as that by PIETRONERO et al.[12] for comparison. Namely, the results recently reported by SAFRAN and HAMANN [4] and by SHIMAMURA [5] are included. On the basis of the tight-binding approximation for the π bands, the former calculated the nonhomogeneous distribution of electrons along the c-axis by

Table 1 Calculated results of the c-axis charge distribution of 3rd and 4th stage GIC with f= 1 by various theoretical method

method	Self-consistent			Thomas-Fermi	Homogeneous
	Whole bands	π bands only			
Layer	OHNO et al [1]	SAFRAN et al [4]	SHIMAMURA [5]	PIETRONERO et al [12]	
(3rd)					
1	0.48	0.438	0.4775	0.42	0.333
2	0.04	0.124	0.045	0.16	0.333
(4th)					
1	0.494	0.425	0.472	0.40	0.25
2	0.006	0.075	0.028	0.10	0.25

adding a self-consistently determined layer-potential term to the tight-binding Hamiltonian where both the graphite and the intercalant layers were treated as charged sheets, while the latter calculated it by taking into account the polarization effect of graphite π orbitals induced by the potential due to the charged sheets of intercalant layers, that is essentially the same approximation as BRINOWSKI and RIGAUX made. Since SAFRAN and HAMANN introduced a dielectric constant ε as a parameter in order to take account of the screening effect by the graphite σ electrons, in Table 1 we refer to their result only for the case of ε= 2 for which the charge distribution along the c-axis is extremely inhomogeneous among their results. On the other hand, SHIMAMURA used the mixing ratio of the graphite anti-σ orbitals to the π orbitals as a parameter and we refer again to their result of extremely inhomogeneous case in Table 1.

By comparing with the extreme case of inhomogeneous distribution by the π band electrons, it can be concluded from Table 1 that the self-consistent non-empirical calculation for the whole energy bands of 1s, σ and π bands by OHNO et al. leads to the strongest inhomogeneous distribution along the c-axis. This means that the polarization effect of the bonding σ bands by the charged intercalant layers is as important as that of the π bands. Furthermore, their result shows that the direction of dipole moments produced by the polarization of bonding σ orbitals along the c-axis at the bounding graphite layer is opposite to that of π orbitals at the same layer. In Fig. 4 the

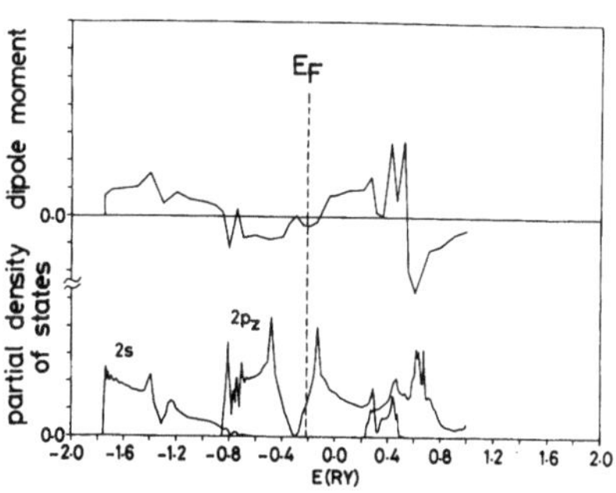

Fig.4 Partial density of states for 2s and $2p_z$ bonding and anti-bonding bands (lower) and the dipole moment induced in the 2s and $2p_z$ bands by the intercalant charged sheets in which the positive sign is taken along the direction towards the interior graphite layers (upper) [1]

partial density of states having the character of 2s and $2p_z(\pi)$ orbitals and the dipole moment at the bounding layer including its sign are shown in the case of 3rd stage. From this figure it is clearly seen that the dipole moment induced along the c-axis by the σ bands whose energy lies in energy region from about -1.8 to -0.8 Ry is directed toward the interior graphite layers, while that by the π bands existing in energy region from about -0.8 Ry to the Fermi energy E_F is directed towards the intercalant layers. The sum of the two types of dipole moment directs towards the interior layers as seen in Fig. 4, but its magnitude is not so large. Thus only the fraction of electrons transferred from the intercalant layers exist in the interior graphite layers and most of electrons are present in the bounding layers.

When only the π bands are taken into account, the dipole moment induced by the mixing of anti-σ orbitals directs towards the interior graphite layers. Because of neglect of σ bands, there exists no cancellation effect, and thus the magnitude of dipole moment in the absence of σ bands is larger than that in the presence of σ bands. As a result, in the former case more electrons exist in the interior graphite layers.

We have shown that the effect of the σ bands gives rise to large screening and as a result the fractional charges in the interior layers are very small. Therefore, if we consider explicitly the potential of intercalant atoms which have been neglected, we may expect that this will cause the significant fluctuation of charge density at carbon atoms within each graphite layer, especially in the interior graphite layers in which only the small amount of charge density exists. Consequently the long-range interaction may occur along the c-axis through the charge fluctuation induced at each carbon atom, and this may be a possible microscopic mechanism for staging in higher stage GIC.

Part II Mechanism of Superconductivity of the First Stage Alkali-Metal GIC

2.1 Introduction

One of the interesting properties of the alkali-metal GIC is the occurrence of superconductivity in the first stage compounds C_8M, because neither graphite nor alkali metals are superconductors. The observed transition temperature T_c is 80 to 550 mK for C_8K [14,15,16], 23 to 151 mK for C_8Rb [14] and 20 to 135 mK for C_8Cs [14], depending on samples rather strongly.

According to the band structure calculations of C_8K carried out by the author's group [11,17,18], there are two kinds of carriers in C_8M, that is the two-dimensional (2D) graphite-like electrons and the nearly isotropic three-dimensional (3D) alkali metal-like electrons. Thus it is very interesting to investigate which kind of carriers, 2D or 3D, contributes to superconductivity. From this point of view much attention has been given to C_8M. The first attack on this problem was initiated by INOSHITA and KAMIMURA [19,20], who calculated the electron-phonon coupling constant between 2D electrons and phonons of carbon and alkali-metal modes, neglecting 3D electrons. They found very large electron-phonon coupling constant because of the existence of two inequivalent pockets in Fermi surfaces which are located at the edges of the hexagonal BZ. For example, the dimensionless electron-phonon coupling constant λ introduced by McMILLAN [21] is 0.31 even for 2nd stage GIC. If the Coulomb interaction between electrons in two different pockets is neglected in the manner of inter-valley Coulomb interaction in the mechanism of superconductivity in multi-valley semiconductors suggested by COHEN [22], the above value of λ yields T_c of 0.23 K even for the 2nd stage GIC. However, this is inconsistent with experimental facts which show no occurrence of superconductivity for higher stage alkali-metal GIC. Thus one may speculate that the interpocket Coulomb interaction is not negligibly small because of higher concentration of electrons in C_8M than in semiconductors and that it possibly cancels the attractive interaction due to the electron-phonon interaction, preventing the 2D electrons from making a substantial contribution to superconductivity.

By this scheme, TAKADA [2] recently developed a theory of superconductivity of C_8M, considering that only the 3D electrons contribute to superconductivity. First he has derived the effective interaction between 3D electrons and developed the method of solving the gap equation in anisotropic systems from the first principles. On the basis of this method, he obtained the proper magnitude of T_c for C_8M. In the next section we would like to review his important theoretical results.

2.2 Mechanism of Superconductivity

Let us explain the microscopic mechanism of superconductivity in C_8M intuitively, according to TAKADA [2]. By the charge transfer from the alkali-metal layers to the graphite layers by the amount of f, an alkali atom M loses its valence electron to become M^+ ion. Since f is 0.5 to 0.6 for C_8M according to the self-consistent band structure calculation [18], the (1-f) electrons move freely in a crystal as the 3D electrons, and the rest of electrons (f electrons) are localized in the graphite layers. Thus each carbon atom is charged as $C^{-f/8}$ on the average. In this situation the 3D electrons can feel a large electric field of the polarization wave associated with the oscillation of $(C_8)^{-f}$ and M^+ ions when the longitudinal optic phonon is created in phase while M^+ ions vibrate out of phase with these carbon atoms. Besides this, the longitudinal acoustic phonon in which $(C_8)^{-f}$ and M^+ ions oscillate in phase can also produce an electric polarization field, because f is less than one. As a result, the 3D electrons in C_8M interact strongly both with the longitudinal acoustic and optic phonons, and this yields the strong phonon-mediated attractive interaction between 3D electrons. On the other hand, the repulsive electron-electron interaction in C_8M is much weaker than in alkali metals, because the concentration of 3D electrons in C_8M is considerably smaller than that in alkali metals. Thus the effective electron-electron interaction between 3D electrons becomes attractive and this results in superconductivity in C_8M. This mechanism of superconductivity is characteristic of C_8M. Therefore, we can clearly understand why superconductivity appears in C_8M but not in alkali metals.

2.3 Calculated Results for T_c

In this subsection the calculated values of T_c are presented for various C_8M according to TAKADA [2]. First let us consider a simple model in which the electrons in the graphite layers are assumed to be localized at the carbon sites. In this case the effective electron-electron interaction is given by the following form by neglecting the anisotropic contributions:

$$V(\vec{q},\omega) = (4\pi e^2/\varepsilon_\infty q^2)/\varepsilon(\vec{q},\omega), \tag{11}$$

where the frequency-wavevector-dependent dielectric function is given by

$$\varepsilon(\vec{q},\omega) = 1 + \frac{\Omega_p^2(f)}{\omega_{LA}^2(f)-\Omega_p^2(f)-\omega^2} + \frac{\tilde{\Omega}_p^2(f)}{\omega_{LO}^2(f)-\tilde{\Omega}_p^2(f)-\omega^2} + \frac{4\pi e^2}{\varepsilon_\infty q^2}\Pi(\vec{q},\omega). \tag{12}$$

Here $\omega_{LA}(f)$ and $\omega_{LO}(f)$ are, respectively, the frequencies of the longitudinal acoustic and optic phonons which depend on f, and $\Pi(\vec{q},\omega)$ is the electronic polarization function of 3D electrons. Further Ω_p and $\tilde{\Omega}_p$ are defined by $\sqrt{4\pi ne^2/\varepsilon_\infty(m_M+8m_C)}\cdot(1-f)$ and $\sqrt{4\pi ne^2/\mu}\cdot(\mu/m_M+f\mu/(8m_C)$, respectively, with $1/\mu$

being $1/m_M + 1/(8m_C)$, and m_M and m_C are the masses of alkali atom M and carbon atom C.

Using the values of the electronic and atomic parameters of C_8M, T_c is calculated by solving the gap equation with (11) and (12) for various C_8M. The result is shown in Fig. 5, where T_c is shown as a function of f for C_8K, C_8Rb and C_8Cs. Although the case of C_6Li is also shown in this figure, it should be noticed that a real Fermi surface of C_6Li is not three dimensional [23,24], and therefore, the present result is not applicable to a real C_6Li. Since the atomic mass increases from Li to Cs, the oscillator strengths of phonons characterized by Ω_p and $\tilde{\Omega}_p$ and thus the attractive part of effective interaction (11) become smaller, as seen from (12). Therefore, T_c decreases with increase of atomic masses, as seen in Fig. 5. According to OHNO et al. [18], f is about 0.6 for C_8M. Thus T_c of C_8M is found from Fig. 5 to be 10 to 100 mK, which is in fair agreement with observed values.

So far we have assumed that 2D electrons are localized at the carbon sites. In reality, however, 2D electrons move freely in graphite layers. Therefore, we have to consider the screening effect of 2D electrons. TAKADA [2] extended his theory to include the screening effect. The calculated T_c in this case are shown in Fig. 6 for C_8K, where ω_t and c_t are the frequency of transverse optic phonons and the velocity of transverse acoustic phonons respectively. In this figure the solid curve shows the case where both s and d wave parts are considered. It can be seen from Fig. 6 that the screening effect of 2D electrons becomes significant when f is larger than 0.6.

Finally a remark is made on the anisotropy of gap function. Since 3D electrons interact with the longitudinal polarization wave along the z-direction, only the Cooper pairs on the Fermi surface near the $k_x - k_y$ plane are mainly scattered into each other. This make the gap function anisotropic.

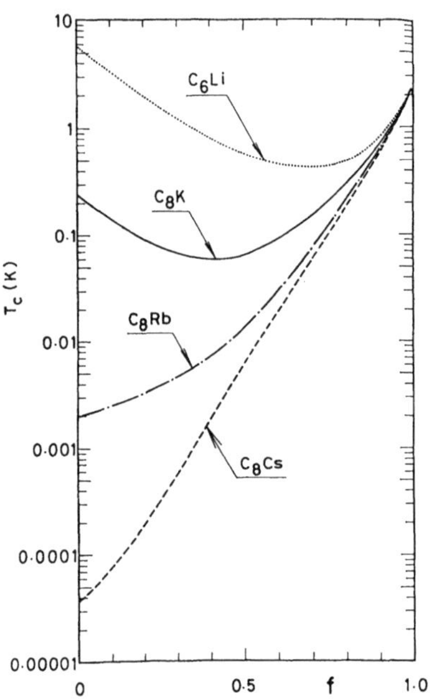

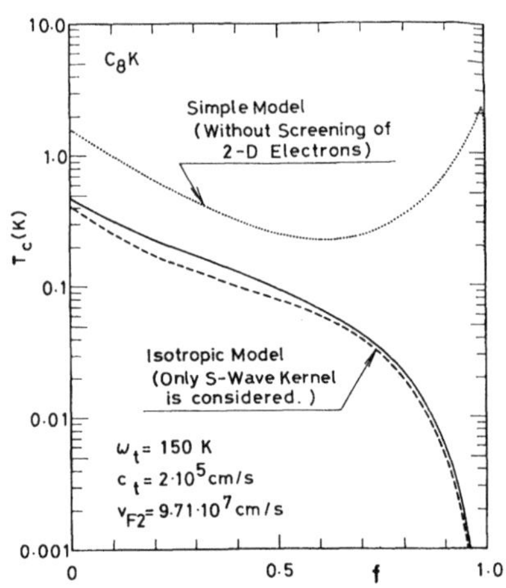

↑ Fig.6 T_c for C_8K calculated by improved and simple models [2]

← Fig.5 T_c for the first stage alkali-metal GIC [2]. Here the velocity of transverse acoustic phonon for C_8K is taken to be 3×10^5 cm/sec

Acknowledgments

It is a pleasure to acknowledge the collaboration of T. Ohno, N. Shima and Y. Takada on the work reported in this review article. I am particularly indebted to T. Ohno who reviewed the final manuscript and made a number of suggestions for its improvment. The research reported here was supported in part by Hōsō-Bunka Foundation.

References

1 T. Ohno, N. Shima and H. Kamimura: to be published.
2 Y. Takada: to be published in J. Phys. Soc. Japan.
3 J. Blinowski and C. Rigaux: J. Physique 41, 667 (1980).
4 S.A. Safran and D.R. Hamann: Phys. Rev. B23, 565 (1981).
5 S. Shimamura: PhD Thesis (Tohoku University, 1981).
6 W. Kohn and L.J. Sham: Phys. Rev. 140, A1133 (1965).
7 H. Conroy: J. Chem. Phys. 47, 5307 (1967).
8 D.E. Ellis and G.S. Painter: Phys. Rev. B2, 2887 (1970).
9 S.A. Safran: Synthetic Metals 2, 1 (1980).
10 U. Mizutani, T.B. Massalski and T. Kondow: Phys. Rev. B17, 3165 (1978).
11 H. Kamimura, K. Nakao, T. Ohno and T. Inoshita: Physica 99B, 401 (1980).
12 L. Pietronero, S. Strässler, H.R. Zeller and M.J. Rice: Phys. Rev. Lett. 41, 763 (1978); Solid State Commun. 30, 399 (1979).
13 S.A. Safran and D.R. Hamann: Phys. Rev. 22, 606 (1980).
14 N.B. Hannay, T.H. Geballe, B.T. Matthias, K. Andres, P. Schmidt and D. MacNair: Phys. Rev. Lett. 14, 225 (1965).
15 M. Kobayashi and I. Tsujikawa: J. Phys. Soc. Jpn. 46, 1945 (1979).
16 Y. Koike, H. Suematsu, K. Higuchi and S. Tanuma: Physica 99B, 503 (1980).
17 T. Inoshita, K. Nakao and H. Kamimura: J. Phys. Soc. Jpn. 43, 1237 (1977).
18 T. Ohno, K. Nakao and H. Kamimura: J. Phys. Soc. Jpn. 47, 1125 (1979).
19 T. Inoshita: PhD Thesis [University of Tokyo, 1979].
20 T. Inoshita and H. Kamimura: Synthetic Metals 3, 223 (1981).
21 W.L. McMillan: Phys. Rev. 167, 331 (1968).
22 M.L. Cohen: Phys. Rev. 134, A511 (1964).
23 N.A.W. Holzwarth, S. Rabii and L.A. Girifarco: Phys. Rev. B18, 5190 (1978).
24 T. Ohno: J. Phys. Soc. Jpn. 49, Suppl.A, 903 (1980).

Experimental Construction of Fermi Surface of $C_{24}SbCl_5$

Sei-ichi Tanuma, Otofumi Takahashi and Yasuhiro Iye

The Institute for Solid State Physics, The University of Tokyo
Roppongi, Minato-ku
Tokyo 106, Japan

1. Introduction

In order to clarify various electronic properties of intercalation compounds, the knowledge of Fermi surface (FS) of the relevant compound is basically demanded. The construction of FS of such complex layered materials as graphite intercalation compounds (GIC's) is interesting by itself. Actually one of the most intensively performed experiments on GIC's is the FS study by means of magneto-quantum oscillatory effects. However, the construction of FS from observed magneto-oscillation frequency spectra is not quite conclusive. The major difficulty is that the observed frequencies are in many cases much more numerous than expected from band models when the frequency spectra are compared to the ones expected from band models.

In cases of donor type GIC's of alkali atom intercalates, the electrical conductivity across the hexagonal plane is much enhanced by the intercalation. The enhancement is even more than the in-plane one, so that the conductivity anisotropy of σ_c/σ_a, where σ_c and σ_a are the conductivities parallel and perpendicular to the c-axis respectively, are larger than the values of pristine graphite in lower stage compounds. This fact together with the superlattice structure along the c-axis as the staging nature gives a good possibility of the construction of FS by c-axis folding of the Sloncewski-Weiss-McClure band of graphite in which the Fermi level is appropriately lifted by the electron transfer from the donor intercalants. The first trial along this way was done on $C_{48}K$ by the group of one of the present authors (S.T.) [1], giving a fair quantitative agreement. DRESSELHAUS et al. [2] developed a phenomenological theoretical model based on graphitic π-bands taking into account the c-axis folding by subtracting intercalate layers as the deplete layers among the periodic carbon layers. Their calculation was compared to the Shubnikov-de Haas oscillation spectra of $C_{60}K$ and the extreme cross sections of FS were mostly explained by adjusting the value of Fermi energy as a single parameter. In order to get good results in each of the above cases, the charge transfer rate from the intercalant atom, f, was not very close to unity; this result may suggest that some larger FS having larger mass was hidden and also may be related to such simplification [1] as the uniform charge distribution over the carbon layers.

On the other hand, σ_c/σ_a values of acceptor GIC's are much less than the value of pristine graphite, and the system of n-carbon layers sandwiched by the neighboring intercalant layers, where n is the stage number, are often regarded as a two-dimensional metal. The electronic nature of such a 2D metal assembly has been calculated by BLINOWSKI and RIGAUX [3] by assuming the interaction of carbon layers being differently charged with holes and

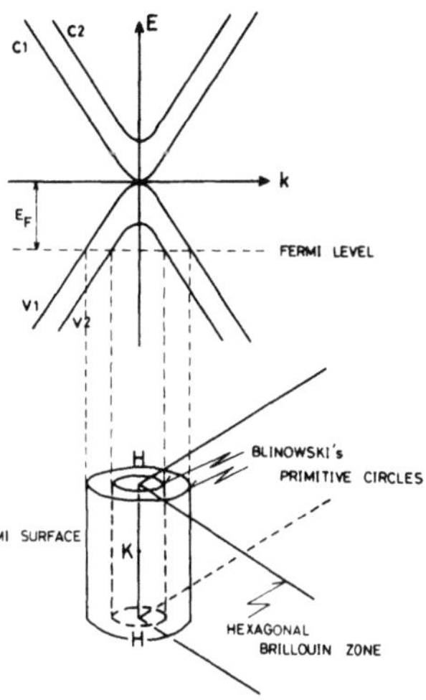

Fig.1 Energy dispersion of the Blinowski-Rigaux band model for stage 2 acceptor GIC

homogeneous distribution of electrons in the intercalated layers. The calculated band structure for the second stage acceptor compounds is shown in Fig.1 where the Fermi level is determined by the f-value of the relevant compound. The charge distribution over the carbon layers for higher stages than the second one is determined self-consistently, taking into account the screening of holes by the negatively charged intercalant layer.

The first experimental construction of FS of acceptor type GIC was done by BATALLAN, BOK et al. [4][5]. They adopted a metallic sandwich model and obtained the FS by folding the free hole circle into the Brillouin zone of the in-plane superlattice due to the intercalant sites. Later, BATALLAN et al. [6] improved their method for NO_3H and other acceptor GIC's in such a way that the Blinowski and Rigaux (BR) band was used instead of a free hole band. FISCHER et al. [7] and MARKIEWICZ et al. [8] tried to explain their de Haas-van Alphen (dHvA) experiments on second stage AsF_5-GIC by using the BR-2D band model. They eliminated the possibility of in-plane band folding because of the incommensurability of AsF_5 intercalant arrangement in the second stage compound. The present authors [9] also studied dHvA effect on AsF_5-GIC's; they tried to adopt the same model as the former 3D graphite rigid band model supplemented by the c-axis folding. As a later criticism by the authors themselves, the result is not as good as the case of alkali atom GIC's, and it seems not reasonable for acceptor GIC's that the c-axis dispersion has to be essential for interpreting all of the observed dHvA branches of frequency versus field angle.

The authors performed the dH-vA experiment on $C_{12n}SbCl_5$ (n=2,3,4) compounds [10]. Differing from AsF_5-GIC, $C_{24}SbCl_5$ (and probably higher stages too) is known to have a commensurate structure, thus the in-plane band folding could be most applicable to this compound. In the following sections, experimental construction of FS due to the in-plane band folding on this compound is described.

2. Intercalant Lattice and Brillouin Zone of $C_{24}SbCl_5$

In the recent study of the authors on $C_{24}SbCl_5$ [10], it was found by the measurements on carefully prepared and treated smaples that the dH-vA frequencies differ from one stage to the other. This fact does not follow the previous paper of BATALLAN et al. [4] which claimed the stage-independent frequency values, and obviously implies that the hole concentration is not fully localized on the carbon layer adjacent to the intercalant layer, but distributed over the n-carbon layers. The distribution is to be calculated appropriately by BR-band model. The n-th stage compound gives each n kinds of conduction and valence bands as being exemplified in Fig.1. In the second stage, no charge distribution exists because of existing only one kind of carbon layers. As Fig.1 shows, the BR-model gives at most two kinds of hole cylinders centered at the hexagonal edge of BZ for the second state compound, while the experimentally obtained number of dH-vA frequencies amounts to eight. This discrepancy could be resolved by considering the in-plane band folding and introducing new energy gaps due to the superlattice Bragg reflection as BATALLAN et al. [6] reported on NO_3H-GIC's. As far as the authors know, a well established commensurability between graphite lattice and intercalant arrangement of acceptor GIC has only been reported for $C_{24}SbCl_5$. For this case only, such FS's are obtainable by the band folding that a simple comparison between the experimental result and the band model could be possible.

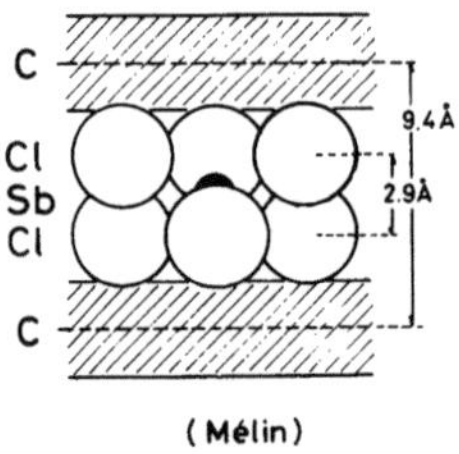

Fig.2 Molecular orientation of $C_{24}SbCl_5$

MÉLIN investigated the structure of $C_{24}SbCl_5$ by electron diffraction [11], and obtained the Cl-Sb-Cl sandwiching layer structure as shown in Fig.2. Each Cl layer makes a triangular lattice and the lattice is rotated ±19.01° from the graphite lattice making the $\sqrt{7}\times\sqrt{7}$ commensuration. The side length of the triangle is 3.27Å which is compared to 2.47Å of the graphite triangle. The unit cell contains six chlorine atoms and fourteen carbon atoms. If the cell contains one antimony atom, the stoichiometry should be $C_{14n}SbCl_6$. On the other hand, the chemically determined stoichiometry is $C_{12n}SbCl_5$ or $C_{14n}Sb_{1.2}Cl_5$, so the lattice site of antimony atom is not simple and not yet determined. As the overall chlorine lattice, there are two possibilities: one is that two Cl layers rotate 19.01° to the same direction, the other is two layers rotating to the opposite direction with each other. In the former case the overall Cl lattice remains as $\sqrt{7}\times\sqrt{7}$, and in the latter case the lattice expands as 7×7. In Fig.3 the 7×7 lattice is shown both in the real space and in the reciprocal space. It is not yet determined which,

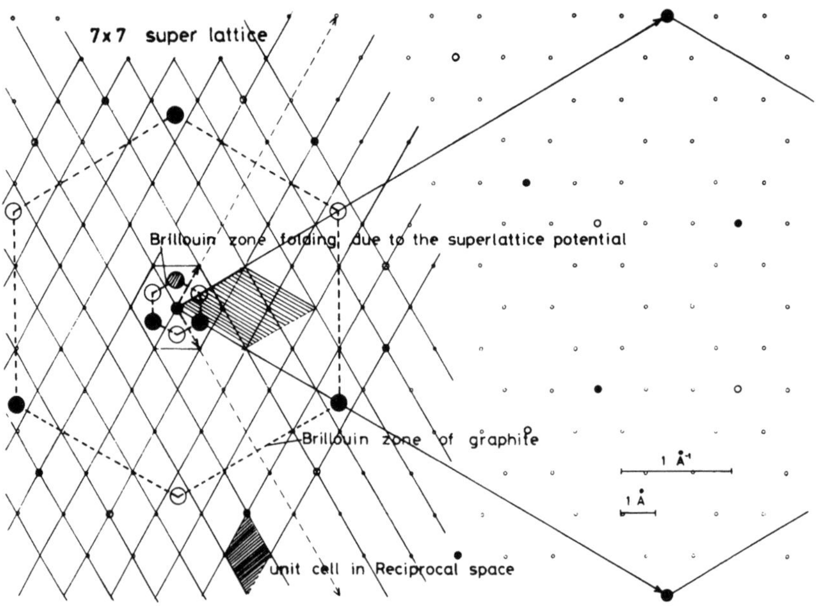

Fig.3 Real space 7×7 superlattice and the Brillouin zone in reciprocal space

$\sqrt{7}\times\sqrt{7}$ and 7×7, is the real one. The authors should give a preference to adopt the latter lattice as the actual one, because the BZ of the former lattice is too large to give any intersection of FS cylinders which are calculated by a BR model using even the largest charge transfer rate, f=1.

3. Fermi Surface Construction

Since the side length of BZ illustrated in Fig.3 can be as small as 1/7 of that of pristine graphite, the Fermi circles can intersect with each other if the Fermi radius is larger than 0.121Å^{-1}, thereafter the number of FS cross sections increases up to three or more. By the BR model, the Fermi radius k_F and energy E_F are expressed as follows:

$$k_F = \frac{1}{3\gamma_0 b}\sqrt{(2E_F \pm \gamma_1)^2 - \gamma_1^2}, \tag{1}$$

$$E_F = \gamma_0\left(\frac{\sqrt{3\pi f}}{2\ell}\right)^{1/2}, \tag{2}$$

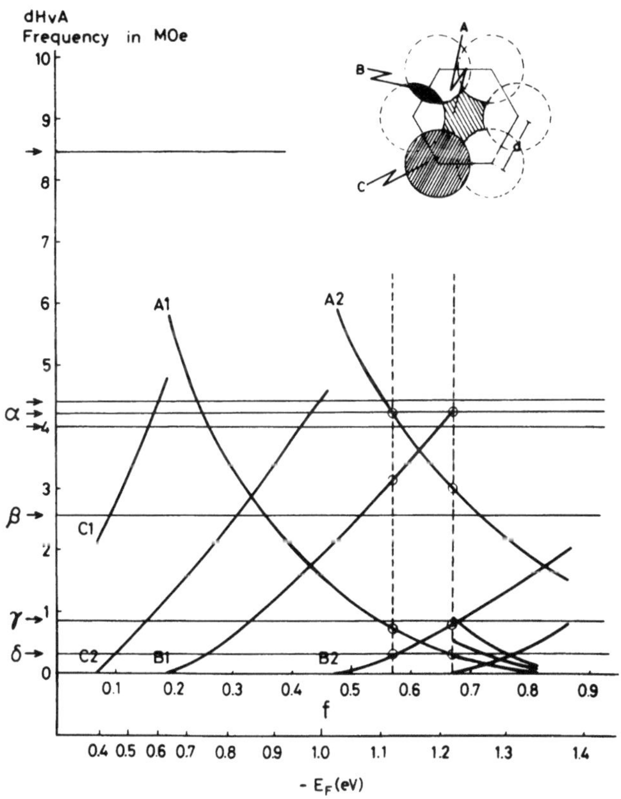

Fig.4. Comparison of the experimental dH-vA frequencies (horizontal lines) and the calculated frequencies by BR-band model (curved lines) for stage 2 $C_{24}SbCl_5$

where γ_0 and γ_1 are overlapping integrals for the nearest atoms of, respectively, the same carbon layer and of adjacent carbon layers, f is the charge transfer rate per intercalant, b=1.42Å and ℓ=12 for this compound. From (1), two primitive cylinders appear corresponding to + and - signs as illustrated in Fig.1.

BLINOWSKI and RIGAUX determined E_F from the optical relectance spectrum as -0.75 eV and then estimated the f-value as 0.45 to 0.25, corresponding to a choice of γ_0 as 2.4 to 3.12eV. The recent values by DRESSELHAUS et al. of γ_0 and γ_1 are 3.12eV and 0.377 eV, respectively. In the present work, the band folding due to the BZ of 1/7 superlattice is calculated by taking E_F as the adjustable parameter, and the FS sections are compared to the dH-vA frequencies as shown in Fig.4. The subscripts 1 and 2 on the curves corresbond to the two primitive cylinders. The f-values for the best fit locate at 0.57 and 0.67. The f-value of 0.57 is chosen for reasons mentioned later. The four kinds of newly appearing orbit are illustrated in Fig.5. The numerical comparison between the experimental values of dH-vA frequencies and the freqeuncies calculated by the FS cross sections are listed in Tab.1, where α, β, γ and δ are the FS sections designated in Fig.5.

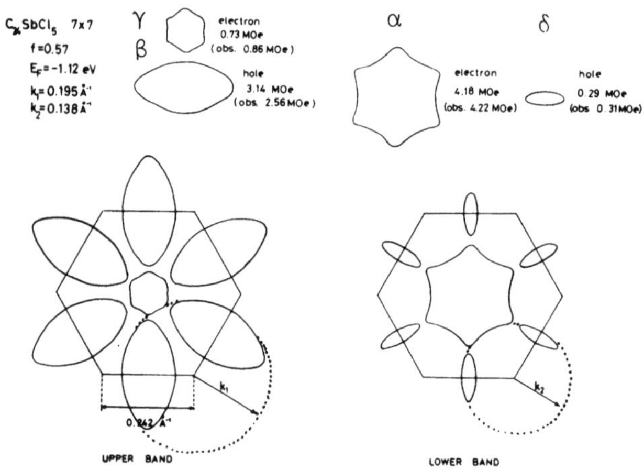

Fig.5. Shape and relative size of Fermi surface sections

The calculated frequencies α, β, γ and δ are fairly well suited to the observed values. The higher harmonics 2α=8.36, 3α=12.54 and 4α=16.72 seem to explain the observed higher frequencies 8.47, 11.9 and 15.2 (15.2 might be an underestimation by a limited accuracy of the field probe used in a field calibration of the 150Oe magnet). The potentials of intercalant charges should not strongly affect the carriers which are moving on the other layers, therefore the band gap energies by the superlattice are rather small, and magnetic breakdowns possibly occur at the magnetic field intensities of the experiment, as BATALLAN et al. claimed. The observed frequency 3.98MOe agrees well the breakdown orbit α-δ of 3.89MOe. It should be mentioned that the frequency 11.9MOe has a large amplitude in the Fourier spectrum as being shown in Tab.1 and this fact is better understood by the orbit of primitive circle of the upper band which gives 11.9MOe. The upper band comes from the symmetrical π-wave function (lower band, from antisymmetrical function) which has the large amplitude between two carbon layers and less affected by

Obs.	Fourier Intensity *	Cal.(I)	f=0.57
0.309		δ 0.29(Lh)	
0.862		γ 0.73(Ue)	
2.56		β 3.14(Uh)	
3.98			α- δ 3.89
4.22	10	α 4.18(Le)	
4.42			
8.47	2	2α 8.36	5.95(Lp)
11.9	8	3α12.54	11.9(Up)
15.2 *	3	4α16.72	6β-γ18.09

Ue ; electoron orbit (Upper band)
Le ; : (Lower band)
Uh ; hole orbit (Upper band)
Lh ; : (Lower band)
Up ; primitive circle orbit (Upper band)
Lp ; : (Lower band)
* Data of 150KOe SCM

Table 1. Observed and calculated de Haas-van Alphen frequencies in MOe in $C_{24}SbCl_5$

the intercalant charges, so that the magnetic breakdown should occur more easily than the primitive circle orbit of lower band 5.95MOe. No such agreement is found in the case of f=0.67.

For purpose of further confirmation, the Hall coefficient, which was measured at ca. 20kOe, is compared with the estimated value by the use of the present FS's. The experimental value is + 2.60×10^{-3} cm^3/C. Under the assumption of negligible magnetic breakdown at the measured field, the Hall coefficient is calculated by the four carrier model with the use of the well-known formula including carrier numbers, masses and relaxation times of α, β, γ and δ orbits. The effective masses of α and δ were measured by dH-vA effect as $0.24m_0$ and $0.07m_0$ and the other two are estimated by the extrapolation using the relation of $m^*\propto\sqrt{F}$, where m* is the effective mass and F is the dH-vA frequency, and, $0.21m_0$ and $0.10m_0$ for β and γ orbit, respectively. The value of Hall coefficient is obtained as + 2.87 10^{-3} cm^3/C, under the assumption of orbit-independent relaxation time. The agreement is fair.

Thus the above described way of FS construction has proved to be successful at least for one of the simplest acceptor-GIC's, $C_{24}SbCl_5$. A remaining problem is that the method is based on the BR-model and the f-value obtained is 0.57, which does not agree with the value 0.45 to 0.25 obtained by BLINOWSKI, RIGAUX from optical experiments. A similar analysis for the 3rd stage is underway.

Acknowledgements

The authors are grateful to Dr. A. W. Moore of Union Carbide Corporation for providing high quality HOPG. They also express thanks Dr. J. Mélin for giving them a copy of his thesis. This work was financially supported by the Mitsubishi Foundation.

References

1. K. Higuchi, H. Suematsu and S. Tanuma: J. Phys. Soc. Jpn 48 (1980) 1532.
2. G. Dresselhaus, S. Y. Leung, M. Shaegan and T. C. Chieu: Synthetic Metals 2 (1980) 321.
3. J. Blinowski and C. Rigaux: J. Physique 41 (1980) 667.
4. F. Batallan, J. Bok, I. Rosenman and J. Mélin: Phys. Rev. Lett. 41 (1978) 330.
5. J. Bok, F. Batallan and I. Rosenman: The Application of High Magnetic Fields in Semiconductor Physics, ed. by J. F. Ryan (Clarendon, Oxford, 1978) p.48.
6. F. Batallan, I. Rosenman and C. Simon: Synthetic Metals 2 (1980) 353. C. Simon, F. Batallan, I. Rosenman, H. Fuzellier: Phys. Rev. B 23 (1981) 2836.
7. J. E. Fischer, M. J. Moran, M. W. Milliken and A. Briggs: Sol. St. Commun. to be published.
8. R. S. Markiewicz, H. R. Hart, Jr., L. V. Interrante and J. S. Kasper: Synthetic Metals 2 (1980) 331.
9. Y. Iye, O. Takahashi and S. Tanuma: Sol. St. Commun. 33 (1980) 1071.
10. O. Takahashi, Y. Iye and S. Tanuma: Sol. St. Commun. 37 (1981) 863.
11. J. Mélin: Thèse, L'université de Nancy (1976).

Note: After the completion of the text, the authors became aware of the work by *Mélin* et al. stating that the 7×7 superlatice of intercalants is realized. J. Mélin, A. Herold: Carbon, 13 (1975) 357.

Effective Mass Studies of Intercalated Graphite

M.S. Dresselhaus and M. Shayegan

Department of Electrical Engineering and Computer Science and
Center for Materials Science and Engineering
Massachusetts Institute of Technology
Cambridge, MA 02139, USA

Abstract

The effective masses for graphite intercalation compounds are discussed in terms of measurement techniques and the relations between masses and the electronic structure. Typical results obtained by the Shubnikov-de Haas and magnetoreflection techniques are presented. The small masses that are observed can be explained on the basis of dispersion relations derived from the graphite π-bands.

1. Introduction

Much recent interest in graphite intercalation compounds (GIC) has focussed on the electronic transport properties, particularly on the large increase in electrical conductivity resulting from intercalation.[1] It is well established that for the common GIC's, the intercalation process increases the carrier density n, but decreases the mobility μ, with the increase in n dominating over the decrease in μ. Considerable attention has been directed toward study of the increase in carrier density or equivalently toward an evaluation of the charge transfer per intercalant, though the problem has not yet been solved.[2] Less attention has however been given to the carrier mobility μ, which depends on both the effective mass m^* and the scattering time τ according to $\mu = e\tau/m^*$. In the present work, we summarize some of the basic issues relating to effective masses, methods used to determine carrier masses and masses for filled and empty bands, and typical results that have been obtained.

When discussing carrier masses in graphite intercalation compounds, we must first recognize that there are multiple carrier types. For example, if we have a stage n compound, c-axis periodicity alone requires that there are n graphitic conduction bands and an equal number of valence bands.[3] Some of these bands are occupied by carriers. Although it is believed that many GIC's have graphite carrier types that are either electrons or holes, there is evidence (as for example from Hall effect measurements on stage 1 alkali metal compounds) that both graphitic electrons and holes may exist in the same sample.[4] In addition there are intercalant levels which may or may not be occupied. At present, there is evidence for intercalant-derived electron pockets for alkali metal donor compounds [4], though no evidence has been presented for intercalant-derived carrier pockets in acceptor compounds.

2. Measurement Techniques

For the occupied carrier pockets or for empty pockets that can be populated by some excitation mechanism, cyclotron resonance provides a direct and accurate determination of effective masses. The only reported cyclotron resonance experiment for GIC's is for desorbed graphite-Br_2 samples.[5] The temperature dependence of de Haas-van Alphen (DHVA) amplitudes provides another method for the determination of effective masses, which though not as accurate as cyclotron resonance, has the advantage of correlating the observed mass with its associated extremal cross sectional area. This method has been applied to obtain effective masses for many donor and acceptor compounds.[2] With the DHVA effect, it is uncertain whether all carrier pockets are being observed, since some low mobility pockets may not satisfy the condition for observation of the DHVA effect ($\omega_c\tau \gg 1$).

To determine effective masses for fully occupied or completely empty bands, interband Landau level transitions using magneto-optic techniques can be exploited. The observation of resonant interband Landau level transitions is not only subject to $\omega_c\tau \gg 1$, but also to the Pauli exclusion condition whereby optically induced transitions are constrained to occur between occupied initial states and empty final states.[6] By using circularly polarized light in the Faraday geometry, a comparison between the magnetoreflection spectra for ± circular polarizations can be analyzed [6] to yield the effective masses for the valence and conduction bands separately.[6] For the strongly coupled bands which characterize both pristine and intercalated graphite, the effective masses may be energy dependent (non-parabolic effects). Thus the DHVA technique which determines effective masses at the Fermi level E_F is complementary to the magnetoreflection technique which measures the masses at the initial and final energies for Landau level transitions. By variation of the photon energy, the energy dependence of the effective masses of each of the bands can be determined separately. Though the magnetoreflection technique permits a more accurate effective mass determination, magnetoreflection has in practice only given well-resolved spectra for magnetic subbands within ± 0.2 eV of E_F. [6]

Since all three methods (cyclotron resonance, DHVA effect, magnetoreflection) measure the cyclotron effective mass, they all can be used to determine the anisotropy of the effective mass tensor by varying the direction of the magnetic field with respect to the c-axis. Such anisotropy measurements, applied to the DHVA technique, can be used to distinguish whether the Fermi surface is open or closed. In particular, these topological symmetery considerations imply that for cylindrical two-dimensional carrier pockets, the effective mass component $m_{zz} \to \infty$.

In addition to these direct methods for effective mass determinations, there are a variety of transport and optical measurements which are sensitive to the carrier masses, but provide information averaged over the various carrier pockets. Since each of these transport phenomena involves a different weighted average of masses over the carrier pockets, it is difficult to deduce quantitative effective mass information from transport experiments. For experiments such as the thermopower, Hall Effect and magnetoresistance, small pockets of low mass (high mobility) carriers make a significant contribution, while for experiments such as the specific heat and the Pauli susceptibility, carrier pockets with high density of states are dominant. Thus such experiments can be used to augment information obtained by direct methods, especially for high mass, low mobility carriers.

3. Relations Between Effective Masses

Using cyclotron resonance, magnetoreflection, or the de Haas-van Alphen technique, the effective mass that is measured is the cyclotron effective mass which is defined directly from the Shockley tube integral or equivalently from the Wannier formula

$$m_c^* = \frac{\hbar}{2\pi}\oint\frac{dk}{v} = \frac{\hbar^2}{2\pi}\frac{(\partial S)}{(\partial E)_{k_H}} \tag{1}$$

where dk is taken along the electron orbit, v is the electron velocity in that orbit, S is the area of the constant energy surface enclosed by the cyclotron orbit, calculated at constant wave vector k_H along the magnetic field. These expressions are applicable to graphite intercalation compounds with arbitrary E(k) dispersion relations, and independent of whether the carriers are three or two dimensional. For a dispersion relation that is isotropic about the HK axis in the Brillouin zone, the cyclotron mass and the transport (or dynamical) effective mass are the same, as can be shown from the Boltzmann equation formulation of the conductivity tensor. A particular case of interest is that for two-dimensional graphitic carriers with a dispersion relation [7]

$$E(k_\rho) = \pm (\sqrt{3}/2)\gamma_0 a_0 k_\rho = \pm \alpha k_\rho \tag{2}$$

where k_ρ is the in-plane wave vector, γ_0 is the nearest-neighbor carbon-carbon overlap integral, a_0 is the in-plane graphite lattice constant $a_0 = 2.46$ Å, and $\alpha = (\sqrt{3}/2)\gamma_0 a_0$. For the dispersion relation given by (2), the effective mass m_c^* obtained from (1) is

$$m_c^* = \hbar^2 k_\rho/\alpha \tag{3}$$

and a derivation of the conductivity formula from the Boltzmann equation in this case also yields the same expression as in (3) for the effective mass. It is of interest to note that the association of an effective mass with a band curvature only makes sense for a parabolic dispersion relation, and not for a linear dispersion relation as in (2).

4. Results and Discussion

Results for the Shubnikov de Haas traces shown in Fig.1 for a stage 5 potassium-intercalated sample give clear evidence that the oscillations can be followed to temperatures as high as ~ 25 K, thereby indicating long mean free paths in the graphitic layers. Furthermore, the observation of the oscillations down to 1 tesla at the lowest temperatures shows that the cyclotron radius $\lambda = (c\hbar/eH)^{1/2}$ can be as large as 250 Å without losing coherence through scattering by defects. From the observed temperature dependence of the amplitudes A, we can determine the cyclotron effective mass through the relation

$$A = T \exp(-2\pi^2 k_B T/\hbar\omega_c) \tag{4}$$

where $\omega_c = eH_0/m_c^* c$ and H_0 is the field value at which the amplitudes are measured. The results for a stage 5 potassium sample are plotted in Fig.2 in terms of $\ln(A/T)$ vs. T. From the slope of these lines, m_c^*/m_0 is determined, and the results for the masses for two of the DHVA frequencies are given in the figure. It is of significance that the masses are small

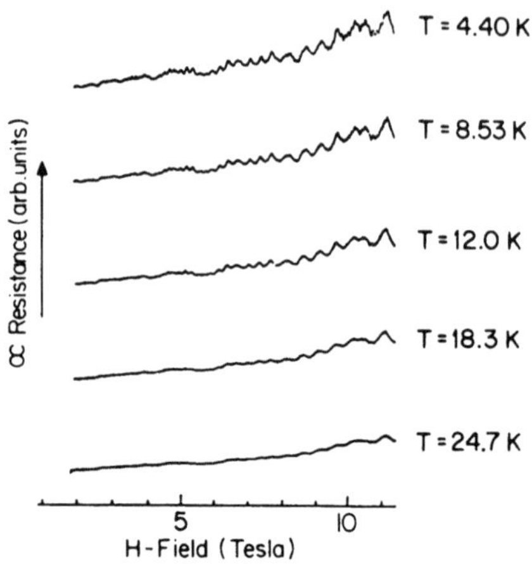

Fig.1 SdH traces taken on a stage 5 sample of potassium intercalated graphite at various temperatures.

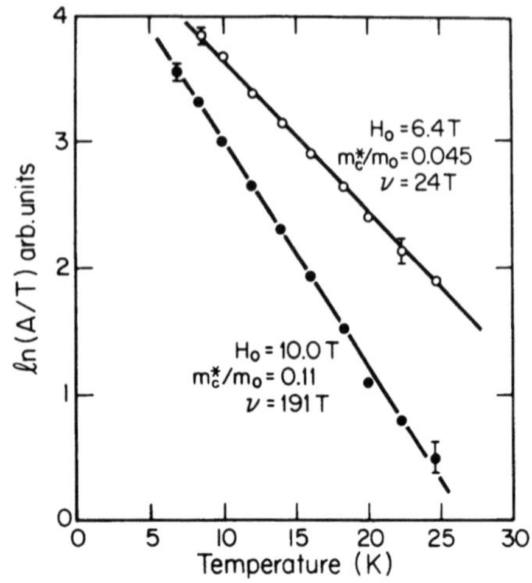

Fig.2 From the slope of $\ell n(A/T)$ vs. T in Fig. 1, m_c^*/m_0 for each SdH cross-section is found from (4).

and are roughly of comparable magnitude to those in graphite. In fact, the effective mass for the smallest cross section is 0.045 m_0, which is smaller than the masses in graphite for the majority electron and hole extremal cross sections. The occurrence of smaller effective masses in intercalated graphite compared to pristine graphite can be understood on the basis of a rigid band argument. Intercalation causes the Fermi surface to grow and the resulting mass for the K-point cross section to increase. The c-axis periodicity associated with staging can be treated by zone folding of the graphitic bands, thereby giving rise to new extremal cross sections away from the K-point. These new cross sections correspond to smaller masses than for the K-point electron orbit, so that there is a greater decrease in m_c^* for the smallest extremal cross section of the folded Fermi surface than the increase in m_c^* resulting from the increased size of the Fermi surface due to charge transfer from the intercalant to the graphite.

To account for the observed SdH frequencies, we have used two models. In model I, the Dilute Limit Model, the Slonczewski-Weiss-McClure (SWMcC) model for the graphite π-bands is used [8] and the Fermi level E_F is raised relative to that for pristine graphite (E_F = − 0.024 eV). Zone folding along the k_z axis is introduced to treat the superlattice periodicity associated with staging, yielding additional extremal cross sectional areas and new SdH frequencies.[9] For a stage 5 compound we have 5 bands that can have carriers. The dependence of the SdH frequencies on E_F calculated from the dilute limit model is given in Fig.3 by the heavy dashed curve. The Fermi level E_F is determined to provide the best fit to the experimental data. In model II, the effect of superlattice periodicity due to staging is also treated by k_z axis zone folding, except that in this case each $(n + 1)^{st}$ layer in the graphite lattice is replaced by an intercalant layer. In the simplest version of model II, the empty intercalate layer model, the intercalate layer is replaced by a "vacuum".[10] The results for the dispersion relations obtained for a stage 5 donor compound within the "empty intercalate layer" model are given in Fig.4, where it is seen

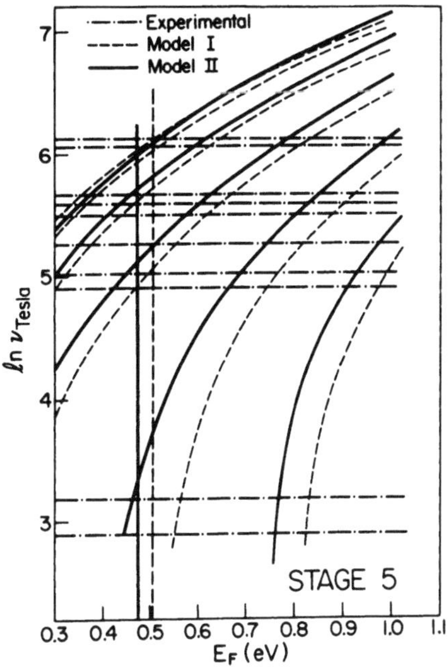

Fig.3 Relation between SdH frequencies (plotted on a log scale) and the Fermi level for stage 5-potassium.

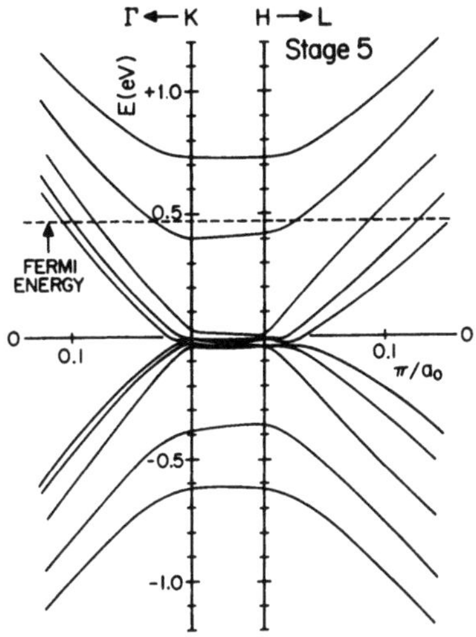

Fig.4 Electronic band structure in the vicinity of the HK axis for a stage 5 compound using the "empty intercalate layer" model. (Model II)

that four of the conduction bands are occupied. The Fermi level for model II also is determined to provide the best fit to the SdH frequencies, as shown in Fig.3 by the heavy solid curves.

The good fit provided by these models to the SdH frequencies has been discussed previously.[11] Of significance is the good fit obtained for the effective masses, as shown in Table 1. In this connection, we note that it is generally more difficult to calculate effective masses than energy levels in crystalline solids.

The mass obtained from the SdH measurements for the smallest cross-sectional area is in excellent agreement with the conduction band mass measured on similar alkali metal compounds using the magnetoreflection technique.[6] As is also the case for pristine graphite, the valence band mass for both donor and acceptor intercalation compounds is found to be greater than for the conduction bands involved in the Landau level transitions. The effective masses determined by the magnetoreflection experiment (for stages $n > 4$ for acceptors and $n > 6$ for donors) are found to be very similar, though always smaller in magnitude, than those for pristine graphite. The magnitude of these valence and conduction band masses decreases with increasing reciprocal stage (intercalate concentration) and this decrease is significantly more rapid for donor compounds than for acceptor compounds of the same stage. A direct calculation of the Landau level transitions using the same k_z axis zone folding technique as for model II yields upon intercalation a decrease in the effective masses

Table 1 Comparison of the calculated and observed SdH frequencies and effective masses for a stage-5 potassium compound.

ν[Tesla]			m_c^*/m		
Model I	Observed	Model II	Model I	Observed	Model II
453	453		0.21		
412	430	401	0.20	0.15 - 0.16	0.14
304	290 267 243	300	0.17	-	0.12
157	191 152 135	163	0.11	0.10 - 0.11	0.083
-	24 18	26.7	-	0.045	0.051
0.505	E_F(eV)	0.475	0.505	E_F(eV)	0.475

associated with the Landau level transitions [12], in agreement with the experimental observations in the magnetoreflection experiment.

Acknowledgments

We thank Dr. G. Dresselhaus, Prof. H. Kamimura and Dr. L. Pietronero for stimulating discussions and AFOSR contract #F49620-81-C-0006 for support of this research.

References

1. G.M.T. Foley, C. Zeller, E.R. Falardeau, and F.L. Vogel, Solid State Commun. 24, 371 (1977).
2. M.S. Dresselhaus and G. Dresselhaus, Adv. Phys. 30, 139 (1981).
3. G. Dresselhaus and S.Y. Leung, Solid State Commun. 35, 819 (1980).
4. T. Ohno, K. Nakao and H. Kamimura, J. Phys. Soc. Jpn. 47, 1125 (1979).
5. S. Tanuma, H. Suematsu, K. Higuchi, R. Inada and Y. Onuki, Proceedings of the Conference on the Application of High Magnetic Fields in Semiconductor Physics, (1978), edited by J.F. Ryan (Oxford: Clarendon Press), p. 85.
6. E. Mendez, T.C. Chieu, N. Kambe and M.S. Dresselhaus, Solid State Commun. 33, 837 (1980).
7. P.R. Wallace, Phys. Rev. 71, 622 (1947).
8. M.S. Dresselhaus, G. Dresselhaus and J.E. Fischer, Phys. Rev. B15, 3180 (1977).
9. K. Higuchi, H. Suematsu and S. Tanuma, J. Phys. Soc. Jpn. 48, 1532 (1980).
10. G. Dresselhaus and S.Y. Leung, Solid State Commun. 35, 819 (1980).
11. G. Dresselhaus, S.Y. Leung, M. Shayegan and T.C. Chieu, Synthetic Metals 2, 321 (1980).
12. T.C. Chieu, G. Dresselhaus and M.S. Dresselhaus, Solid State Commun. 37, 561 (1981).

Analysis of Physical Properties of Potassium-Tetraydrofuran-Graphite Derivatives $KC_{24s}(THF)_m$

L. Facchini and A.P. Legrand

Laboratoire de Physique Quantique ERA 676, E.S.P.I. - 10, rue Vauguelin
F-75231 Paris Cedex 05, France

F. Beguin and R. Setton

C.R.S.O.C.I. - C.N.R.S. - 16, rue la Ferollerie
F-45045 Orléans Cedex, France

G. Hermann

Bruker Spectrospin, F-67160 Wissembourg, France

1. Introduction

Graphite can form intercalation compounds (GIC) containing along with an alkali metal an organic molecule [1,2] which by its presence, completely modifies the structural, magnetic and electronic properties of the parent binary derivatives. Crystallographic characteristics of the four KC_{24s} $(THF)_m$ ternary GIC which are studied in the present work were given in [1,4]. From ^1H NMR [4] we have shown that the THF molecule is more bound in first stage than in second stage compounds. The physical properties of the ternary compounds were first studied in [5]. Modifications of the specific heat and static magnetic susceptibility appeared relatively to the initial binary compound. In the first stage compound KC_{24} $(THF)_1$, the apparent density of states at the Fermi level (in state/eV. C atom) deduced from the measured specific heat at low temperature is about four times that of the initial binary compound. From the evolution of the static magnetic susceptibility with temperature, AMIELL et al. deduced that KC_{24} $(THF)_1$ does not contain localized paramagnetic centers, whereas a Curie law is observed for KC_{24} $(THF)_2$ showing a partial localization of the electrons (0.2 e/K atom) [1]. Unfortunately the measured static magnetic susceptibility is the sum of several terms : the first three terms which contain the physical information are the Pauli susceptibility proportional to $N(E_F)$, the orbital magnetic susceptibility [6] and the paramagnetic susceptibility of the localized centers ; the fourth term, which is the core diamagnetism χ_o, is very ambiguous. In previous papers [1,5] this term was determined from the known values of χ_o for K^+ and carbon (as in [8] for binary compounds) ; the THF contribution was obtained from magnetic susceptibility measurements on liquid THF. The latter evaluation implies two assumptions :

- a full ionization of the potassium atom (i.e. the charge transfer is one) ;

- a very weak bond between the THF and the graphitic planes or K^+ (no perturbation in the valence orbitals of THF).

The EPR spectrum of the ternary [1,4,5] and acceptor compounds are very similar : narrow line, relatively small A/B ratio and a value of g which is very close to the free electron g_o value.

It implies that the orbital momentum of the electron is quenched. By comparison with the EPR of binary compounds [8], we may think that the diffusion time along the $\vec{c}$ axis is longer than that of the initial binary compound. This remark is consistent with the low value of the Debye temperature (∿ 80 K) for KC_{24} $(THF)_1$, deduced from the measured high temperature specific heat, showing a weak coupling between graphite and intercalant.

A local probe of electronic properties is the ^{13}C NMR [9,12]. The ternary compounds contain two types of ^{13}C : the carbon of graphite and the carbon of the organic molecule. The first problem is to resolve two sorts of lines in order to obtain a selective information :

- a first line, characteristic of the THF molecules, which gives information on the interaction between the organic molecule and the substratum ;

- a second line, characteristic of the graphite planes, which gives information on the bond between the graphitic plane and the intercalant layer, and on the electronic properties like $N(E_F)$ and the charge transfer as shown by CONARD et al. [9,10].

2. Experimental

This work was carried out on a Bruker WM 250 cryomagnet system (about 6T), because the use of a high magnetic field turned out to be essential to the resolution of the spectrum. Because of the low S/N ratio, a large number of spectra has been accumulated ; in order to remain within reasonable duration the cycling rate has been fixed to 2 seconds. As this rate is certainly smaller (or of the same order) than the spin-lattice relaxation time, the surface of the lines is not correlated to the number of ^{13}C in graphitic planes and THF molecules.

In Fig.1 results are reported for the first and second stage compounds. The position of the lowfield part of the lines was obtained from [10] for KC_{24} and from [9] for C_8Br.

3. Discussion

3.1 Attribution of the Lines

The spectrum is clearly composed of two distinct lines :

- the high field line is shifted by about + 50 ppm. This value is roughly the one obtained in KC_8, which has the most shifted line among the binary compounds [9] ;

- in first stage ternary compounds the greater value of the charge transfer is 1/24 per carbon atom, so the shift of the graphite line must be of the same order as the shift of KC_{24} ;

- the low field line of KC_{24} $(THF)_1$ is very similar to the low-field part of the line in KC_{24} ;

- for KC_{24s} $(THF)_2$ the lowfield line is shifted from stage 1 to 2 whereas the highfield line is not shifted.

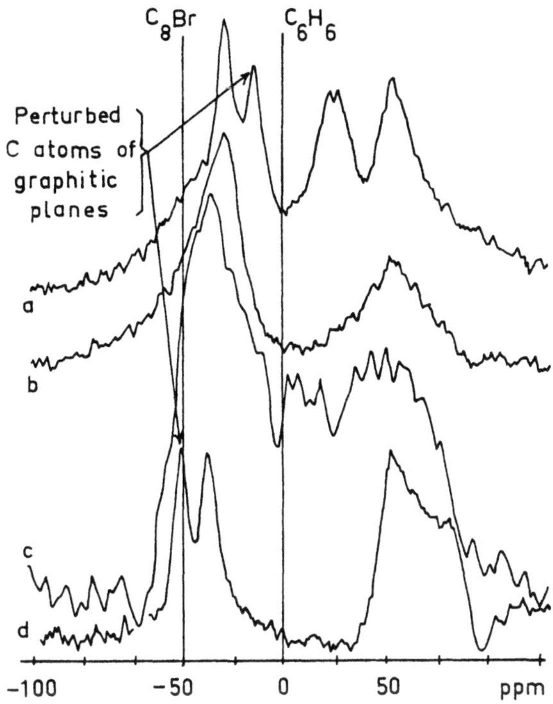

Fig. 1 : High field ^{13}C NMR spectra of :
(a) KC_{24} $(THF)_1$, (b) KC_{24} $(THF)_2$,
(c) KC_{48} $(THF)_1$, (d) KC_{48} $(THF)_2$

It implies a ^{13}C line of the graphitic planes shifted towards the lowfields and we assign the highfield line to the ^{13}C of the THF molecule.

3.2 Analysis of the Lines

In the present work, we shall only consider the graphitic line.

For KC_{24} $(THF)_1$ and its parent binary compound KC_{24}, the low-field parts of the graphite line are very similar. Following CONARD'S analysis for second stage compounds; we shall consider that two types of carbon exist : one perturbed by the intercalant (chemical bond ?), the other one not perturbed (no bond with the intercalant).

For KC_{24} $(THF)_2$ the graphite line is not resolved, so we have two possibilities : an anisotropic shift or two mixed lines, characteristic of two sites as in KC_{24} and KC_{24} $(THF)_1$.

We observe a reverse situation for the second stage compounds. The spectrum of KC_{48} $(THF)_2$ is resolved in two lines whereas the spectrum of the second stage compound with $\frac{THF}{K} = 1$ is not resolved.

As there is no reason to assume a large modification of the chemical bond between the intercalant layer and the graphitic plane from stage 1 to 2, we must admit a two ^{13}C sites system (resolved or not resolved) for the four compounds.

Relative to C_8Br, one of the graphite lines has the same shift in KC_{24} (THF)$_1$ and KC_{24} (about + 22 ppm). A similar shift was observed for the ternary compound KC_{24} $(C_6H_6)_2$ [13]. Since we have no reason to consider a similar perturbation of the ^{13}C in these four compounds (with very different structural properties), we deduce that the lines with the common shift for KC_{24}, KC_{24} (THF)$_1$ and KC_{24} $(C_6H_6)_2$ are characteristic of <u>non-perturbed carbon of graphitic planes</u>.

In order to understand this observation, we must analyse the causes of the shift. Since ^{13}C has a spin 1/2 and is in very poor natural abundance (1%), the structure of the graphitic lines is only due to the interaction of the nucleus with the electrons. This interaction is very different if the electrons are localized (chemical shift) or delocalized (KNIGH shift). We consider that the ^{13}C which are not directly perturbed by intercalant are only in an environment of π electrons and we take this intrinsic chemical shift as a reference (chemical shift characteristic of intrinsic isolated graphitic plane). Since it is known that the change transfer in C_8Br is about zero, in this compound the ^{13}C line shift is only due to intrinsic chemical shift in isolated graphitic plane ; so we take the C_8Br spectrum as a reference. In our analysis the shift of the common line is relative to C_8Br is only due to the effect of delocalized electrons.

According to ABRAGAM [14] the shift can be interpreted as resulting from :

- an interaction between the orbital momentum of electrons and the spin of nuclei. But as shown before orbital momentum is quenched and this term can be neglected.

- a dipolar and indirect contact interaction between electrons and nuclei. This term is proportional to $N(E_F)$ and contains <u>the same information as the Pauli susceptibility.</u>

- a coupling of the magnetic moment of nucleus with the magnetic field resulting from the electronic current due to the action of static magnetic field on the delocalized electrons. It contains the same information as the orbital susceptibility discussed by SAFRAN [6] for 2D graphitic planes. Since the graphitic plane is weakly coupling with the intercalant, we assume that only the contribution along the $\vec{c}$ axis is of interest (2D system).

- an indirect coupling of the nuclei via the electron which must be neglected because of the poor natural abundance of ^{13}C in the compounds.

It seems that only the component $\delta_\perp$ (shift of the component perpendicular to the $\vec{c}$ axis) is observed. The component $\delta_{//}$ may be masked by the line of the organic molecule. Since $\delta_\perp$ is very dependent on $N(E_F)$ [10], it implies that the value of $N(E_F)$ by carbon unit, for the three compounds KC_{24}, KC_{24} (THF)$_1$ and

KC_{24} $(THF)_2$, is the same for the unperturbed carbon. The density of states at the Fermi level of graphitic plane is not changed by the intercalation of the organic molecule ($N(E_F) \sim 0.1$ state/eV. C or 2.5 states/eV.molecule) [7].

The greater value of $N(E_F)$ deduced from specific heat measurements is certainly due to an increase of the electron-phonon interaction.

From stage 1 to 2 the line is shifted towards the lowfields. Unfortunately, for KC_{48} $(THF)_1$ the graphitic line is not resolved and we cannot deduce quantitatively the shift of the common line as a function of stage. For KC_{48} $(THF)_2$ the common line is towards the high field (Fig.1), the new shift is 11 ppm, which is exactly half of the shift of the common line for the first stage compound. We deduce that the value of $N(E_F)$ from carbon atoms is twice as small, i.e. the density of states at the Fermi level for KC_{24} $(THF)_2$ and KC_{48} $(THF)_2$ is the same if $N(E_F)$ is defined by a molecule unit.

From the two resolved lines of KC_{24} $(THF)_1$ and KC_{48} $(THF)_2$, with the assumption of no evolution of chemical shift with the stage, we deduce the number of localized electrons around the perturbed carbon atoms. We note that the line of perturbed carbon is shifted to the highfield relatively to the unperturbed carbon for $\frac{THF}{K} = 1$, whereas the line is shifted to the lowfield for compounds with $\frac{THF}{K} = 2$. Like CONARD [9,10] we used the standard shift of 160 ppm/e and we obtain :

KC_{24} $(THF)_1$ $\longrightarrow$ + 0.1 e/perturbed carbon

KC_{48} $(THF)_2$ $\longrightarrow$ - 0.07 e/perturbed carbon

We note :

- an excess of electrons around the perturbed carbon for the compounds with $\frac{THF}{K} = 1$.

- a deficiency of charge around the perturbed carbon for the KC_{24}s $(THF)_2$ compounds. From the partial localization of the electrons (0.2 e/K) obtained from the static magnetic susceptibility measurements, we deduce that three carbons are perturbed in KC_{24} $(THF)_2$ molecule.

4. Conclusions

^{13}C NMR appears to be a very powerful method to study the physical properties of ternary lamellar compounds . We show that :

- the density of states at the Fermi level does not depend on stage if $N(E_F)$ is defined by molecule unit, and is about the same as KC_{24}. By comparison with the specific heat measurements in KC_{24} $(THF)_1$, it implies a strong electron-phonon interaction.

- the number of perturbed atoms in KC_{24} $(THF)_2$ is small.

- the density of charge in a graphitic plane is not uniform and depends on the $\frac{THF}{K}$ ratio.

Intercalation compounds containing along with an alkali metal an organic molecule are probably a good way for synthesizing compounds with the characteristics of an acceptor (decoupling of the graphitic planes and weak interaction between intercalant and the graphitic plane) but with high value of charge transfer, i.e. a high value of the density of states at the Fermi level.

REFERENCES

1 - F. BEGUIN; Thesis, University of Orleans France (1980)
2 - F. BEGUIN et al.: Mat. Sci. Eng. 40, 167 (1979)
3 - L. FACCHINI et al.: Physica 99B, 525 (1980)
4 - L. FACCHINI et al.: 3rd International Carbon Conference Baden-Baden 100, (1980)
5 - J. AMIELL et al.: Mat. Sci. Eng. 31, 243 (1977)
6 - S.A. SAFRAN et al.: Phys. Rev. B 20, 4889 (1979)
7 - P. DELHAES: Mat. Sci. Eng. 31, 225 (1977)
8 - P. LAUGINIE Physica 99 B, 514 (1980)
9 - J. CONARD et al.: Physica 99 B, 521 (1980)
10 - J. CONARD et al.: Synthetic Metals 2, 227 (1980)
11 - SUGANUMA et al.: Phys. Rev. B 22, 5079 (1980)
12 - D.T. HOWORTH et al.: J. Sol. State Chem. 31, 343 (1980)
13 - F. BEGUIN et al.: 15th Biennal Conference on Carbon Philadelphia (1981)
14 - A. ABRAGAM: Les Principes du Magnetisme Nucléaire PUF (1961)

Inelastic Electron Scattering Spectroscopy of Intercalated Graphite and Polyacetylene

John J. Ritsko

Xerox Corporation, Webster Research Center
Webster, NY 14580, USA

Synthetic metals can be made by intercalating a wide variety of atoms and molecules between the layers of graphite. The increase in conductivity over that of pristine graphite is 1-2 orders of magnitude. Some highly conjugated polymers, such as polyacetylene, will react with some of the graphite intercalants to produce synthetic metals also, but in these materials the increase in conductivity is 12 orders of magnitude (1). Although the structure of doped polyacetylene is not nearly as crystalline and well understood as that of the graphite compounds, there is evidence that the dopants occupy planes similar to graphite intercalants (2). In addition, the increase in conductivity in these two classes of synthetic metals is certainly associated with the charge transfer which accompanies the intercalation reaction. Thus, the nature of the donor-acceptor interaction in these materials is of prime importance. In a number of recent experiments (3-6), we have attempted to quantitatively measure the change in the electronic structure of graphite when it is reacted with acceptor intercalants such as $FeCl_3$, AsF_5 and $SbCl_5$. The results showed that in these materials the graphite electronic structure is not significantly altered and that the amount of charge transfer and the Fermi level location could be determined in several ways. The results described here will extend these measurements to a donor intercalant, potassium, where significant changes are observed in the graphite electronic structure. In addition, measurements on pristine and doped polyacetylene will show that the intercalants which produce little change in the graphite electronic structure strongly affect polyacetylene.

The electronic excitation spectra of the materials under study have been measured by electron energy loss spectroscopy utilizing an 80 KeV electron beam with an energy loss resolution of 0.1 eV and a momentum resolution of $0.05Å^{-1}$. From these spectra the momentum dependent dielectric functions have been determined by Kramers-Kronig analyses (6-8) and the results directly compared with theoretical calculations of the dielectric function based on models of the graphite band structure (3,5,7,8).

The energy loss spectrum for stage-1 potassium intercalated graphite, KC_8, for momentum $q = 0.1Å^{-1}$ parallel to the graphite planes is shown in figure 1. It differs from previous measurements (9) in having higher resolution and much lower

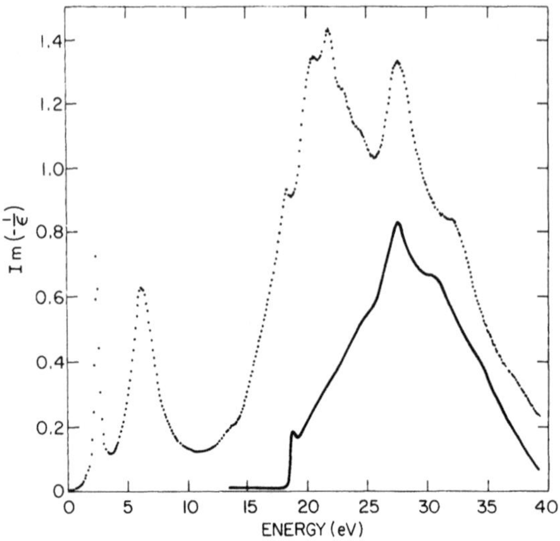

Fig. 1 Electron energy loss spectrum of KC_8; dots show actual data points. The solid curve is the unnormalized shape of the energy loss spectrum of metallic potassium.

statistical noise. There is an intraband plasmon at 2.38 eV and an interband plasmon at 6.23 eV as well as structure at 18.4, 20.6, 21.9, 23.2, 24.5, 27.7 and 32.3 eV. Also shown in figure 1 is the spectrum for potassium metal which allows us to identify the features at 18.4, 27.7 and 32.3 eV as being due to the potassium intercalant (8). The nature of the remaining fine structure becomes evident from the results of the Kramers-Kronig analysis of the data which is given in figure 2. The three rather weak peaks in ϵ_2 at 21.3, 22.9, and 24.3 eV are shown more clearly in the inset to figure 2. These features occur near the zero crossing of ϵ_1 and hence are relatively prominent in the energy loss spectrum. This fine structure must be associated with the potassium superlattice in KC_8 because in KC_{24} (stage-2) where the potassium superlattice does not exist, these features are not observed, as can be seen in figure 3. Thus, the fine structure in KC_8 between 19 and 25 eV is most likely due to graphite interband transitions in the backfolded graphite bands imposed by the two-dimensional potassium superlattice (8).

We have constructed a theoretical model of the backfolded graphite band excitations (8). These transitions are between occupied 2p wavefunctions at the Brillouin zone center and unoccupied 3p states which were originally at the zone boundary in pristine graphite but are folded back to the zone center by the interaction with the intercalant potential. The model specifically takes account of the splitting and symmetry of conduction band states as well as symmetry mixing of valence band wavefunctions. Interplanar as well as intraplanar interactions are included. The results indicate that the carbon-intercalant

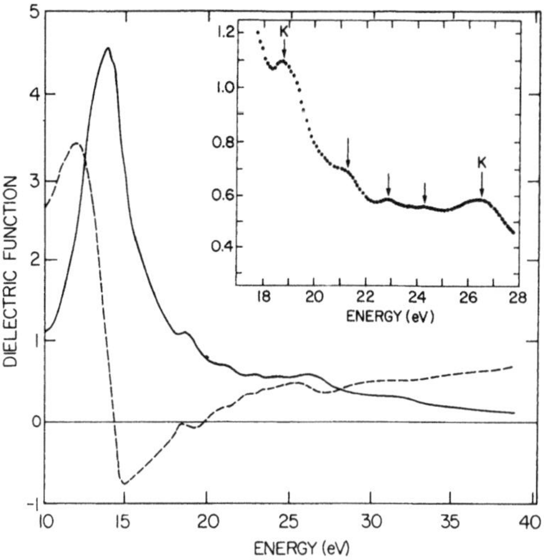

Fig.2 Dielectric function of KC_8; ϵ_1-dashed curve, ϵ_2-solid curve. Inset shows ϵ_2 on expanded scale with actual data points given. K denotes potassium excitation.

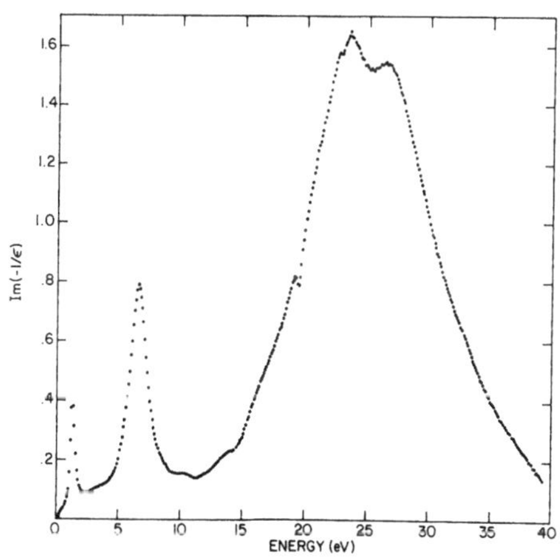

Fig.3 Electron energy loss spectrum of KC_{24} at $q=0.1 Å^{-1}$; dots show actual data points.

interaction for the carbon 3p orbitals modulates the carbon-carbon nearest neighbor transfer integral by about 20% and the intralayer interaction is about 2.6 times as strong as the interlayer interaction. Despite this strong modulation the oscillator strength of the new interband excitations is measured and calculated to be only 2% of the total π electron oscillator strength (8).

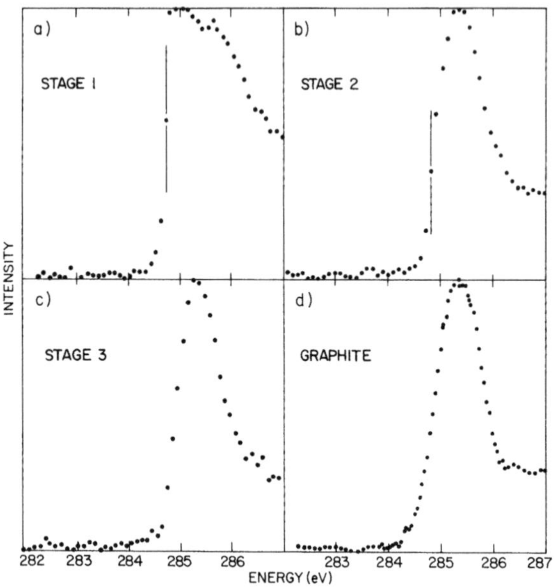

Fig.4 Carbon 1s excitation spectra (raw data) for potassium intercalated graphite compounds. Vertical lines indicate Fermi level.

The direct influence of the intercalant on states near the Fermi level is shown in figure 4. Each panel shows the excitation spectra of carbon 1s core electrons into the empty states just above the Fermi level. In acceptor intercalated compounds the Fermi level is simply lowered and the states above the pristine graphite Fermi level are quite unperturbed (3,4). This clearly is not the case for potassium compounds. Figure 4 shows that above a very sharp Fermi level there are two peaks in KC_8 and the spectrum is much broader than the absorption edge shape of pristine graphite. In KC_{24} the Fermi edge is less prominent and the overall shape just slightly broader than in graphite, while in stage-3 (KC_{36}), the spectrum looks a great deal like pristine graphite although minor differences persist. The splitting of the two peaks above the Fermi level in KC_8 agrees reasonably well with features in the density of states calculated in a semiempirical tight-binding model (10). Of course the measured spectral shape is strongly influenced by the electron-hole interaction (3). Nevertheless, the double peaked spectrum of KC_8 indicates that there is definite hybridization between carbon and potassium wavefunctions. In fact, the spectrum is so different from that of graphite that a simple analysis of the lineshape to determine the Fermi level position relative to graphite as was done in the acceptor compounds (3,4,6) cannot be done here. A new theoretical calculation of the core electron absorption lineshape using hybridized carbon and potassium bands must be carried out to obtain a measure of the Fermi level position and the charge transfer in KC_8 from the present data.

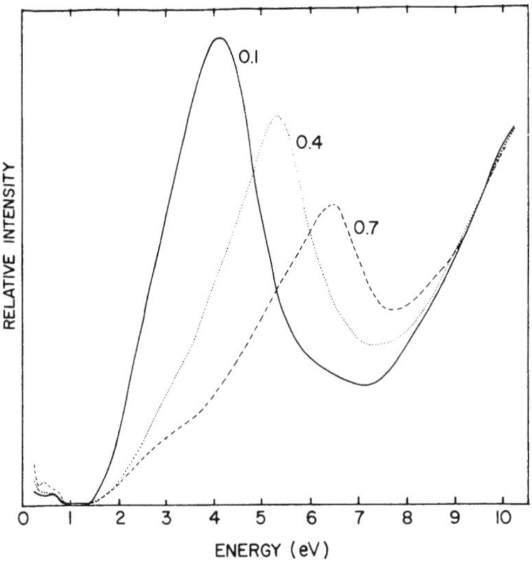

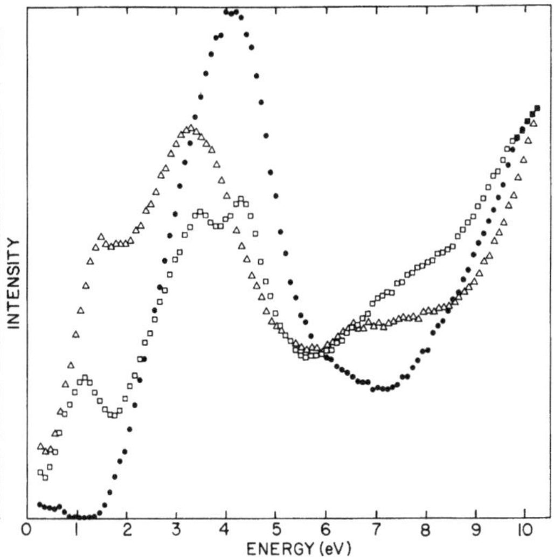

Fig.5 Polyacetylene valence excitations: energy-loss spectra for momentum indicated (in $Å^{-1}$).

Fig.6 Energy-loss probability at q = 0.1 $Å^{-1}$: dots, pure $(CH)_x$; squares, iodine-doped $(CH)_x$; triangles, AsF_5-doped $(CH)_x$.

Valence excitations of pristine (predominantly trans) polyacetylene are shown in figure 5. Apart from minor impurity absorption below 1 eV the absorption onset occurs at 1.4 eV followed by a peak at 4.1 eV which is a plasmon associated with the π electronic excitations similar to that observed in graphite at 7 eV (3). The 1.4 eV threshold is consistent with the semiconducting state of this material which is due to bondlength alternation. In the absence of such bondlength alternation the electrons would form a simple one-dimensional metallic band. As illustrated in figure 5 the interband plasmon shows positive dispersion with increasing momentum indicating that the energy bands in $(CH)_x$ are about as wide as in inorganic semiconductors (11). The absorption threshold itself, however, does not change with increasing momentum consistent with an excitonic or indirect absorption edge (11).

The effects of doping polyacetylene to saturation with AsF_5 or iodine are shown in figure 6. When heavily doped with AsF_5 the 4.1 eV plasmon shifts to 3.25 eV and continuous absorption begins at very low energy. Iodine doping results in a more inhomogeneous mixture of doped and undoped regions as evidenced by an unshifted peak near 4 eV as well as a downshifted peak near 3.3 eV (11). A Kramers-Kronig analysis of the energy loss data shows that in pristine $(CH)_x$ there is a large peak in ϵ_2 near 2 eV which shifts to very low energy in heavily doped samples. This behavior is consistent with the closing of the semiconducting gap and a simple metallic state (11).

The changes in the density of states in $(CH)_x$ which would come about when the semiconducting gap is closed are illustrated in figure 7. The solid lines show the density of states for a simple one-dimensional semiconductor model of $(CH)_x$ (12). The large peak in ϵ_2 in the pristine material would be associated with interband transitions or excitons at the band edges. In the absence of bondlength alternation a simple-metallic density of states indicated by the dot-dashed curve would result. In this case Drude-like intraband transitions at the Fermi level would result in a metallic dielectric function in which ϵ_2 is very large at low energy. This simple description seems to fit both pristine and doped samples if there is no bondlength alternation in the heavily doped samples.

If in addition to closing the gap, the heavily doped samples also involved charge transfer, the Fermi level in the metallic state would be lowered to E_m in figure 7. An alternative model for the very high conductivity which is observed (1) would be a degenerate semiconductor in which bond alternation persisted but with the Fermi level lowered below the conduction band edge to position E_s (12). The spectrum of core level excitations can easily distinguish between these two models since it is sensitive to the density of states near the Fermi level. The results of these measurements are shown in figure 8. Pristine polyacetylene shows a strong peak at the absorption edge which is due to the excitonic enhancement of the peak at the conduction band edge (12). In heavily doped AsF_5 samples the absorption edge is

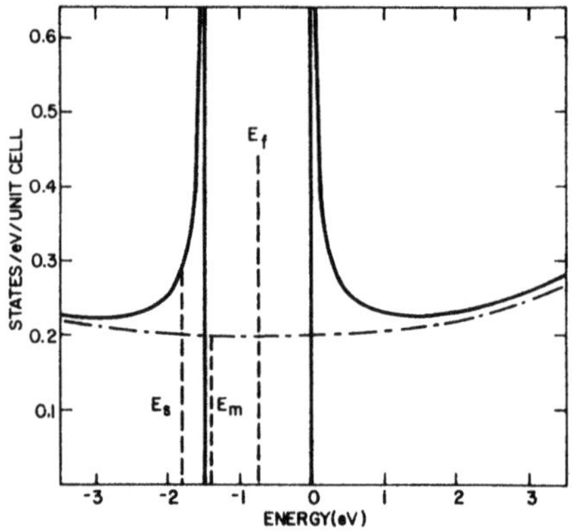

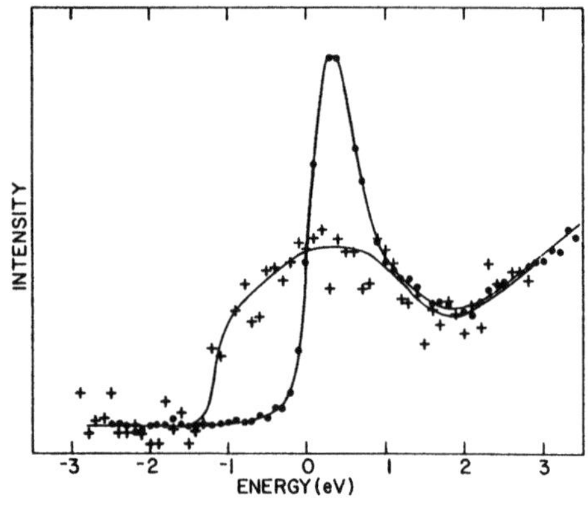

Fig.7 Model density of states near the Fermi level, E_F, of polyacetylene: Solid curves, alternating C-C bond lengths; dot-dashed curve, uniform bond lengths. E_s (E_m), shown as vertical dashed lines, indicate Fermi level of degenerate semiconductor (simple metal) model for polyacetylene doped to saturation with AsF_5.

Fig.8 Carbon 1s excitation spectra (raw data): dots, $(CH)_x$; crosses, $(CH)_x$ doped to saturation with AsF_5.

shifted to lower energy and the strong excitonic peak is missing. The new spectrum more closely resembles the simple metallic step-like edge than the shape expected for a degenerate semiconductor. Although iodine doped samples show evidence of inhomogeneous doping in their core excitation spectra as well as their valence excitation spectra, they also show a step-like core absorption threshold. Hence, the simple metallic state of heavily doped samples does not depend on the specific dopant.

Therefore, electron energy loss spectroscopy has shown that acceptor intercalants produce charge transfer but little change in the electronic structure of graphite (3-7). However, the same or similar intercalants cause not only charge transfer but also remove the bondlength alternation in heavily doped polyacetylene. In graphite, donor intercalants, specifically potassium, can hybridize with carbon wavefunctions near the Fermi level and when the intercalant forms a superlattice the graphite bandstructure is backfolded into a reduced Brillouin zone generating new interband transitions which yield information concerning the carbon-intercalant interaction.

References

1. C.K. Chiang, C.R. Fincher, Y.W. Park, A.J. Heeger, H. Shirakawa, E.J. Louis, S.C. Gau, A.G. MacDiarmid, Phys. Rev. Lett. 39, 1098 (1977).
2. S.L. Hsu, A.J. Signorelli, G.P. Pez, R.H. Baughman, J. Chem. Phys. 69, 106 (1978).
3. J.J. Ritsko and E.J. Mele, Physica (Utrecht) 99B, 425 (1980).
4. J.J. Ritsko and E.J. Mele, Syn. Met. 3, 73 (1981).
5. E.J. Mele and J.J. Ritsko, Syn. Met. 3, 89 (1981).
6. J.J. Ritsko, E.J. Mele and P.C. Eklund, to be published.
7. J.J. Ritsko, E.J. Mele, Phys. Rev. B 21, 730 (1980).
8. J.J. Ritsko, E.J. Mele and I.P. Gates, to be published.
9. D.M. Hwang, M. Utlaut, M.S. Isaacson and S.A. Solin, Phys. Rev. Lett. 43, 882 (1979) and Physica (Utrecht) 99B, 435 (1980).
10. T. Inoshita, K. Nakao and H. Kamimura, J. Phys. Soc. Japan 43, 1237 (1977).
11. J.J. Ritsko, E.J. Mele, A.J. Heeger, A.G. MacDiarmid and M. Ozaki, Phys. Rev. Lett. 44, 1351 (1980).
12. J.J. Ritsko, Phys. Rev. Lett. 46, 849 (1981).

On the Theoretical Models for Polyacetylene

D. Baeriswyl

Laboratories RCA Ltd.
CH-8048 Zürich, Switzerland

1. Introduction

The theoretical work on the electronic structure of hydrocarbon chains dates back to the early days of quantum mechanics when the molecular orbital approach was introduced to deal with the extended wavefunctions of the π -electrons [1]. But only in 1959 was it clearly demonstrated that this scheme led to a ground state with alternating bond lengths and an electronic gap at the Fermi surface [2]. It was also recognized that simple defect structures, "bond-alternation defects", could exist as kinds of domain-walls between the two degenerate dimerization patterns [3]. The singly occupied local orbital associated with such a defect was proposed to give rise to the Curie-type magnetic susceptibility observed experimentally [4].

The simple molecular orbital approach is a one-particle theory of the Hartree-Fock type and thus, at least formally, neglects correlation effects. This has led to criticism against the bond-alternation picture and it was pointed out that by including correlation a spin-density wave was more likely to represent the stable ground state [5]. However, one has to keep in mind that, on the one hand, the parameters involved in molecular-orbital theories are not determined self-consistently but adjusted to experimental data and therefore correlation is taken into account in an averaged way. On the other hand, the criticism raised against these theories [5] was also based on a set of empirical parameters and in fact depended quite sensitively on their specific values. One expects that only involved ab initio calculations will finally decide about the relevance of correlation and, hopefully, the nature of the ground state. Recent self-consistent Hartree-Fock calculations, which take all s- and p-electrons into account, indeed suggest that correlation does play a major role since both the bandwidth and the bandgap (arising from bond alternation) are found to be larger by a factor of at least two than the values believed to be appropriate for polyacetylene [6].

The successful synthesis of highly crystalline $(CH)_x$ films by SHIRAKAWA and coworkers [7] and the subsequent discovery by the Penn group that the conductivity of these films could be increased from the semiconducting to the metallic regime upon exposure to various donor and acceptor species [8] produced great activity in the field [9], and an enormous amount of papers has been published since then, revealing many novel but sometimes also controversial results [10].

2. Polyacetylene

Polyacetylene is usually obtained as a film of randomly oriented fibers with a diameter of 200 Å and indefinite length which fill about 1/3 of the volume and show a high degree of crystallinity. If the material is synthesized below room temperature, the $(CH)_x$ chains appear to have mostly the cis-transoidal form, whereas subsequent heat treatment transforms them into the trans-transoidal form (Fig. 1). The crystal structure has not been determined unambiguously [11] and it has been suggested that, depending on the preparation, different modifications may be obtained [12].

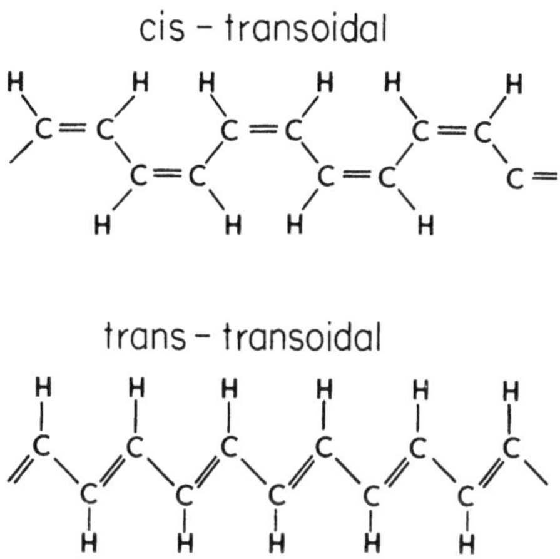

Fig. 1. The two isomeric forms of polyacetylene.

Due to the random orientation of the fibers (and thus also of the $(CH)_x$ chains) the electrical conductivity and optical absorption are found to be isotropic. Consequently the one-dimensional character of the chains can only be examined by means of specific spectral dependences reflecting the quasi one-dimensional density of states. On the other hand partially oriented material obtained either by stretch alignment [13] or by shear flow polymerization [14] does show some anisotropy in electrical and optical properties.

The mean chain length or, more precisely, the conjugation length can be estimated from Raman scattering experiments which suggest that the cis form consists mainly of long segments with more than 100 CH groups [15] whereas trans polyacetylene has a peak in the distribution of chain lengths at 60 CH groups [16]. Therefore one has to conclude that "conjugation defects", e.g. cross-links, are introduced during cis-trans isomerization.

3. Hückel-type Model for trans $(CH)_x$

Due to the complex morphology of $(CH)_x$, any theoretical description has to make drastic simplifications such as to consider a single idealized chain and to assume that the σ-electrons merely affect the elastic properties of the chain. Such a Hückel-type model has been introduced by SU, SCHRIEFFER and HEEGER in the form of the following Hamiltonian [17]

$$H = \frac{1}{2}K_0 \sum_n (u_{n+1}-u_n)^2 - \sum_{ns} t_{n,n+1} (c^+_{ns} c_{n+1s} + c^+_{n+1s} c_{ns}) . \qquad (1)$$

Here u_n denotes the displacement of the n-th CH group parallel to the chain axis, K_0 is a screened elastic constant [18], c^+_{ns} and c_{ns} are creation and annihilation operators, respectively, for π-electrons in the n-th Wannier state and with spin projection $s = \pm 1$, and the resonance integral $t_{n,n+1}$ is expanded in powers of the displacement coordinates

$$t_{n,n+1} = t_0 - \alpha(u_{n+1} - u_n) . \qquad (2)$$

For an infinite system this model implies bond alternation. In fact the Ansatz $u_n = (-1)^n u$ leads to a gap $2\Delta_0 = 8\alpha u$ at the Fermi level and a lowering of electronic energy proportional to $u^2 \ln u$, whereas the elastic energy increases quadratically as a function of u. The minimalization of the total energy yields the gap equation

$$2\Delta_0 \approx \frac{4}{e} W \exp\left(-\frac{1}{2\lambda}\right) , \qquad (3)$$

where $W = 4t_0$ is the π-bandwidth and the parameter $\lambda \equiv 2\alpha^2/(\pi t_0 K_0)$ has been assumed to be small, i.e. $\lambda \ll 1$. The second derivative of the ground state energy with respect to u is found to be $K_{eff} \approx 2\lambda K_0$ and corresponds to the effective force constant for oscillations around the new equilibrium positions. The values believed to be appropriate for polyacetylene are $W \approx 12$ eV, $2\Delta_0 \approx 1.4$ eV, $K_{eff} \approx 20$ eV Å^{-2}. They fix the three parameters of the theory $t_0 \approx 3$ eV, $K_0 \approx 50$ eV Å^{-2}, $\alpha \approx 6.9$ eV Å^{-1} [10, 19].

4. Bond Alternation and Bond Alternation Defects in Finite Chains

Consider a finite trans chain with N CH groups and the boundary condition $u_1 = -u_N$. For even N the Ansatz $u_n = (-1)^n u$ is compatible with this equation, whereas in an odd-numbered chain the bond-alternation sequence would have to change somewhere.

In an even-numbered chain and with cyclic boundary conditions an interesting difference exists between N = 4M and N = 4M+2 where M is an integer. In the former case the electronic ground state is degenerate for u = 0 and due to the Jahn-Teller theorem, bond alternation occurs for all values of M. In the latter case bond alternation is expected to occur only as a pseudo Jahn-Teller effect for long enough chains [1] ($M \geq 4$ for $\lambda \approx 0.2$). This difference is reflected in the particular dependences on chain length of the electronic gap or the effective elastic constant [10].

For odd-numbered chains there always exists a localized state at mid-gap, i.e. at the zero-energy level [20, 10]. In fact, seeking this state in the form $\sum_n \lambda_n c_{ns}$, the coefficients λ_n are obtained as

$$\lambda_{2m} = 0$$
$$\lambda_{2m+1} = (-1)^m \lambda_1 \prod_{\ell=1}^{m} (t_{2\ell-1,2\ell}/t_{2\ell,2\ell+1}) . \tag{4}$$

In the situation illustrated in Fig. 2 it is clear that $t_{2\ell-1,2\ell}/t_{2\ell,2\ell+1} > 1$ (< 1) at the left (right) hand side of the point where the bond-alternation sequence changes. Thus λ_{2m+1} increases by approaching this point and the state represented by Eq. (4) is localized. This result is independent of

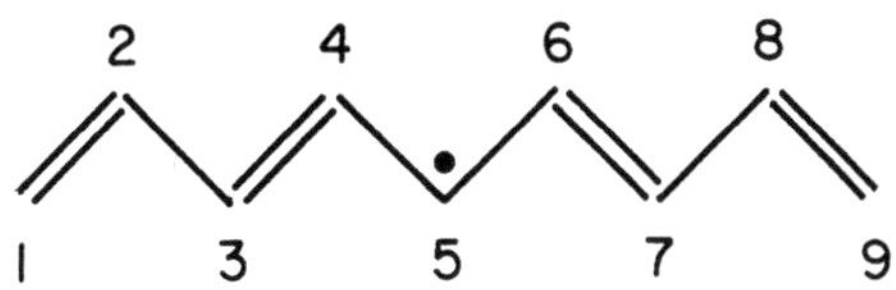

Fig. 2. Bond alternation defect in an odd-numbered chain.

the detailed form of the dimerization pattern and merely due to the odd number of lattice points. A natural form representing such a bond-alternation defect is [17, 21, 22]

$$u_n = (-1)^n u \tanh (n/n_0) . \tag{5}$$

With Eqs. (2) and (4) this pattern leads to a mid-gap state with coefficients

$$\lambda_{2m+1} \sim (-1)^m \left(\operatorname{sech} \frac{2m+1}{n_0}\right)^{n_0 \delta} , \tag{6}$$

where $\delta \equiv 2\alpha u/t_0 \approx 0.12$. RICE [21] and BRAZOVSKII [22] propose the relation $n_0\delta = 1$, which turns out to become exact in the continuum limit (below). Therefore the extent of the defect is of the order of 15 CH groups, as also found by numerical computation [17]. Due to this large extent, such a defect is only weakly pinned to the lattice and the continuum limit of the discrete model should be appropriate.

5. Kinks, Polarons and the Soliton Lattice

In the continuum limit, the slowly varying staggered displacement $(-1)^n u_n$ is replaced by a field $\varphi(x)$ and, instead of operators c_{ns} related to the Wannier function at site n, one uses two types of field operators representing π-electrons close to $+k_F$ and $-k_F$, respectively. The wavefunctions $v_\lambda(x)$ and $w_\lambda(x)$ associated with these field operators are determined by the Bogoliubov-de Gennes equations [23, 24]

$$\begin{aligned} -i\hbar v_F \, \partial_x v_\lambda + \Delta(x) w_\lambda &= \epsilon_\lambda v_\lambda \\ i\hbar v_F \, \partial_x w_\lambda + \Delta(x) v_\lambda &= \epsilon_\lambda w_\lambda , \end{aligned} \tag{7}$$

and the gap parameter $\Delta(x) \equiv 4\alpha\sqrt{a}\,\varphi(x)$ (a = lattice parameter) has to satisfy the self-consistency relation

$$\Delta(x) = -4\alpha^2 a/K_0 \sum_{\lambda s} n_\lambda \left[v_\lambda^* (x) w_\lambda (x) + v_\lambda (x) w_\lambda^* (x)\right], \tag{8}$$

where n_λ measures the occupation of the level λ.

It is remarkable that, besides the homogeneous solution $\Delta(x) = \Delta_0 =$ constant, exact inhomogeneous solutions of Eqs. (7) and (8) have been found (Fig. 3). The simplest case is the kink (or the anti-kink) which is nothing other than the continuum version of Eq. (5) with a single localized level at mid-gap, a characteristic length $\xi = \hbar v_F/\Delta_0$ and an energy $E_K = 2\Delta_0/\pi$ (relative to the homogeneous ground state) [24]. Due to its stability and particle-like properties it is often called a "soliton". It plays an important role in the discussion of experiments on undoped and lightly doped polyacetylene. In fact, depending on the occupation of the localized level, the soliton may be charged ($\pm e$) and spinless or charge-neutral with spin (Fig. 3a). Neutral solitons are thought to be present in undoped $(CH)_x$; charged solitons may be created upon doping.

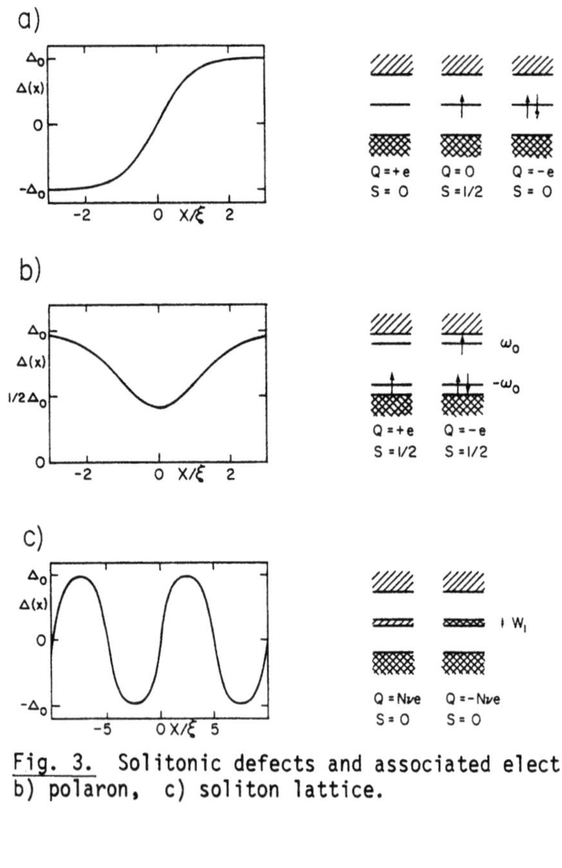

Fig. 3. Solitonic defects and associated electronic states: a) kink, b) polaron, c) soliton lattice.

The addition of a single electron to an even chain cannot generate a kink due to topological reasons, instead a "polaron" would be produced as found numerically for the discrete model [25] and analytically as an exact solution of Eqs. (7) and (8) [26]. The change in the displacement pattern leads to two localized electronic states with levels at $\pm\omega_0$, where $\omega_0 = \Delta_0/\sqrt{2}$. The energy of a polaron, with a level-filling as shown in Fig. 3b, would be $2\sqrt{2}\Delta_0/\pi \approx 0.9\,\Delta_0$.

Adding (or removing) more electrons leads to the problem of finding the ground state for $1\pm\nu$ π-electrons per CH group. For $\nu \ll 1$ it is expected to correspond to a "soliton lattice" [27, 28]. Indeed, as shown by HOROVITZ [29], such a solution of Eqs. (7) and (8) exists in the form of equally spaced kinks and anti-kinks (Fig. 3c). A very narrow electronic band at mid-gap is associated with such a configuration. The energy of the soliton lattice would have a minimum for $\nu_0 \approx 4\%$ (for the characteristic parameters of $(CH)_x$ [10]) and therefore at a light doping level ($\nu < \nu_0$) a phase separation is predicted to occur into a region with a soliton lattice of period ν_0 and one without solitonic defects.

6. Solitons in Undoped and Lightly Doped Polyacetylene

The defect structures described in the previous section are very natural states of the Hückel-type model Hamiltonian. If the model is appropriate for trans-polyacetylene, these defects should be observable although disorder, quantum and correlation effects may influence their appearance in a particular experiment. Moreover, it makes sense to call them solitons provided that the term "soliton" is not used in a mathematical sense but in a physical sense as a localized, quite stable and mobile intrinsic defect structure.

Notice that the cis form of $(CH)_x$ would exhibit quite different defect structures to the trans form since its ground state is non-degenerate (Fig. 1). Therefore a kink would be either pinned to a chain end or tightly bound to an anti-kink, forming sort of a "bi-polaron" [30]. This difference between the two isomers has been used to explain why photoconduction is observed in the trans but not in the cis form while photoluminescence is seen in cis- but not in trans-polyacetylene [31]. Quantitative and sometimes even qualitative differences also show up in magnetic experiments. A strong increase in the number of Curie spins is observed upon cis-trans isomerisation [32] in agreement with the Raman data, which indicate that in trans $(CH)_x$ the chains are shorter than in the cis form. In addition, the ESR line associated with these spins becomes very narrow. This has been interpreted as evidence for highly mobile spins in trans $(CH)_x$ [33] and attributed to neutral kinks [34]. NMR experiments support this interpretation [35] although a second species of immobile spins (possibly localized at cross-links) has to be invoked in order to reconcile the ESR and NMR data [36]. On the other hand, this picture has been questioned on the basis of spin-echo measurements [37].

Doping has been proposed as a mechanism leading to the formation of charged solitons [17, 21]. Characteristic effects observed after introducing various types of donor or acceptor molecules into $(CH)_x$ films are quite independent of the particular dopant used, indicating that doping merely consists in adding or removing π-electrons. The subsequent structural relaxation is then expected to produce charged solitonic defects.

Charged kinks are predicted to give rise to strong optical absorption in the visible (transitions involving the mid-gap level) [38, 39] and in the infrared (due to local vibrations of the defects)[40]. Additional optical absorption upon doping is indeed observed at wavelengths close to the predicted values, although appreciably broadened, yielding support to the soliton picture [10].

An interesting question is if solitons will be involved in the charge transport. The simplest mechanism, based on depinning of charged solitons from the ions of the dopant, requires an activation energy of about 0.3 eV. KIVELSON has proposed phonon-assisted hopping of electrons between bound, charged kinks and mobile neutral kinks as an alternative mechanism, at least at a low doping level [41]. The competing anisotropies due to preferential overlap of π-electrons along one chain and the larger probability of finding two solitons close to each other on different chains leads to an essentially three-dimensional transport. The electrical conductivity of this model depends in a complicated way on the dopant concentration and varies strongly with temperature, $\sigma \sim T^{13.7}$, and frequency, $\sigma(\omega) \sim \omega(\ln \omega)^4$. It appears that experiments on trans $(CH)_x$ can be nicely fitted to this theory [42].

The interpretation of magnetic, optical and electrical experiments on pristine and lightly doped trans $(CH)_x$ in terms of the soliton concept is indeed quite satisfactory in view of the simplicity of the model. However, it is not yet clear if it represents not only a useful, but a necessary framework to deal with the experimental observations. Also, details like the origin for the broad linewidths found in IR absorption [43] or the different constants for spin diffusion as obtained from ESR, NMR and spin-echo measurements remain to be clarified.

7. Heavily Doped Polyacetylene and the Nature of the Semiconductor-Metal Transition

The self-consistent set of equations (7) and (8) has been derived to deal with a half-filled band where the order parameter can be taken as real, or more precisely, the phase of the complex order parameter is completely pinned and the excitations are amplitude solitons. For a nearly half-filled band, the fluctuations in the phase become relevant and finally, in the "incommensurate limit" where the (Peierls) distortion of the ground state has a period $2k_F = \pi(1 \pm \nu)/a$, the phase of the order parameter and the associated electronic charge become free to move (at least in the ideal system [44]; in the real system the phase is pinned by impurities [45]). In this limit the electronic gap is strongly reduced, $\Delta_{inc} \approx 0.1\,\Delta_0$, and correspondingly the distortion amplitude is much smaller than for a half-filled band. It is in fact so small that quantum fluctuations (and possibly also disorder [46]) are expected to wipe out the lattice distortion and to fill the electronic gap. (Quantum effects should also be appreciable in the case of a half-filled band[10] although numerical calculations indicate that the probability distribution for the displacement amplitudes remains doubly peaked [47]).

The Pauli-like susceptibility observed in heavily doped polyacetylene is indeed consistent with a filled gap, and the density of states deduced from these measurements corresponds to an effective transfer integral $t \approx 2.25$ eV [48]. In addition, the optical absorption becomes Drude-like for most dopants, at least in the visible and near infrared [49, 50]. Thus, it is tempting to associate heavily doped $(CH)_x$ with a one-dimensional metal, although the temperature dependence of the electrical conductivity is in general not metal-like (this may be related to the fibrillar nature of the material [51]). The question as to how the transition to this state occurs is controversial. Some authors claim that the Pauli susceptibility remains zero up to a dopant concentration of 6%, above which it suddenly increases towards its asymptotic value, whereas the transition to the highly conducting state takes place around 1% [52, 53]. The nature of the spinless charge carriers which have to be postulated in the intermediate region is not clear, in particular since a similar effect has been observed in polyparaphenylene [54]. Other authors did not find this step-like change in susceptibility but a more gradual increase consistent with the formation of metallic clusters [37]. A similar gradual increase found with iodine doped material has been attributed to inhomogeneous doping of the fibrils [55] which seems to be particularly pronounced with this dopant [56].

8. Concluding Remarks

The experimental data for undoped and weakly doped (trans) polyacetylene are quite consistently explained in terms of bond alternation which leads to an electronic gap at the Fermi level and bond-alternation defects ("solitons") which are responsible for the motionally narrowed ESR line, certain peaks in the optical absorption spectrum and unusual electrical transport. Heavily doped $(CH)_x$ can be described in terms of an incommensurate Peierls distortion which is quenched due to quantum fluctuations and disorder leading to a closed-gap "metallic" state. However, the nature of the transition has to be clarified, in particular with respect to the possible existence of an intermediate phase with spinless charge carriers.

Acknowledgement

The author wishes to thank R. Morf for discussions and a critical reading of the manuscript.

References

1. See L. Salem, The Molecular Orbital Theory of Conjugated Systems, Benjamin, London, 1966.
2. H.C. Longuet-Higgins and L. Salem, Proc. Roy. Soc. A, 251, 172 (1959).
3. J.A. Pople and S.H. Walmsley, Molec. Phys. 5, 15 (1962).
4. E.g. G.M. Holob, P. Ehrlich and R.D. Allendoerfer, Macromolecules 5, 569 (1972).
5. R.A. Harris and L.M. Falicov, J. Chem. Phys. 51, 5034 (1969); A.A. Ovchinnikov, I.I. Ukrainskii and G.V. Kventsel, Usp. Fiz. Nauk 108, 81 (1972) (Sov. Phys. Usp. 15, 575 (1973)). See also the contribution of P. Horsch in this volume.
6. For recent work see S. Suhai, J. Chem. Phys. 73, 3843 (1980); A. Karpfen and R. Höller, Solid State Commun. 37, 179 (1981).
7. T. Ito, H. Shirakawa and S. Ikeda, J. Polym. Sci., Polym. Chem. Ed. 12, 11 (1974).
8. C.K. Chiang, C.R. Fincher, Y.W. Park, A.J. Heeger, H. Shirakawa, E.J. Louis, S.C. Gau and A.G. MacDiarmid, Phys. Rev. Lett. 39, 1098 (1977).

9. See e.g. the proceedings of two recent conferences, Physics in One Dimension , eds. J. Bernasconi and T. Schneider, Springer Series in Solid State Sciences, Vol. 23, Berlin 1981 and "Low-Dimensional Synthetic Metals" (Helsingor, 10-15 August 1980) to be published in Chemica Scripta.
10. For a recent review see D. Baeriswyl, G. Harbeke, H. Kiess and W. Meyer, to be published in Electronic Properties of Polymers , eds. J. Mort and G. Pfister, Wiley, New York.
11. R.H. Baughman, S.L. Hsu, G.P. Pez and A.J. Signorelli, J. Chem. Phys. 68, 5405 (1978).
12. G. Lieser, G. Wegner, W. Müller, V. Enkelmann and W.H. Meyer, Makromol. Chem. Rap. Comm. 1, 627 (1980).
13. H. Shirakawa and S. Ikeda, Synth. Met. 1, 175 (1980).
14. W.H. Meyer, Synth. Met. 4, 81 (1981).
15. L.S. Lichtman and D.B. Fitchen, Synth. Met. 1, 139 (1980).
16. H. Kuzmany, phys. stat. sol. (b) 97, 521 (1980).
17. W.P. Su, J.R. Schrieffer and A.J. Heeger, Phys. Rev. Lett. 42, 1698 (1979) and Phys. Rev. B 22, 2099 (1980).
18. See Ref. 10, Section 3.2 for a discussion of the origin of the screening.
19. D. Vanderbilt and E.J. Mele, Phys. Rev. B 22, 3939 (1980).
20. G.V. Kventsel and Y.A. Kruglyak, Theor. Chim. Acta 12, 1 (1968).
21. M.J. Rice, Phys. Lett. 71A, 152 (1979).
22. S.A. Brazovskii, Pis'ma Zh. Eksp. Teor.Fiz. 28, 656 (1978) (Sov.Phys. JETP Lett. 28, 606 (1979)).
23. P.G. de Gennes, Superconductivity of Metals and Alloys , Benjamin, New York, 1966, Chapter 5.
24. H. Takayama, Y.R. Lin-Liu and K. Maki, Phys. Rev. B 21, 2388 (1980).
25. W.P. Su and J.R. Schrieffer, Proc. Natl. Acad. Sci. USA 77, 5626 (1980).
26. D.K. Campbell and A.R. Bishop, preprint 1981.
27. A. Kotani, J. Phys. Soc. Japan 42, 416 (1977).
28. M.J. Rice and J. Timonen, Phys. Lett. 73A, 368 (1979).
29. B. Horovitz, Phys. Rev. Lett. 46, 742 (1981).
30. K. Dransfeld, private communication.
31. L. Lauchlan, S. Etemad, T.C. Chung, A.J. Heeger and A.G. MacDiarmid, preprint 1981.
32. H. Shirakawa, T. Ito and S. Ikeda, Makromol. Chem. 179, 1565 (1978).
33. I.B. Goldberg, H.R. Crowe, P.R. Newman, A.J. Heeger and A.G. MacDiarmid, J.Chem.Phys. 70, 1132 (1979).
34. B.R. Weinberger, J. Kaufer, A.J. Heeger, A. Pron and A.G. MacDiarmid, Phys. Rev. B 20, 223 (1979).
35. M. Nechtschein, F. Devreux, R.L. Greene, T.C. Clarke and G.B. Street, Phys. Rev. Lett. 44, 356 (1980).
36. K. Holczer, J.P. Boucher, F. Devreux and M. Nechtschein, Phys. Rev. B 23, 1051 (1981).
37. Y. Tomkiewicz et al. in Physics in One Dimension , eds. J. Bernasconi and T. Schneider, Springer Series in Solid State Sciences, Vol. 23, Berlin, 1981.
38. N. Suzuki, M. Ozuki, S. Etemad, A.J. Heeger and A.G. MacDiarmid, Phys. Rev. Lett. 45, 1209 (1980), Erratum in Phys. Rev. Lett. 45, 1463 (1980).
39. J.T. Gammel and J.A. Krumhansl, preprint 1980.
40. E.J. Mele and M.J. Rice, Phys. Rev. Lett. 45, 926 (1980).
41. S. Kivelson, Phys. Rev. Lett. 46, 1344 (1981) and Preprint 1981.
42. A.J. Epstein, private communication.
43. S. Etemad et al. preprint 1980.
44. H. Fröhlich, Proc. Roy. Soc. A 223, 296 (1954).

45. P.A. Lee, T.M. Rice and P.W. Anderson, Solid State Commun. 14, 703 (1974).
46. M.J. Rice and E.J. Mele, Proc. Int. Conf. on Low-Dimensional Synthetic Metals (Helsingor, 10-15 Aug. 1980), to be published in Chemica Scripta.
47. W.P. Su, unpublished.
48. B.R. Weinberger, J. Kaufer, A.J. Heeger, A. Pron and A.G. MacDiarmid, Phys. Rev. B 20, 223 (1979).
49. R. Fincher Jr. et al., Phys. Rev. B 20, 1589 (1979).
50. M. Tanaka, A. Watanabe and J. Tanaka, Bull. Chem. Soc. Japan 53, 645 (1980).
51. P. Sheng, Phys. Rev. B 21, 2180 (1979).
52. S. Ikehata et al., Phys. Rev. Lett. 45, 1123 (1980).
53. M. Peo et al., Solid State Commun. 38, 467 (1981).
54. M. Peo et al., Solid State Commun. 35, 119 (1980).
55. A.J. Epstein, H. Rommelmann, M.A. Druy, A.J. Heeger, and A.G. MacDiarmid, preprint 1981.
56. See the contribution by J.J. Ritsko in this volume.

Electronic Correlation Effects in Cyclic Polyenes and in Polyacetylene

Peter Horsch

Max-Planck-Institut für Festkörperforschung
D-7000 Stuttgart 80, Fed. Rep. of Germany

1. Introduction

Most papers written nowadays on polyacetylene [1,2] seem to accept as a well-established fact that the origin of the optical gap in trans-$(CH)_x$ is due to bond alternation arising from the Peierls instability. Yet a completely different explanation for the origin of the optical gap has been given by OVCHINNIKOV and coworkers [3], who suggested that the gap results from electronic correlations and not from bond alternation. Estimating the relevant physical parameters for polyacetylene from substances such as ethane, benzene, ethylene and graphite [4], they concluded that the electron-phonon coupling is much too small to produce a bond alternation large enough to explain the measured gap.

Recently we have studied the role of electronic correlations on the bond alternation in polyacetylene [5]. These calculations have shown that electronic correlations can induce a substantial bond alternation even when the electron-phonon coupling is weak. Here we present new results for cyclic polyenes of size 4N + 2 and a comparison is made with the results found for trans-$(CH)_x$. Finally we discuss the discrepancy found for the electron-phonon coupling constant when estimated either from the Raman frequency [6] or from the bond order-bond length relation [4,7], which can be resolved when considering correlation effects.

2. Bond Alternation in Closed Shell Cyclic Polyenes

The electronic states of polyenes close to the Fermi level can be well approximated by a tight-binding Hamiltonian [8] containing only π-electrons explicitly.

$$H = H_0 + H', \tag{1}$$

$$H_0 = - \sum_{n,\sigma} t_{n+1,\,n}(c^+_{n+1\sigma}c_{n\sigma} + \text{H.c.}) + \frac{1}{2}\sum_n K\rho^2{}_n, \tag{2}$$

$$H' = U\sum_n n_{n\uparrow}n_{n\downarrow} \tag{3}$$

Here $c^+_{n\sigma}$ and $c_{n\sigma}$ are fermion creation and annihilation operators, respectively, referring to atomic $2p_z$-orbitals at site n and with spin σ. The π-electron number operator is denoted by $n_{n\sigma} = c^+_{n\sigma}c_{n\sigma}$. The first part of the Hamiltonian (1) is equivalent to the Hamiltonian used by SU et al. [1]. The resonance integral $t_{n+1,n}$ is changed when the bond length of the n'th bond is changed by ρ_n. To first order in ρ_n the resonance integral is given by

$$t_{n+1,n} = t_o - \alpha\rho_n, \quad (4)$$

where α is the electron-lattice coupling constant, K is the bond stretching force constant. Note that here the definitions of the constants K and α refer to changes of the bond length and differ from those of [1] by factors of 4/3 and $2/\sqrt{3}$, respectively. The electronic interaction H' is simplified by a Hubbard interaction, modifications due to the long range character of the Coulomb interaction have been discussed elsewhere [5,8].

Within the Born-Oppenheimer approximation it poses no problem to solve for the electronic structure for a given dimerization of the polyene

$$\rho_n = (-1)^n \rho_o , \quad (5)$$

while treating the Hubbard interaction within mean-field theory [1,7]. For illustration we give here the resulting total energy for the case of $(CH)_x$ in the limit where the bond alternation is small.

$$E_{MF} \simeq -\frac{2to}{\pi} \left[2+(\ln\frac{4}{\delta} - \frac{1}{2})\delta^2 - \frac{1}{2\lambda}\delta^2\right] + \frac{U}{4} . \quad (6)$$

Here the parameter $\delta = \alpha\rho_o/t_o$ characterizes the distortion according to (4) and the coupling constant λ, defined as

$$\lambda = \frac{2\alpha^2}{\pi K t_o} , \quad (7)$$

determines the strength of the Peierls distortion. The Peierls instability has its origin in the $\delta^2\ln\delta$ term which arises from the kinetic energy. The minimum of the total energy is at $\delta = (4/e)\exp(-1/2\lambda)$ and is not affected by the Hubbard interaction within the mean-field approximation. The explanation for this is that the bond alternation is characterized by alternating bond orders [7] and not by alternating atomic charges.

2.1 Correlation Energy

The treatment of the electronic interaction within the mean-field approximation implies the complete neglect of correlations between electrons with antiparallel spin. The correlation energy is defined as the difference of the mean-field or Hartree-Fock ground state energy and the exact one. Here we employ a variational method to study the correlation problem where the correlated many body ground state is constructed from the ground state within the independent particle picture by means of a certain set of local operators. The method and applications of it have been outlined elsewhere [5,9,10].

The most important correlations are those which prevent two electrons with opposite spins to be at the same site. This occurs with a probability of 1/4 in the mean-field ground state and gives rise to the energy contribution U/4 in (6).

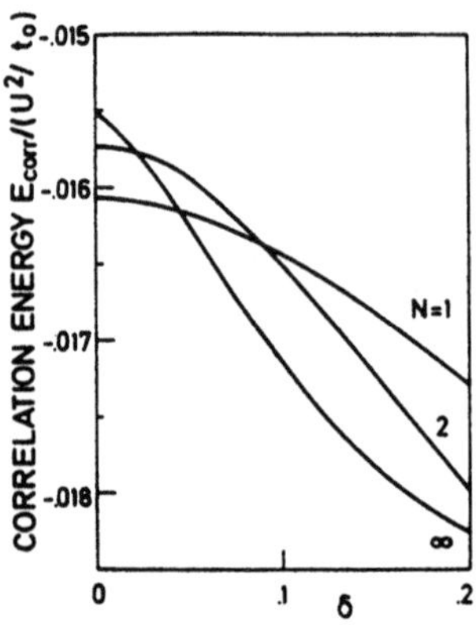

Fig. 1 Correlation energy of closed shell polyenes of size 4N+2 versus the distortion parameter δ. The correlation energy of the N = 10 cyclic polyene is rather close to that of polyacetylene, $N=\infty$. Note that the correlation energy curves have a minimum at about $\delta=.25$.

The leading term of the correlation energy which is proportional to U^2 is shown in Fig. 1. The correlation energy increases with increasing δ and therefore involves an enhancement of the bond alternation. This effect is particularly strong in the case of $(CH)_x$, where the numerical result for the correlation energy can be fitted, for small δ, by a linear function of δ

$$E_{corr} \simeq - \frac{U^2}{t_o} (c_o + c_1\delta) \text{ for } 0 < \delta \leq 0.1 \ , \tag{8}$$

where $c_0 = 0.0157$ and $c_1 = 0.015$. For very small δ the linear dependence changes into δ^μ, where μ is numerically found to be 1.15. The U^2 expansion of the correlation energy is known to give reliable results for $U/4 \cdot t_o < 1$.

These results change the accepted picture for the Peierls instability substantially, since the linear term in the correlation energy (8) will replace the $\delta^2 \ln\delta$-term of the kinetic energy (6) as driving term of the bond alternation.

2.2 Total Energy and Raman Force Constant

The total energy for a large cyclic polyene is shown in Fig. 2, the effect of the correlation energy is to enhance: (i) the

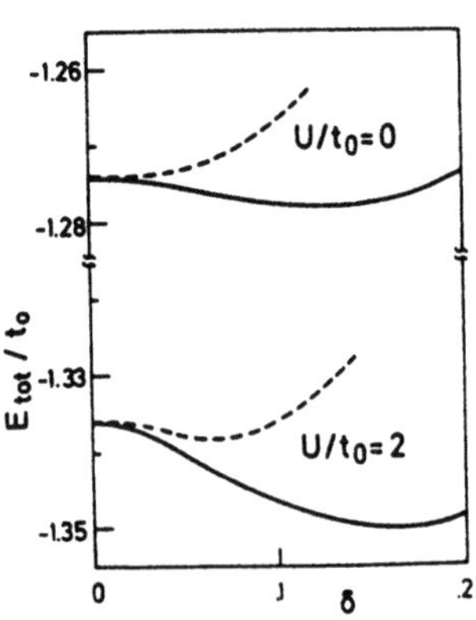

Fig. 2 Total energies for the N = 12 cyclic polyene for $\lambda = 0.12$ (dashed line) and 0.20 (solid line) for two different values of U/t_o. In the plot of $E_{tot} = E_{MF}+E_{corr}$ versus δ the constant mean-field interaction $U/4(6)$ has been left out.

bond alternation, (ii) the condensation energy of the dimerized structure and (iii) the force constant corresponding to the antisymmetric vibrational mode. The effect is particularly pronounced when the coupling constant λ is small, in this case the dimerization is almost completely due to the correlation energy.

The total energies obtained within the variational approach employed here are much lower than those computed within an unrestricted Hartree-Fock treatment, i.e. assuming a SDW-ground state [5,13,14].

The second derivative of the total energy at the minimum δ is related to the force constant K_T of the antisymmetric Raman active phonon mode [6] which has been measured in $(CH)_x$ at $1470\ cm^{-1}$.

From (6) and (8) we obtain for $(CH)_x$

$$K_T = \frac{4\alpha^2}{\pi t_o} Z_R , \qquad (9)$$

where the renormalization constant Z_R,defined as

$$Z_R = [\ 1 + \frac{\pi U^2}{4t_0^2} \quad \frac{c_1}{\delta}\ ], \qquad (10)$$

describes the enhancement of K_T due to correlations.

For the cyclic polyenes K_T depends sensitively on the ring size since the antisymmetric vibrational mode goes to zero at a particular N_c above which the polyene rings are dimerized. We define the size dependent renormalization constant Z_R^N according to (9) via

$$K_T^N = \frac{4\alpha^2}{\pi t_o} Z_R^N , \qquad (11)$$

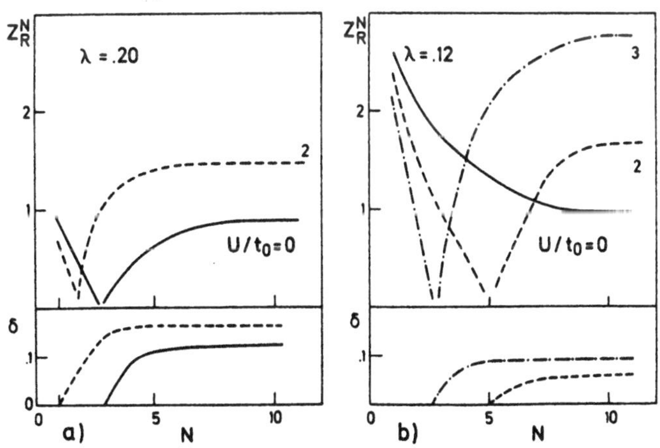

Fig. 3 Numerical results for the renormalization constant Z_R^N(11) corresponding to the antisymmetric phonon mode in (4N+2) cyclic polyenes versus N. In the lower part of the figures the corresponding distortion parameters δ are shown.

which is also strongly dependent on U as can be seen from Fig. 3. The instability in cyclic polyenes is supposed to occur at N_c larger than 4 [11]. If correlations are neglected, i.e. $U/t_o = 0$, alternation exists for $\lambda = 0.2$ [1,6] for $N > 3$, wheras for $\lambda = 0.12$ [4] there is no alternation for the polyenes considered in Fig. 3(b). However, a rather small interaction of $U/t_o = 2$ is sufficient to produce an alternation of $\delta = 0.06$.

2.3 Determination of α and U from Elastic Constants of Benzene and Polyacetylene.

Using the elastic constant characterizing the B_{2u}-mode of benzene $K_T^1 = 24.6$ [eV/Å] [7] and the corresponding constant of polyacetylene $K_T = 25.9$ [eV/Å^2] [6] we can derive a $U(\lambda)$ curve from the requirement that according to (9) the equation

$$K_T^1/K_T = Z_R^1(\lambda, U/t_o)/Z_R(\lambda, U/t_o) \tag{11}$$

must be satisfied. Furthermore combining (7) and (9) we obtain a second equation

$$K_T/K = 2\lambda Z_R \ , \tag{12}$$

which determines a particular λ and U. Assuming for K the bond-stretching force constant of benzene, i.e. $K = 47.6$[eV/Å^2] [7,4], we find $\lambda = 0.12$ and $U/t_o = 3$. Using $t_o = 2.5$[eV] we obtain for the electron-phonon coupling constant $\alpha = 4.7$ [eV], which is close to $\alpha = 4.6$ [eV] derived from the bond order - bond length relation [4,11], whereas without considering correlation effects one finds a coupling constant approximately twice as large [6]. Yet it has been pointed out that the α derived from bond length changes of different molecules, such as ethane and benzene, could be masked by atomic potential effects [12], and therefore should not be used for polyacetylene.

3. Conclusions

We have examined the influence of the electronic interaction on the bond alternation in cyclic polyenes using the framework of a recently developed local approach to the computation of correlation energies. For a Hubbard interaction which has been studied here, electronic correlations strongly favour bond alternation. For realistic values of U/t_o the gain of correlation energy due to dimerisation can be much larger than the gain of kinetic energy, which is the driving force for the Peierls transition. This result implies a significant change in the accepted picture of a Peierls transition and it clearly shows that electronic correlation effects have to be adequately taken into account.

Finally we want to mention that the results obtained above are modified if one considers the case of a slowly decreasing interatomic interaction [5]. Then one finds that long-range correlation effects, which are particularly large in the metallic regime, turn out to stabilize the non-alternating structure.

Acknowledgments

The author wants to express his sincere thanks to Dr. S.Strässler and Prof. P. Fulde for their continuous interest in this work and for many stimulating discussions.

References

1. W.P.Su, J.R. Schrieffer, and A.J. Heeger, Phys.Rev.Lett. 42, 1698 (1979).
2. A. Karpfen and R. Höller, Solid State Commun. 37, 179 (1981).
3. A.A. Ovchinnikov, I.I. Ukrainskii, and G.V. Kventsel, Sov. Phys. Usp. 15, 575 (1973).
4. T. Kakitani, Prog. Theor. Phys. 51, 656 (1974).
5. P. Horsch, submitted to Phys. Rev. B
6. D. Vanderbilt and E.J. Mele, Phys. Rev. B 22, 3939 (1980).
7. H.C. Longuet-Higgins and L. Salem, Proc. Roy. Soc. A 251, 172 (1959).
8. I. Ohmine, M. Karplus, and K. Schulten, J. Chem. Phys. 68, 2298(1978).
9. G. Stollhoff and P. Fulde, Z. Phys. B 26, 251(1977); J. Chem. Phys. 73, 4548 (1980).
10. P. Horsch and P. Fulde, Z. Phys. B 36, 23 (1979).
11. L. Salem, Molecular Orbital Theory of Conjugated Systems (W.A. Benjamin, New York, 1966) p. 514.
12. S. Strässler and L. Pietronero, Festkörperprobleme 21, 313 (1981).
13. R.A. Harris and L.M. Falicov, J. Chem. Phys. 51,5034 (1969).
14. K.R. Subbaswamy and M. Grabowski, submitted to Phys. Rev. B.

X-Ray and Shubnikov-De Haas Studies of $C_{8n}AsF_5(n=1,2,3)$

R.S. Markiewicz

Northeastern University, Boston, MA 02115, USA and

J.S. Kasper, H.R. Hart, Jr., and L.V. Interrante

General Electric Research & Development Center
Schenectady, NY 12301, USA

$(0,0,\ell)$-X-ray-reflection studies of single crystal and HOPG graphites intercalated with AsF_5 have provided maps of the c-axis charge density distribution of these intercalates. Analysis of this distribution allows an estimate of the ratio of AsF_6^-:AsF_5 in the intercalant layers. This in turn yields the charge transfer f (how far the reaction $3\ AsF_5 + 2\ e^- \rightarrow 2\ AsF_6^- + AsF_3$ has proceeded to the right). Presence of significant AsF_5 shows that in stage 2, $f \approx 0.4 - 0.45$, considerably below the maximum value of 2/3.

Shubnikov-de Haas oscillations provide an independent means of measuring this charge transfer, giving the electronic density directly from the total area of the (two-dimensional) Fermi surfaces. We report the observation of large area Fermi surfaces in all three stages, which yield values of f in substantial agreement with the chemical measure described above, f = 0.37, 0.41, 0.45 for stages (n =) 1, 2, 3. In addition, the areas and effective masses are in good agreement with BLINOWSKI and RIGAUX's theoretical interpretation of optical spectra. The stage 3 samples show significant non-uniformities in the c-axis distribution of charge. The values of f are in substantial agreement with earlier estimates of charge transfer.

1. Introduction

For a fundamental understanding of the process of graphite intercalation and of the high electrical conductivities of several of the resulting compounds, it is necessary to determine the mechanism and degree of charge transfer between intercalant and graphite. The reaction which controls intercalation of AsF_5 is believed to be [1]:

$$3\ AsF_5 + 2\ e^- \rightarrow 2\ AsF_6^- + AsF_3\ . \tag{1}$$

Chemically, the degree of charge transfer is specified by f_{chem}, the fraction of intercalated molecules converted to AsF_6^-. If (1) goes to completion, $f_{chem} = 2/3$, but there is evidence that some unreacted AsF_5 enters the graphite. The conductivity and other electronic properties are determined not by f_{chem} but by f_{elec}, the fraction of charge in the graphite layer which is delocalized. In general, it is possible that $f_{elec} < f_{chem}$, with the difference going to produce localized C-F bonds [2].

Determination of f_{chem} and f_{elec} for the graphite-AsF_5 system is important not only in understanding the properties of these highly conducting compounds [3,4], but also because similar chemical reactions may govern the intercalation of many other acceptor compounds [5]. In sections 2 and 3 we shall describe our experimental studies of the graphite-AsF_5 system, and the information these provide about f_{chem} and f_{elec}. In section 4 we shall compare our results with other studies.

2. X-ray Structure Studies

Our studies have concentrated on two complementary techniques, one of which samples f_{chem}, the other f_{elec}. By x-ray structure analysis [6] we attempt to determine what chemicals are actually present in the intercalant layers, while measurements of Shubnikov-de Haas (SdH) and de Haas-van Alphen (dHvA) oscillations [7] are used to find the Fermi surfaces of the materials.

The amplitudes of the $(0,0,\ell)$ x-ray reflection peaks are Fourier transformed to produce a charge density distribution as shown in Fig.1 (top). From the observed distribution, the following results can be established. (1) The atomic positions are consistent with the presence of undistorted molecules of AsF_3, AsF_5, and AsF_6^-. (2) The trigonal-bipyramid AsF_5 must have its axis tilted by an angle $\theta \sim 35\text{-}65°$ (Fig.1) with respect to the graphite c-axis. (3) The well-defined maxima associated with flourine show that the molecules are not freely rotating about an axis perpendicular to the c-axis. (4) The intercalant layer density is not constant in a given stage. For stage 2, nominally $C_{16}AsF_5$, the As:C ratio can vary from the stoichiometric 1:16 (single crystal sealed with excess AsF_5) to about 1:20 (HOPG stored under N_2 vapor). X-rays showed that both samples were of stage 2, but the latter had a smaller c-axis spacing. (5) An attempt to separate the flourine peak into separate contributions from AsF_5, AsF_6^-, and AsF_3 suggests that $f_{chem} \simeq 0.4$ for both samples, with the intercalated molecules being mostly neutral AsF_5 or AsF_3.

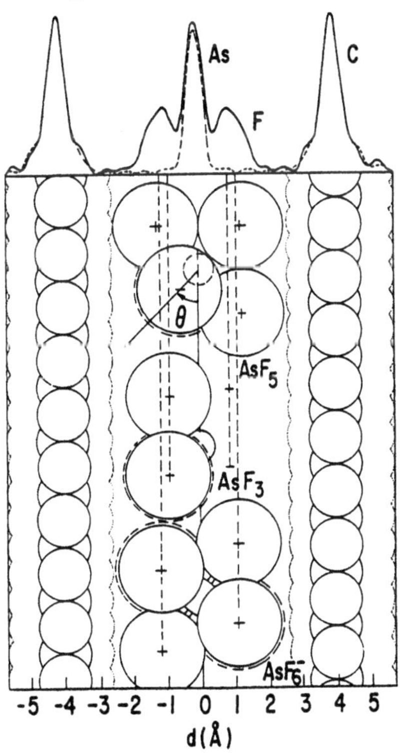

Fig.1 Top. Observed charge density distribution (solid line) in $C_{16}AsF_5$, with fit (dashed line) to the C and As peaks. Bottom. Molecules which may be present in intercalate layer, drawn to scale (dotted lines = limits of single layer, 3.353 Å across).

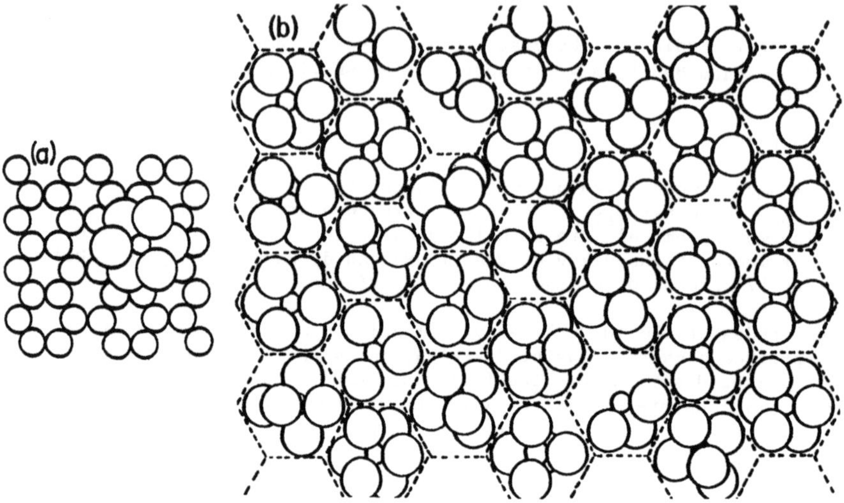

Fig.2a Possible positioning of AsF_6^- molecule on C-layer.
Fig.2b Possible arrangement of (disordered) AsF_5, AsF_3, AsF_6^- in intercalant plane, corresponding to stoichiometric As:C ratio and $f_{chem} \simeq 0.4$.

Figure 2 shows a possible in-plane arrangement of the molecules. McCARRON and BARTLETT [8] suggest that the As in AsF_6^- lies preferentially above a C-atom, so that the three F-ions can rest over open spaces in the C-network (Fig.2a). Such an arrangement requires an A-B stacking of graphite planes across the intercalant layer, which we observe [6]. The tilting of the AsF_5 allows three F to assume a triangular shape similar to one face of AsF_6^-. The resulting structure in Fig.2 suggests that vacancies in the fluorine layer can allow F-ions to hop between cells. This could explain the rapid F-exchange observed in NMR experiments [9].

3. The Fermi Surface of $C_{8n}AsF_5$

Several groups have now observed S-dH and dHvA oscillations in low stage AsF_5-intercalated graphite [7,10,11], and the results, where they overlap in stages 2 and 3, seem to be approximately consistent [see the discussions in Refs. 7 and 11]. Fig.3 shows the Fourier transforms of SdH oscillations for stages 1-3. The low frequency oscillations are often variable from sample to sample, while the high frequencies are much more constant. We interpret these spectra on the basis of model 2-d Fermi surfaces proposed by BLINOWSKI and RIGAUX [12]. For stage n, there are n cylindrical sections of Fermi surface, with their splitting determined by intralayer (γ_0) and interlayer (γ_1) interactions. For stage $n \geq 3$, the hole distribution is nonuniform in the various graphite layers (this is determined by the parameter δ). We find this screening effect to be strong, with only 11% of the holes on the interior graphite layer (stage 3). For stage 2 we interpret the SdH frequencies A and B to be the two fundamental sections of Fermi surface with the other periods being mixing frequencies of the form A ± B, A ± 2B. Making similar identifications for stages 1 and 3, we derive the band parameters listed in Table 1. Once the band structure is known, the effective masses can be calculated and, as seen from the Table, they are in general agreement with the masses measured from the

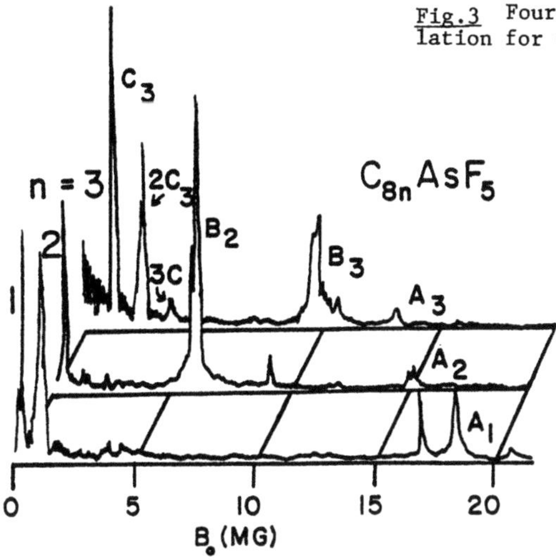

Fig.3 Fourier transform of SdH oscillation for $C_{8n}AsF_5$, stages 1-3.

Table 1 Properties of $C_{8n}AsF_5$

Property	Stage	Present Paper (SdH Oscillations)	Blinowski, et al [12] (Optical Reflection Spectra)	Ritsko and Mele [22] Electron Energy Loss
f	1	0.37	0.37[+]	
	2	0.41	0.43[+]	0.4
	3	0.45		
E_F	1		1.3 eV	
	2		1.0 eV	0.9 eV
	3	0.77 eV[+]		
γ_0	1	2.60 eV		
	2	2.68 eV	3.12-2.4 eV	2.8 eV
	3	(2.60 eV)[+]		
γ_1	2	0.43 eV	0.38 eV	
	3	0.43 eV[+]	0.38 eV	
δ	3	0.15[+]	~0.20[+]	
		expt. / thy.		
m^*/m_0	1	0.38 / 0.32		
	2	0.18 (α2) / 0.19		
		0.47 (α5) / 0.29		

[+] Assumes γ_0 = 2.60 eV

T-dependence of the SdH amplitude. The charge transfers found this way are $f_{elec} \simeq 0.37$ (stage 1), 0.41 (2), and 0.45 (3). As the samples age, f seems to increase slightly. This is probably related to the decrease of c-axis sandwich thickness and strongly suggests that only inert molecules (AsF_5, AsF_3) are lost from the lattice.

The above results show that a consistent picture of charge transfer can be formulated for the compounds $C_{8n}AsF_5$, with $f_{chem} \simeq f_{elec} \simeq 0.4$. A more careful examination of the SdH experiments, however, suggests that this picture is only a first step in understanding the band structure of these compounds. The presence of mixing frequencies and of additional very low frequencies is not explained by the present model. Also, the non-saturation of the magnetoresistance [7] may be indicative of magnetic breakdown effects. It may be that the actual low-field Fermi surfaces are more complex, due to the presence of superlattices. In high fields magnetic breakdown occurs across the minigaps, reproducing the simpler underlying structure. Further study is necessary to confirm this.

4. Comparison with Earlier Estimates of f_{chem}, f_{elec}

Extreme care must be taken in handling these samples, because of the ease with which they deintercalate. Thus, while the stoichiometric formula is $C_{8n}AsF_5$ for stage n, stage 2 compounds stored under N_2 have the composition $C_{20}AsF_5$ [6]. Additional material is lost upon exposure to moist air or vacuum. Thus stage 1 samples can be vacuum pumped to a mixture of stage 1 and 2, with a composition $C_{20}AsF_5$ [2] or $C_{26}AsF_5$ [13], while stage 2 samples can be pumped to stage 3, $C_{34}AsF_5$ [2]. Moreover, not all of the loss need be neutral molecules. Thus dHvA studies [7,10,11] show a range of f $\sim$ 0.4 - 0.47 for stage 2, but this decreases to $\sim$ 0.32 in air exposed samples. Many of the early studies were done on a stage 1 compound $C_{10}AsF_5$, indicating that $\sim$ 20% of the AsF_5 had already deintercalated. X-ray studies [13,14] show this is a mixture of stages 1 and 2.

The presence of AsF_3 has been detected via x-ray absorption edge studies [1] on $C_{10}AsF_5$ and by NMR on a ternary compound ($C_{16}AsF_5SbF_5$) [9], although it was not detected in an earlier NMR study of $C_{10}AsF_5$ [15]. These probes cannot readily distinguish AsF_5 and AsF_6^-. The presence of AsF_5 is shown by the reactivity of the compounds with several chemicals (N_2F_4, IOF_5) which do not react with AsF_6^- or AsF_3 [16]. Also, the gas which is pumped off of the compounds in vacuum is found to be initially AsF_5 [11], with significant amounts of AsF_3 released in the later stages [8]. Finally, other species such as AsF_5^- are rules out by the absence of a Curie-type susceptibility [18].

There have been few quantitative measures of f_{chem}. The x-ray absorption edges [1] were interpreted as showing f_{chem} = 2/3, but they were done on $C_{10}AsF_5$, and the presence of as much as 20% extra AsF_5 could probably not be ruled out. The NMR study [9] found $f_{chem} \sim 0.3$. Weight loss measurements [2] were interpreted as showing $f_{chem} \simeq 0.27$ (stage 1) and 0.35 (stage 2). However, this analysis assumes that all the gas pumped off is AsF_5, and no AsF_5 is produced by the back-reaction of Eq. 1. Neither of these assumptions is likely to be rigorously correct.

Early estimates of $f_{elec} \simeq 0.5 - 0.15$ [19,20] assumed a hole mass appropriate to pure graphite. Since the correct mass [7,11] is about 6 times larger, the values of f_{elec} should be enhanced by a like factor. BLINOWSKI and RIGAUX's [12] reanalysis of the optical data [19] is entirely consistent with the dHvA results (Table 1). Spin susceptibility [21] gives f_{elec} = 0.24 (stage 1) and 0.48 (stage 2). Electron-energy loss studies [22] give f_{elec} = 0.4 for $C_{16}AsF_5$ (Table 1). The position of the C(15) core electron excitation peak directly measures the net charge on the carbon, but since the change in position is small, the accuracy is limited. Electron-energy loss [22] gives a shift of 0.2 eV, for $f_{elec} \simeq 0.56$ in stage 2. The additional shift of 0.1 eV in going to stage 1 [2] gives $f_{elec} \simeq 0.41$ (stage 1).

In summary, most experiments are approximately consistent with the dHvA results for f_{elec}. Data on f_{chem}, while not as quantitative, suggest $f_{chem} \simeq f_{elec}$. It should be pointed out that BARTLETT [23] has synthesized $C_8AsF_6^-$, as well as stage 1 compounds of several other hexafluorides. For these compounds $f_{chem} = 1$, but the low conductivities suggest $f_{elec} \ll f_{chem}$.

References

1. N. Bartlett, R.N. Biagioni, B.W. McQuillan, A.S. Robertson, and A.C. Thompson, J. Chem. Soc. Chem. Comm., 200 (1978).
2. M.J. Moran, J.E. Fischer, and W.R. Salaneck, J. Chem. Phys. 73, 629 (1980).
3. G.M.T. Foley, C. Zeller, E.R. Falardeau, and F.L. Vogel, Sol. St. Commun. 24, 371 (1977). The high conductivity values reported by these authors have not been reproduced in more recent studies [4].
4. L.V. Interrante, R.S. Markiewicz, and D.W. McKee, Synth. Met. 1, 287 (1979/80); S.C. Singhal and A. Kernick, Synth. Met. 3, 247 (1981); T.E. Thompson, E.M. McCarron, and N. Bartlett, Synth. Met. 3, 255 (1981); M.J. Moran, J.W. Milliken, C. Zeller, R.A. Grayeski, and J.E. Fischer, Synth. Met. 3, 269 (1981).
5. H.A. Resing, F.L. Vogel, and T.C. Wu, Mat. Sci. & Eng. 41, 113 (1979); S. Laughlin, R.A. Grayeski, and J.E. Fischer, J. Chem. Phys. 69, 3740 (1978); J.E. Fischer, A. Metrot, P.J. Flanders, W.R. Salaneck, and C.F. Brucker, Phys. Rev. B23, 5576 (1981); M.L. Dzurus and G.R. Hennig, Journ. Am. Chem. Soc. 79, 1051 (1957).
6. R.S. Markiewicz, J.S. Kasper, and L.V. Interrante, Synth. Met. 2, 363 (1980).
7. R.S. Markiewicz, H.R. Hart, Jr., L.V. Interrante, and J.S. Kasper, Sol. St. Commun. 35, 513 (1980) and Synth. Met. 2, 331 (1980).
8. E.M. McCarron and N. Bartlett, J. Chem. Soc. Chem. Comm., 404 (1980).
9. G.R. Miller, H.A. Resing, P. Brant, M.J. Moran, F.L. Vogel, T.C. Wu, D. Billaud, and A. Pron, Synth. Met. 2, 237 (1980).
10. Y. Iye, O. Takahashi, and S. Tanuma, Sol. St. Commun. 33, 1071 (1980) and S. Tanuma, Y. Iye, O. Takahashi, and Y. Koike, Synth. Met. 2, 341 (1980).
11. J.E. Fischer, M.J. Moran, J.W. Milliken, and A. Briggs, to be published.
12. J. Blinowski, N.H. Hau, C. Rigaux, J.P. Vieren, F. LeToullec, G. Furdin, A. Herold, and J. Melin, J. Physique 41, 47 (1980), and J. Blinowski and C. Rigaux, Synth. Met. 2, 297 (1980).
13. E.R. Falardeau, L.R. Hanlon, and T.E. Thompson, Inorg. Chem. 17, 301 (1978).
14. L.B. Ebert, D.R. Mills, and J.C. Scanlon, Mat. Res. Bull. 14, 1369 (1979).
15. L. Ebert and H. Selig, Mater. Sci. Eng. 31, 177 (1977).
16. V. Munch and H. Selig, Synth. Met. 1, 407 (1979/80).
17. H. Selig, M.J. Vasile, F.A. Stevie, and W.A. Sunder, J. Fluor. Chem. 10, 299 (1980).
18. B.R. Weinberger, J. Kaufer, A.J. Heeger, E.R. Falardeau, and J.E. Fischer, Sol. St. Commun. 27, 163 (1978).
19. L.R. Hanlon, E.R. Falardeau, and J.E. Fischer, Sol. St. Commun. 24, 377 (1977).
20. C. Zeller, L.A. Pendrys, and F.L. Vogel, J. Mat. Sci. 14, 2241 (1979).
21. B.R. Weinberger, J. Kaufer, A.J. Heeger, J.E. Fischer, M.J. Moran, and N.A.W. Holzwarth, Phys. Rev. Lett. 41, 1417 (1978).
22. J.J. Ritsko and E.J. Mele, Synth. Met. 3, 73 (1980).
23. N. Bartlett, B. McQuillan, and A.S. Robertson, Mat. Res. Bull. 13, 1259 (1978).

Model Self-Consistent Bands for Graphite Intercalation Compounds

S.A. Safran and N.A.W. Holzwarth

Corporate Research-Science Laboratories
Exxon Research and Engineering Company
P.O. Box 45, Linden, NJ 07036, USA and

D.R. Hamann

Bell Laboratories, Murray Hill, NJ 07974, USA

Abstract

The electronic energy levels and charge distribution of the π electrons in high stage graphite intercalation compounds are determined within a self-consistent model. The model for a stage n compound treats the graphite π electrons in terms of a three-dimensional LCAO Hamiltonian. The effects of the inhomogeneous distribution of electrons in the n carbon layers (screening) are taken into account by adding a self-consistently determined layer potential term to the LCAO Hamiltonian, while the σ electrons are treated in terms of a background dielectric constant. The model energy bands, potentials, and charge distributions for a third stage compound are compared with the results of a first-principles, self-consistent calculation for LiC_{18}.

Introduction

The stage dependence of the electronic properties of graphite intercalation compounds has been an area of recent activity for both theorists [1-5] and experimentalists [6,7]. The variation of the energy band structure and the electronic charge distribution as a function of stage is important not only for an understanding of electronic properties, but also for an analysis of the cohesive properties and the stability of the staged structures themselves. This paper reviews a model for a self-consistent band structure for graphite intercalation compounds [3] which has been used to study the stage dependence of the electronic energy of these compounds. The results of the model for a third stage compound XC_{18} are compared with the results of a first principles calculation for LiC_{18} [8]. The comparison indicates that the simple model is a reasonable first approximation for the band structure and charge density for a staged intercalation compound.

I. Model Self-Consistent Band Structure

A. Model Hamiltonian

Several models of the electronic structure of graphite intercalation compounds have recently been proposed which consider only the π bands of the graphite layers. In these calculations, the hybridization of the carbon and intercalant orbitals is ignored and a parameter, f, is introduced to define the charge transfer per intercalant to the carbon layers. HOLZWARTH [4] and BLINOWSKI et al. [2] introduced models that took into account the effects of the change in layer stacking upon intercalation. Interactions between graphite layers on either side of an intercalate layer were assumed to be small, and the stage-n compound was modeled by an n-layer, thin-film of carbon layers. Non-self-consistent calculations of the band structure of intercalation compounds were also performed by DRESSELHAUS et al. [5] who took into account the changes in the band structure due to the symmetry of the intercalation compound. BLINOWSKI et al. [2] demonstrated the importance of the effects of including the electrostatics of the charged graphite layers, within their simplified tight-binding model, which they treated analytically. In [3], we introduced self-consistent electrostatic layer potentials into the thin-film models and presented a calculation of the band structures, layer potentials and charge densities, and total energies for stages n=2-8.

The self-consistent model is derived from an expression for the total energy of n graphite layers bounded by two intercalant layers. The main assumptions of the model are:

(i) Hybridization between the graphite and intercalant orbitals is negligible so that an effective charge transfer per intercalant (f) of electrons to the π bands of graphite can be defined.
(ii) An LCAO Hamiltonian is used to describe the π bands of graphite and changes in the intersite LCAO parameters due to charge transfer are neglected.
(iii) Any spatial variations of the electrostatic potential within a plane are neglected. The electrostatic potential for both the graphite and intercalant layers are calculated by treating them as charged sheets.
(iv) Hopping of electrons between graphite sheets separated by an intercalant layer is neglected, as is hopping between next-nearest-neighbor graphite layers. The n-layer sandwiches in a stage-n compound are thus decoupled and the thin-film model is applicable.

The theoretical justification for these approximations is discussed in [3]. The relevance of these assumptions to the case of LiC_{18} is discussed in Section II, where the model calculation is compared with the results of a first-principles, self-consistent band structure. With these approximations, the exact wave functions of the interacting system are written $\psi_p^k(r)$, where k is a two-dimensional (in-plane) wave vector defined in the two-dimensional, hexagonal, Brillouin zone whose sides have length $4\pi/3a$, where a is the graphite in-plane lattice constant. The wave function is indexed by p (p=1.....2n) which labels the band for stage n. The model Hamiltonian is constructed from an expansion of the wave function in a basis of orthonormal orbitals $\phi(\vec{r}-\vec{R}_{\alpha i})$ with π symmetry, localized on the carbon atom in layer i (i=1...n) and the index α=a,b denotes the two atoms in the unit cell of the single graphite layer. Minimizing the total energy of the n-layer film in the charged sheet approximation, with respect to the expansion coefficients $c_{pi}^{k\alpha}$ with the constraint that the wave function be orthonormal, yields the equation,

$$\sum_{\beta j} [E_{\alpha\beta}^{ij}(\vec{k}) + V_i\, \delta_{\alpha,\beta}\, \delta_{i,j}]\; c_{pj}^{k\beta} = E_p(\vec{k})\; c_{pi}^{k\alpha} \tag{1}$$

where $E_{\alpha\beta}^{ij}(\vec{k})$ are the LCAO matrix elements [3] and where V_i is the self-consistent potential at layer i given by [3]

$$V_i = -V_o \sum_j q_j\, |i-j| \tag{2}$$

In Eq. (2), q_i is the average charge per carbon atom in the ith layer and

$V \simeq 116$ eV $/\varepsilon$, where ε is an appropriate dielectric constant representing the screening due to the polarizability of the σ electrons (see Section II and Appendix B of [3]). With a starting guess for the q_i and hence the V_i, the eigenvalues $E_p(k)$ and eigenvectors $c_{pi}^{k\alpha}$ are determined. After determining the Fermi energy ε_F, the new charge distribution q_i and potential V_i are recomputed and the process is iterated to self-consistency.

B. Energy Bands, Potentials and Charge Densities for LiC

Self-consistent charge densities, layer potentials, and energy bands for the model Hamiltonian described above were presented in [3] for compounds with stoichiometry $C_{12n}X$, where n is the stage number and n=2...8. Fig. 1 shows the self-consistent, model bands calculated for LiC_{18}(n=3), assuming a charge transfer per intercalant of f=1. Table I lists the charge distribution in the two inequivalent graphite layers and the potential difference between those layers for values of ε=1 and 3. The lower-energy conduction bands have wave functions mostly localized in the bounding graphite layers (those layers adjacent to the intercalant layers), while the higher-lying bands have wave function mostly localized in the interior graphite layer. The screening of the intercalant layer is shorter-ranged in LiC_{18} than in the stage-3 compounds XC_{36}, discussed in [3], due to the higher Li concentration and larger total charge transfer in these compounds. The trend towards longer-ranged screening with increasing values of the effective dielectric constant, can be understood in terms of the polarizability of the σ electrons obviating the need for strong screening of the intercalant by the π electrons.

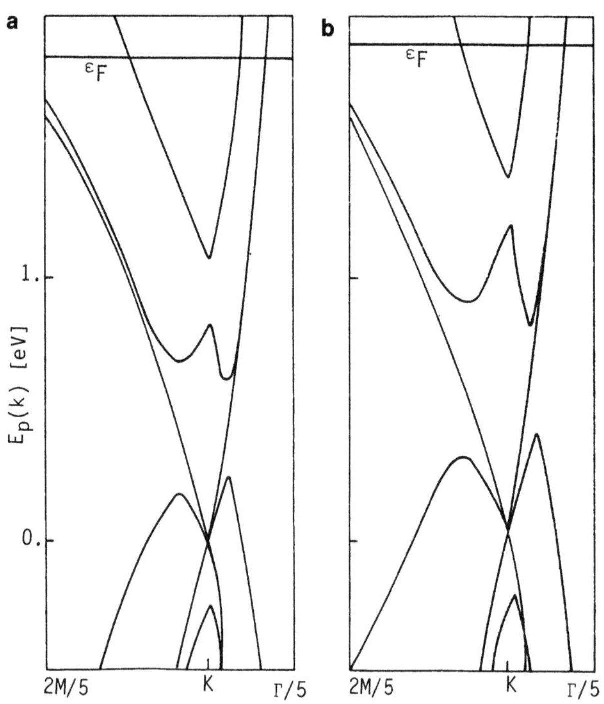

Fig. 1 Energy bands $E_p(\vec{k})$ for stage n=3 calculated from the model described in Section I. ε_F is the Fermi energy and the Γ, M, and K points are located

at (0,0), $\frac{2\pi}{a}$ $(\frac{2}{3}, 0)$, and $\frac{\pi}{a}$ $(1, \frac{1}{\sqrt{3}})$ where a is the graphite in-plane lattice constant. The charge transfer per Li, f=1.
(a) Bands calculated for effective dielectric constant [3] $\varepsilon = 3$.
(b) Bands calculated for effective dielectric constant $\varepsilon = 1$.

Table I Layer Charges and Potential Difference

	Model		First Principles Calculation
	ε=3	ε=1	
Bounding Charge[(a)]			
Total	0.44	0.47	0.48
Conduction	0.42	0.44	0.40
Interior Charge[(a)]			
Total	0.13	0.06	0.06
Conduction	0.17	0.13	0.19
Potential Difference[(b)]			
(eV)	0.8	1.2	0.8

(a) Charge is electrons (charge transfer f=1 for the model). See Figure 2 for the definition of regions a, b, and c. The bounding charge is the charge in the volume a+b, while the interior charge is in the volume 2c. Up to roundoff errors, the total charge in volume 2(a+b) + 2c = 1 electron.
(b) Layer-plane average.

As discussed in detail in [3], the stability of the high stage compounds is directly related to the relatively strong dependence of the total electronic energy upon stage. The range of stability in chemical potential (proportional to the log of the intercalant vapor pressure) for a given stage, $\Delta\mu_n$, is related to the electronic energy per intercalant, E_n, by [3]

$$\Delta\mu_n = n \; [E_{n+1} + E_{n-1} - 2E_n] \qquad (3)$$

In [3], the results for $\Delta\mu_n$ calculated from the self-consistent bands were compared with experiment for stages 3-5 K,Rb, and Cs intercalation compounds and good general agreement was obtained. Since the screening of the intercalant layer is shorter-ranged in the Li intercalation compounds (LiC_{18}) due to the higher Li concentration, the relative ranges of stability (e.g. the ratio $\Delta\mu_3/\Delta\mu_2$ in the high stage Li compounds is smaller than those of the heavier alkali metals. This could explain the difficulty in growing high stage Li intercalation compounds [9].

II. Comparison with Pseudopotential Calculations

First principles, self-consistent, band structure calculations have been performed for LiC_{18} for stages n=1-3. These calculations, whose details are reported in [8], were carried out in the local density approximation using pseudopotentials [10] and a mixed basis set [11]. For simplicity, the structure studied for the stage-three compound was ABC α CAB β BCA γ ABC....[12]. In order to establish a basis of comparison, the electronic structure of graphite, modified to this structure was also determined.

The charge density difference of LiC_{18} minus that of modified graphite gives an indication of the charge distribution in this material, and its layer-plane average is plotted in Fig. 2. The corresponding electrostatic potential, including the Li^+ ionic contribution, is also shown in this figure. The charge and potential were integrated around the bounding (regions a+b in Fig. 2) and interior (region c in Fig. 2) layers to define layer charges and potential differences that are compared with the model calculations in Table I. The main result of the first principles calculation is that the charge is localized on the bounding layer with only 6% of the total density difference being associated with the interior layer. On the other hand, the conduction electron contribution (as defined by the density of electrons in partially occupied bands) is much more delocalized, with 19% of the total conduction charge being associated with the interior layer.

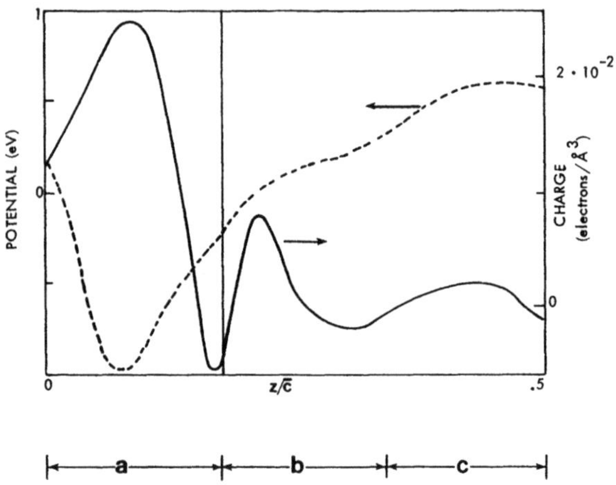

Fig. 2 Electronic charge density and potential averaged in the layer plane and plotted along the c-axis, where is the unit cell dimension in the c-axis. The Li layer is located at z=0 and the interior layer at $z=\bar{c}$. The position of the bounding graphite layer is shown by the vertical line. The potential and charge density were offset by the pure graphite values and the zero of the potential is set to be its volume average.

A comparison of the first principles and model calculation band structures is difficult because of in-plane zone folding effects which are neglected in the model calculation and because of the different layer geometries of the two calculations. However, integrated quantities, such as the c-axis charge and potential distribution, lend themselves to a meaningful comparison as shown in Table I. Allowing for the differences in the graphite structure used in the two calculations, the charge distribution and potential differences between the interior and bounding layers are in good agreement. Both the model and the first principles calculation show the delocalization of the conduction band charge as compared with the total charge. For the model calculation, this effect arises from the screening contribution of the valence

π electrons to the self-consistent potential. However, the response of the σ electrons to the potential is included only as an effective dielectric constant in the model. In the first principles calculation, it is seen that in addition to the valence electron screening, there is also a distortion polarization of the valence electrons so that the net redistributed charge is highly localized (89%) in the sandwich region between the two graphite planes surrounding an intercalant layer as seen in Fig. 2. The exact relationship of this distortion polarization to the effective dielectric constant of the model calculation has yet to be determined.

The authors wish to acknowledge useful discussions with S. G. Louie.

References

1. L. Pietronero, S. Strassler, H. R. Zeller and M. J. Rice, Phys. Rev. Lett. 41, 763 (1978), Solid State Comm. 30, 399 (1979).
2. J. Blinowski, N. Hau, C. Rigaux, J. P. Vicren, R. Le Toullec, G. Furdin, A. Herold, and C. Melin, J. Phys. (Paris) 41, 47, 667 (1980).
3. S. A. Safran and D. R. Hamman, Phys. Rev. B. 22, 606 (1980) and Phys. Rev. B 23, 565 (1981).
4. N. A. W. Holzwarth, Phys. Rev. B 21, 3665 (1980).
5. G. Dresselhaus, S. Shayegan, T. Chieu and S. Y. Leung, Solid State Comm. 35, 819 (1980).
6. J. E. Fischer and T. E. Thompson, Phys. Today 31 (7), 36 (1977).
7. Proceedings of the Second International Conference on Intercalation Compounds of Graphite, Provincetown, Mass. 1980, J. Synth. Metals, 3, (1980).
8. N. A. W. Holzwarth, S. G. Louie, and S. Rabbi (to be published).
9. D. Billaud, E. McRae, J. F. Mareche, and A. Herold, Synth. Metals, 3, 21, (1981).
10. D. R. Hamann, M. Schluter, and C. Chiang, Phys. Rev. Lett. 43, 1494 (1979).
11. S. G. Louie, K-M Ho, and M. L. Cohen, Phys. Rev. B19, 1774 (1979).
12. A, B, C denote graphite layer stacking, while α, β, γ denote intercalant layer stacking. Intercalant layers form a $\sqrt{3} \times \sqrt{3}$ lattice with respect to the graphite layers. The separation between graphite layers surrounding an intercalant layer was chosen to be that of the first and second stage compounds, D. Guerard and A. Herold, Carbon 13, 337 (1975).

Charge Superlattice Effects on the Electronic Structure of a Model Acceptor Graphite Intercalation Compound

G. Campagnoli

GNSM-CNR, Istituto di Fisica Teorica, Università di Pavia, Italy and

E. Tosatti

Scuola Internazionale Superiore di Studi Avanzati, Trieste, and International Centre for Theoretical Physics, Trieste, Italy

Many existent electronic structure calculations of graphite intercalation compounds (GIC's) deal with situations where all carbons in each plane are equivalent or treat the intercalate layer as a uniform sheet of charge [1,2,3]. The discussion has focused so far on the non-uniformity of the charge distribution across the planes, which is by now very well established [4,2,5]. We have previously discussed how in fact a large in-plane charge non-uniformity between inequivalent carbons is also to be expected, since the typical decay length of a charge disturbance is estimated to be below 1Å [6]. Recent ^{13}C NMR data in donor GIC's provide clear evidence of inequivalent carbon sites, with a large charge difference [7]. No such evidence has been so far found in acceptor GIC's. Nevertheless, several facts lead to the suspicion that the in-plane charge inhomogeneity must be at least as large, possibly much larger, than that of the donors. In fact, while the donor level usually gives rise to a band, this will generally not be the case in acceptors, because of the poor overlap, both direct and carbon-mediated, between the acceptor molecules. This is indicated by large spacings, both parallel and perpendicular to the planes, by the extremely poor c-axis conductivity, and by the presence of chemical, as well as positional, disorder [8,9]. If a band does not exist, then the state of charge of a molecule must be integer - either neutral or negative - unless it is in a dynamical mixed-valence state. Setting aside for the time being this latter possibility, for which there is no evidence, we conclude that, if f is the average charge-transfer per molecule, then a fraction f of all molecules carries one electron, and all others are neutral. In the case of AsF_5 GIC's and similar intercalates, there is further evidence of outright chemical disproportionation into distinct charged (AsF_6^-) and neutral species (AsF_3, AsF_5) [10,9].

We have thus a picture of charge being concentrated into well-separated islands, between which only neutral molecules and essentially neutral carbons - a majority - are to be found. For example, recent X-ray diffraction experiments on stage-1 SbF_5 GIC have been interpreted in terms of strings of intercalant molecules [11]. The average spacing between two molecules in the string is about 5.7 Å but that given between two charged molecules is as large as 23 Å! Even if the electron depletion is actually shared by a whole carbon ring or more, this still leaves - with a charge decay length of less than 1Å - plenty of neutral carbons in between.

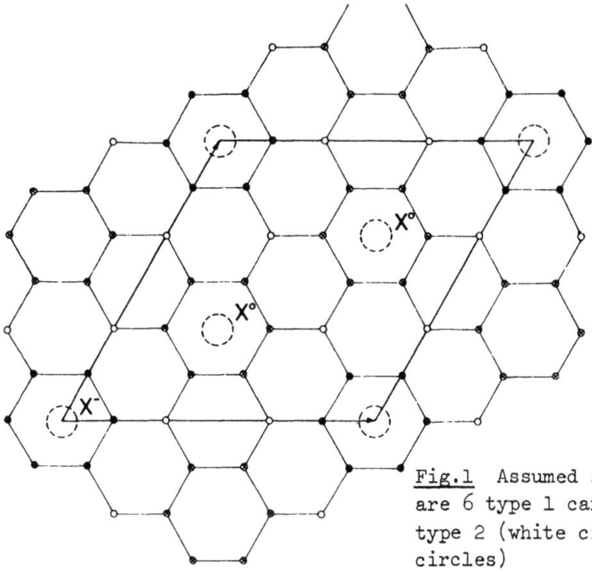

Fig.1 Assumed structure for $C_{24}X_2^oX^-$. There are 6 type 1 carbon sites (black circles), 6 type 2 (white circles) and 12 type 3 (crossed circles)

The purpose of our work is to pursue this picture of acceptor GIC's in the attempt to bring out features that could be experimentally recognized or refuted. In the present attempt, we have chosen to disregard disorder effects, and consider a model ordered situation (a super-superlattice) where, starting from a basic stoichiometry C_8X, a fraction 1/3 of the molecules acquire one electron, the remaining 2/3 being left neutral. This super-lattice is simple enough, with a unit cell $C_{24}X_2^oX^-$ shown in Fig. 1 (X stands for the intercalant molecule) still of manageable size. The average charge transfer $f = 1/3$ is also rather close to the value 0.37 found in stage-1 AsF_5 GIC by magnetooscillations [12], in close agreement with optical determinations [1, 6]. Thus, even if the real intercalate is undoubtedly much more complicated, we may still hope to catch with this model at least some elements of reality. We have performed an electronic structure calculation for this situation, using the same simple model introduced earlier for C_8X^{f-} [6]. Only carbon π-electrons are treated using tight-binding plus electrostatic (Hartree) self-consistency. The neutral molecules are supposed to be both electrically and chemically inactive. The assumption of uniform stacking along the c-axis has been made for simplicity and could be relaxed if desirable, for example to consider ABC stacking.

2. Charge Population

There are in this model three types of carbons, called 1, 2 and 3 as in Fig.1. At self-consistency, we obtain a charge imbalance given by $\rho_1 = -.109$, $\rho_2 = -.014$ and $\rho_3 = -.021$ electrons (so that $6\rho_1 + 6\rho_2 + 12\rho_3 = -1$). In comparison with C_8X with similar f [6], we note that the concentration of one full electron on one molecule has further enhanced the non-uniformity.

However, this is likely to be over-estimated by the present model, because of our assumption of point-like charge distribution of X^- [6] and of AAA c-axis stacking.

We can fit ρ_1 and ρ_2 to an exponentially decaying hole charge about the X^- sites, as done previously. We obtain a decay length $\lambda \simeq 0.7$Å, very close to our value of 0.65Å for C_8X [6]. This suggests that this decay length is somewhat characteristic of graphite layers and could be extrapolated to other situations as well. The order-of-magnitude difference between λ and the intermolecular distance is our main result, which implies that in-plane charge inhomogeneity must be a general phenomenon beyond the limitations of the present model.

We also note that $|\rho_3|$ is about one order of magnitude larger than predicted by the exponential decay, indicating in fact an oscillatory behaviour of the charge away from the acceptor site. It is not clear at the moment whether this oscillation is a Fermi surface effect that would be present even for a single molecule or a constructive interference due to having a lattice of molecules.

3. Band Structure

Our calculated π-band structure and density of states are shown on Fig.2. The size of the super-superlattice mini-gaps at the new zone boundaries is in the order of 0.1 - 0.2 eV, close to that of the C_8X superlattice gaps [6]. The density of states has new features at these gaps. However, the gross

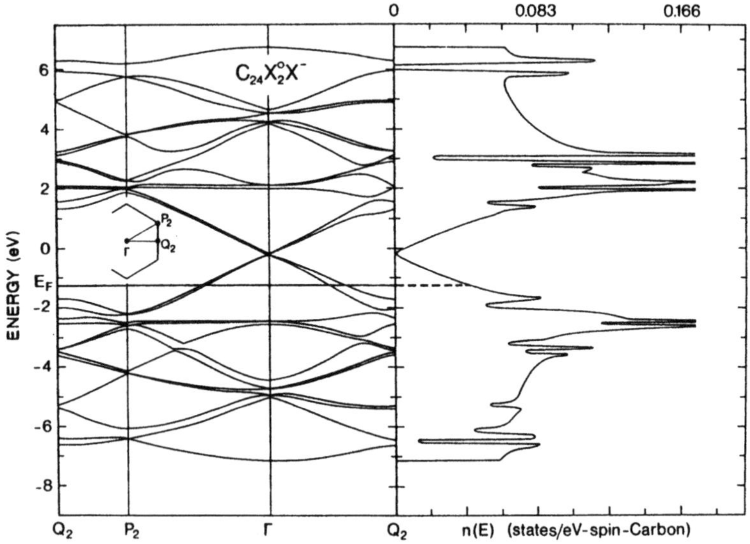

Fig.2 π-band structure and density of states of $C_{24}X_2^oX^-$. The energy zero is that of the unperturbed π-bands

features of n(E) remain close to the rigid band model. This contrasts with the charge population, which is distinctly non-uniform. The non-uniformity is an integrated effect over all Bloch functions, a piece of information the bands themselves do not convey.

4. Fermi Surface

It consists of two hole cylinders, whose cross-sections are in the shape of a rounded hexagon (S_1) and of a rosette (S_2). They originate from zone-folding of the two rounded triangular cyliners of C_8X, as shown in Fig.3.

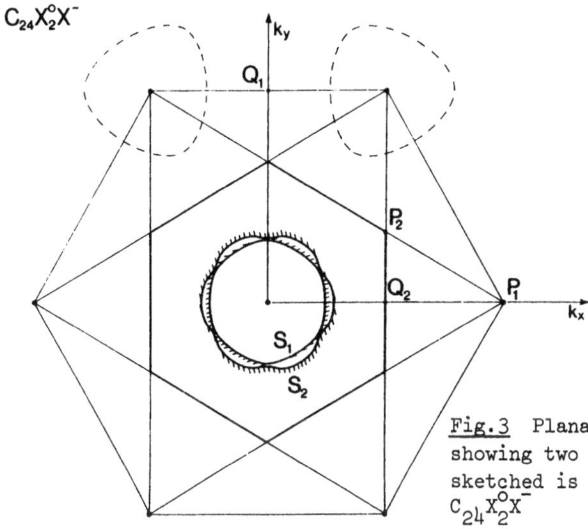

Fig.3 Planar section of the Fermi surface showing two hole orbits S_1 and S_2. Also sketched is the zone folding from C_8X^{f-} to $C_{24}X_2^oX^-$

Their area ratio is $S_2/S_1 \approx 1.25$. This can be compared with magneto-oscillatory data in stage-1 AsF_5 GIC [12] that show three large orbits of relative size roughly 1, 1.08, 1.23. Thus, our assumed super-superlattice seems to account well for the largest and the smallest but not the middle one. We find however that the unperturbed orbit of C_8X before folding (dashed on Fig 3) has area ratio $S/S_1 \sim 1.1$. The possibility arises, therefore, that in this GIC, the slight excess of charged molecules (0.37 rather than 1/3) and their disorder might allow breakdown between the two orbits. That would explain the middle orbit as well, with surprisingly good numerical accuracy.

5. Optical and Energy Loss Properties

The $\vec{q} = 0$ complex dielectric function $\varepsilon(\omega)$ has been calculated from the band structure of Fig.2, with an accurate procedure described earlier [6]. From this, the normal-incidence reflectivity $R(\omega)$ and energy loss function $-\mathrm{Im}\,\varepsilon^{-1}(\omega)$ are easily obtained and are shown in Fig.4. All the same gross

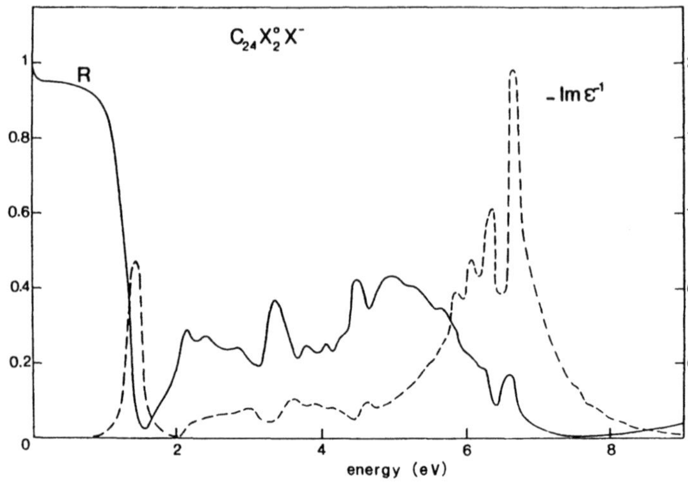

Fig.4 Calculated opitcal reflectivity (solid line) and energy loss function (dashed line)

features found without the super-superlattice [6], namely a) the Drude-like behaviour at low frequency and corresponding intra-band plasmon; b) the Fermi edge around 2 eV; c) the broad $\pi \rightarrow \pi^*$ peak centered at 4.6 eV; d) the broad π-plasmon peak around 6.5 eV are still present. Some of the mini-gaps show up as narrow peaks and dips, but that is not likely to survive non-stoichiometry and disorder. In fact, all the above broad features can be seen in reflectivity data of $SbCl_5$ GIC [13] and also in energy loss in various acceptor GIC's [14], with no evidence of mini-gaps.

6. Conclusions

We have studied a simple model acceptor GIC, where the charged X^- molecules form a $\sqrt{3} \times \sqrt{3}$ super-superlattice with respect to the basic C_8X structure. Our main conclusions are:

(i) the self-consistent hole charge in the graphite layers is strongly concentrated around the X^- molecules, with a decay length of about 0.7Å. Thus, the charge distribution in the carbon plane is very non-uniform;

(ii) the Fermi surface consists of two hole pockets, with an area ratio of about 1.25. This result can be directly compared with the magneto-oscillation data of MARKIEWICZ et. al. [12];

(iii) the π-band structure and the interband optical properties, though exhibiting mini-gaps and other distinctive superlattice features, are roughly not too far from those obtained in the uniform case.

References

1. J. Blinowski et al.: J. Physique 41. 47 (1980); J. Blinowski and C. Rigaux: J. Physique 41, 667 (1980)
2. H. Kamimura, K. Nakao, T. Ohno and T. Inoshita: Physica 99B, 401 (1980)
3. G. Dresselhaus and S.Y. Leung: Solid State Comm. 35, 819 (1980)
4. L. Pietronero, S. Strässler, H. Zeller and M.J. Rice: Phys. Rev. Lett. 41, 763 (1978)
5. S. Safran and D.R. Hamann: Phys. Rev. B23, 565 (1981)
6. G. Campagnoli and E. Tosatti: J. Phys. Soc. Japan, Suppl. A49, 919 (1980); J. Phys. C., to appear
7. J. Conard et al.: Synth. Metals 2, 227 (1980)
8. J. Fischer: Physica 99B, 383 (1980)
9. R.S. Markiewicz et al.: Synth. Metals 2, 363 (1980), and this conference
10. N. Bartlett, B. McQuillan and A.S. Robertson: Mat. Res. Bull. 13, 1259 (1978)
11. T.C. Wu and F.L. Vogel: to be published.
12. R.S. Markiewicz, H.R. Hart, L.V. Interrante and J.S. Kasper: Synth. Metals 2, 331 (1980), and this conference
13. P. Pflüger et al.: Synth. Metals 2, 285 (1980)
14. J.J. Ritsko and E.J. Mele: Physica 99B, 425 (1980), and this conference

Knight Shift and Nuclear Quadrupolar Interaction in Alkali-Graphite Intercalation Compounds

G. Roth and K. Lüders

Institut für Atom- und Festkörperphysik, Freie Universität Berlin
D-1000 Berlin 33, Fed. Rep. of Germany

and

P. Pfluger[1] and H.-J. Güntherodt

Institut für Physik, Universität Basel
CH-4056 Basel, Switzerland

1. Introduction

In order to get more information about the electronic structure of Alkali-Graphite Intercalation Compounds, a number of experiments have recently been performed such as electron spectroscopy, NMR on the carbon nucleus, electronic specific heat or positron annihilation [1-4]. In most cases, the experimental results have been discussed with respect to the well-known properties of pristine graphite. Application of the NMR method to the intercalant nuclei, however, offers complementary information by comparison with the properties of the respective alkali metals. Due to the anisotropy of these compounds it is of particular interest to investigate the orientation dependence. In this paper NMR measurements on powder samples as well as on samples of Highly Oriented Pyrolytic Graphite (HOPG) intercalated with Cs and Li are reported. Earlier measurements using powder samples have already been reported by CARVER [5] and by CONRAD and ESTRADE [6].

2. Experimental

The C_8Cs samples were prepared by gas-phase reaction by the classical two temperature method [7], while the C_6Li samples were obtained by immersion into a liquid lithium pool. The resulting compounds were subjected to x-ray (00ℓ) and (hk0) analysis.

The NMR measurements were made using a cw-spectrometer at fields up to 4.8 T and in the temperature range 4 - 300 K. For the orientation dependence measurements, intercalated HOPG sandwiches of several slabs of 14x9x0.5 mm, isolated by mica sheets, were sealed under vacuum into pyrex tubes of 9 mm inner diameter (Fig.1). The orientation of the c-axis is normal to the slabs and could be varied with respect to the static magnetic field B_0. For signal detection a modified Pound-Knight-Watkins detector is used in combination with field modulation and lock-in technique [8].

3. Powder Samples

Figure 2 shows the ^{133}Cs NMR signal in a C_8Cs powder sample, obtained at a temperature of 3.8 K and a magnetic field of 3.2 T. The resonance is shifted by 0.22 % compared to an aqueous solution of CsCl as reference. At both sides of the central line, satellite lines appear, which are due to nuclear elec-

---

[1] present address: P. Pfluger, IBM Research Laboratory, 5600 Cottle Road, San José, CA 95193, U.S.A.

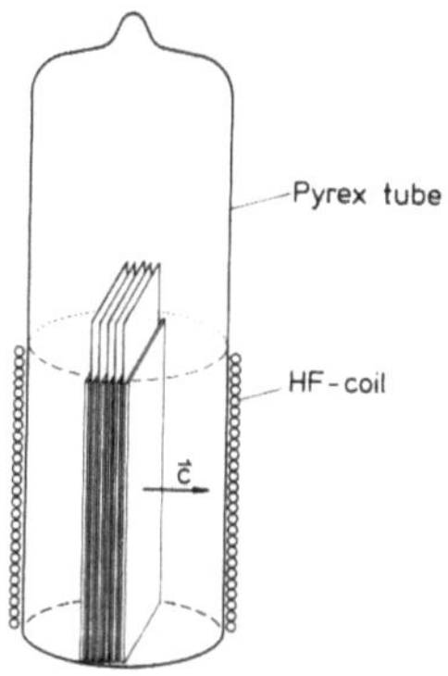

Fig.1. NMR sample, consisting of several slabs of intercalated HOPG isolated by mica sheets and sealed under vacuum.

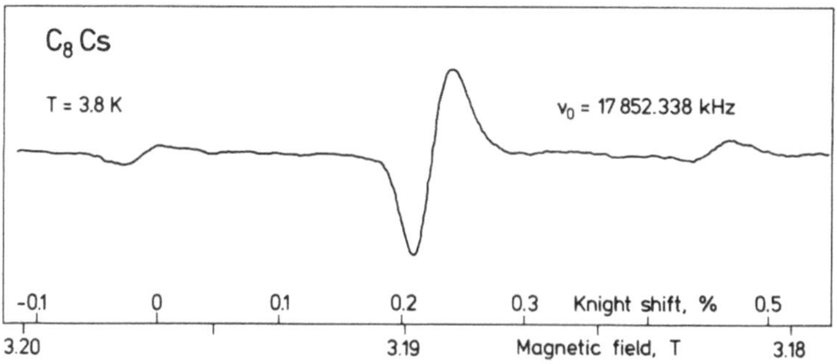

Fig.2. ^{133}Cs NMR signal of a C_8Cs powder sample. The satellite lines are due to nuclear electric quadrupolar interaction.

tric quadrupolar interaction. Their distance is equal to the quadrupolar frequency ν_Q.

The central resonance line is influenced by two effects. First, due to second-order nuclear quadrupolar interaction the linewidth is frequency dependent and decreases with increasing resonance frequency ν_0, as given by

$$\delta\nu = \frac{25}{144}\frac{\nu_Q^2}{\nu_0}\left[I(I+1) - \frac{3}{4}\right] \quad . \tag{1}$$

Second, due to the non-cubic structure of these compounds an anisotropic contribution to the Knight shift is present. The influence of this anisotropic Knight shift has the opposite frequency dependence and is given by

$$\delta\nu = 3\, K_{ax}\, \nu_0 \quad , \tag{2}$$

where $K_{ax} = 1/3(K_{\parallel} - K_{\perp})$ is the axial part of the shift. Fig.3 shows the measured frequency dependence of the ^{133}Cs linewidth. At first $\delta\nu$ decreases with increasing frequency, whereas at higher frequencies it increases again. From the low-frequency branch the quadrupolar frequency ν_Q and from the high frequency branch the axial part of the Knight shift are determined to be approximately 89 kHz and 0.012 %, respectively.

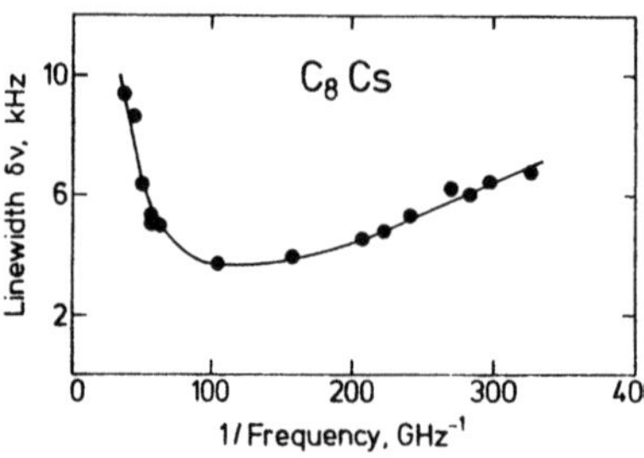

Fig.3. Frequency dependence of the ^{133}Cs line width in a C_8Cs powser sample at 4.2 K

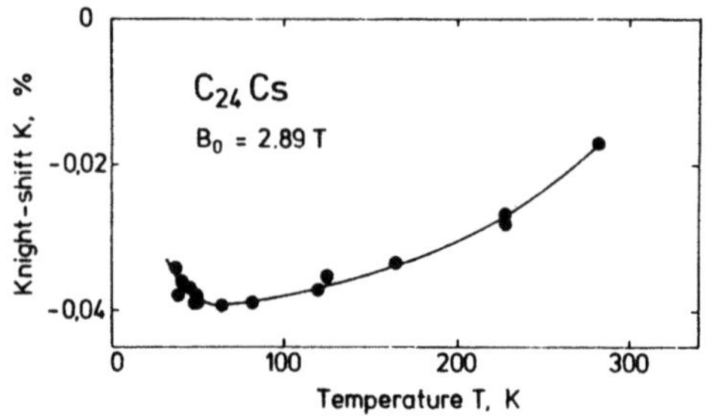

Fig.4. Temperature dependence of the ^{133}Cs Knight shift in $C_{24}Cs$

For C_8Cs and $C_{24}Cs$ the temperature dependence of the isotropic part of the Knight shift has been determined in the temperature range between 4.2 K and room temperature. For C_8Cs the Knight shift increases linearly from 0.22 % at 4.2 K to 0.255 % at room temperature. The results for $C_{24}Cs$ are shown in Fig.4. The Knight shift is negative in the whole temperature range. Again, the Knight shift is temperature dependent and decreases with decreasing temperature, until it reaches a minimum at approximately 65 K. As the total Knight shift is the sum of several contributions, this negative shift may result from the influence of the diamagnetic part of the susceptibility.

4. Oriented Samples

For the orientation-dependent measurements, samples of C_8Cs and C_6Li were prepared starting from highly oriented pyrolytic graphite. The arrangement shown in Fig.1 allows the study of the orientation dependence of the quadrupolar splitting and of the Knight shift as well by varying the direction of the magnetic field with respect to the c-axis.

As an example the orientation dependence of the quadrupolar splitting Δ of the ^{133}Cs resonance in C_8Cs is shown in Fig.5. The full line is the result of a fit to the data points using the theoretical expression given in the figure. The resulting quadrupolar frequency ν_Q is shown in Tab.1. The same orientation dependence of the quadrupolar splitting is obtained for the ^7Li resonance in C_6Li [9]. The quadrupolar frequency ν_Q is lower than in the case of C_8Cs (Tab.1). The resulting electric field gradients for both compounds

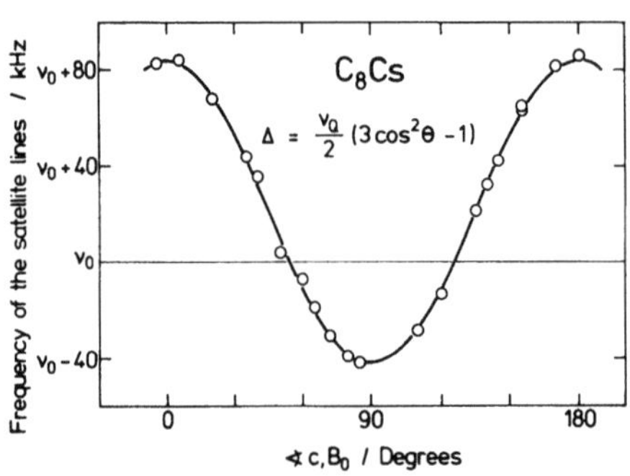

Fig.5. Orientation dependence of the quadrupolar satellite lines in C_8Cs with respect to the resonance frequency of the central line (4.2 K).

Table 1 Quadrupolar frequency ν_Q, electric field gradient eq and Sternheimer factor γ_∞ for C_8Cs and C_6Li

	ν_Q [kHz]	eq [Vcm^{-2}]	γ_∞
C_8Cs	83.8 ± 0.2	8.1×10^{17}	- 121
C_6Li	28.3 ± 0.8	2.9×10^{15}	+ 0.26

are also given in Tab.1. The great difference in the electric field gradients is mainly due to the very different Sternheimer antishielding factors.

For a determination of the Knight shifts this quadrupolar interaction must be taken into account. Fig.6 shows the measured orientation dependence of the ^{133}Cs resonance in C_8Cs (curve 1). The influence of the second-order quadrupolar interaction can be calculated using the theoretical expression $\Delta\nu_{quadr} = -15/16(1-\cos^2\theta)(9\cos^2\theta-1)\nu_Q^2/\nu_0$, which leads to curve 2. From the difference of these two curves the real orientation dependence of the Knight shift, described by $\Delta\nu_{aniso} = K_{ax}\,\nu_0(3\cos^2\theta-1)$, can be separated leading to curve 3. This curve results as best fit using the axial and isotropic part of the Knight shift as fit parameters. The resulting isotropic and axial Knight shift values are given in Tab.2.

In the case of C_6Li, quadrupolar effects are of first order only. Therefore, the Knight shift values $K_\parallel$ and $K_\perp$ can be measured directly by choosing the appropriate sample orientation. From these values the isotropic and axial part of the Knight shift can be calculated (Tab.2). Besides the usual reference LiCl, there exists the additional possibility of direct comparison with Li metal. Due to the liquid phase preparation method, applied in this case, small amounts of metallic Li have penetrated into microcracks of the sample

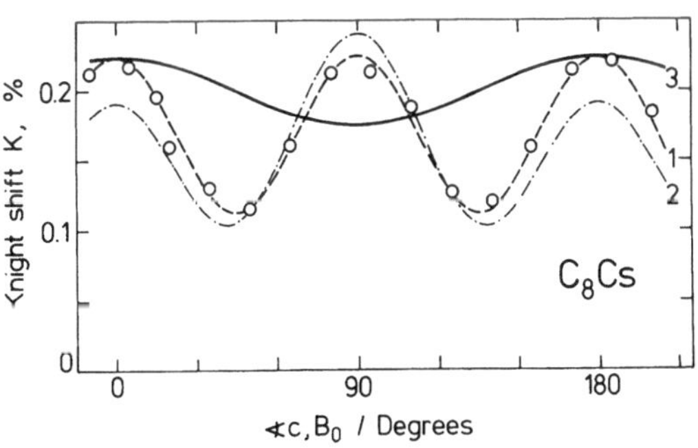

Fig.6. Orientation dependence of the ^{133}Cs resonance frequency in C_8Cs at 4.2 K (curve 1). The orientation dependence of K (curve 3) is obtained after separating the influence of second order quadrupolar interaction (curve 2).

Table 2 Knight shift values $K_\parallel$, $K_\perp$, K_{iso} and K_{ax} in the intercalation compounds C_8Cs and C_6Li. $K_{iso} = 1/3(K_\parallel + 2K_\perp)$ and $K_{ax} = 1/3(K_\parallel - K_\perp)$ are calculated from the measured values $K_\parallel$ and $K_\perp$

sample	$K_\parallel$ [%]	$K_\perp$ [%]	K_{iso} [%]	K_{ax} [%]	K_{met} [%]	K_{iso}/K_{met}
C_8Cs	0.223 ± 0.005	0.175 ± 0.005	0.191	0.016	1.57	0.122
C_6Li	0.0037 ± 0.0004	0.0021 ± 0.0004	0.0026	0.0005	0.026	0.10

and the NMR signal of this metallic Li can also be observed. The absolute values of the isotropic Knight shift are very different, but if one compares these values with the Knight shift of the respective metals,one finds that the Knight shift ratios are similar.

5. Discussion

In order to discuss the results, the Knight shift equation based on the Fermi contact interaction is used:

$$K = \frac{8\pi}{3} 2\mu_0 \mu_B^2 \langle|\Psi(0)|^2\rangle_{FS} N(E_F) \quad . \tag{3}$$

The Knight shift is proportional to the density of states at the Fermi level, $N(E_F)$, and to the electron probability density at the alkali metal nuclear site, averaged over the Fermi surface. There are two possible contributions to the Knight shift, arising from s- and π-electrons, at the site of the alkali atoms.

In the case of C_8Cs, the total density of states is known from measurements of the electronic specific heat [3]: $N_{tot} = 0.19$ states/eV atom. Subtraction of $N_\pi = 0.04$ states/eV atom, which is determined from the graphite-π-conduction band structure [10] and from the Fermi level shift known from photoemission spectra [1] leads to $N_s = 0.15$ states/eV atom. This splitting of N_{tot} into s- and π-states is consistent with a charge transfer of 0.3 electrons per alkali atom into the graphite π-conduction band. Assuming that the matrix elements for the s-like electrons remain unchanged, a Knight shift ratio of 0.11 is calculated, which is slightly smaller than the experimental one (0.122, Tab.2). The difference may be due to the neglected influence of the π-electrons. Thus in this case, the reduction in the Knight shift is mainly due to the reduction in the density of states.

In the case of C_6Li it is known from theoretical band structure calculations [11] and from positron annihilation experiments [4] that the charge transfer into the graphite-π-conduction band is almost complete. This means that the Knight shift is predominantly caused by π-electrons and therefore it can be no longer assumed that the matrix element remains unchanged. Now, approximating the matrix element by the charge density at the Li nuclear site, as calculated by HOLZWARTH et al. [11], and knowing the ratio of the density of states from electronic specific heat measurements [3], a Knight shift ratio of $K_{C_6Li}/K_{Li} = 0.106$ results. This ratio agrees very well with the experimental ratio of 0.10.

In summary, the measured Knight shift ratio in the case of C_6Li is in very good agreement with the assumption of a total charge transfer whereas in the case of C_8Cs, a partial charge transfer of about 0.3 electrons per Alkali atom agrees with the measured Knight shift ratio of 0.122.

Acknowledgements

We would like to thank Prof. Dr. K.H. Bennemann for helpful discussions. We are also indebted to V. Geiser for careful sample preparation. This work was supported by the Swiss National Science Foundation and by the Deutsche Forschungsgemeinschaft.

References

1. P. Oelhafen, P. Pfluger, E. Hauser, H.-J. Güntherodt, Phys. Rev. Lett. 44, 197 (1980)
2. J. Conard, M. Gutierrez-Le Brun, P. Lauginie, H. Estrade-Szwarckopf, G. Hermann, Synth. Met. 2, 227 (1980)
3. M.G. Alexander, D.P. Goshorn, D.G. Onn, Phys. Rev. B22, 4535 (1980)
4. E. Cartier, F. Heinrich, P. Pfluger, H.-J. Güntherodt, Sol. St. Comm., in press
5. G.P. Carver, Phys. Rev. B2, 2284 (1970)
6. J. Conard, H. Estrade, Mat. Sci. Eng. 31, 173 (1977)
7. K. Fredenhagen, G. Cadenbach, Z. anorg. allg. Chemie 158, 249 (1926)
8. R. Karcher, U. Köbler, K. Lüders, Z. Szücs, phys. stat. sol. (b) 84, 189 (1977)
9. G. Roth, K. Lüders, P. Pfluger, H.-J. Güntherodt, Sol. St. Comm. 39, 423 (1981)
10. B.R. Weinberger, J. Kaufer, A.J. Heeger, J.E. Fischer, M. Moran, N.A.W. Holzwarth, Phys. Rev. Lett. 41, 1417 (1978)
11. N.A.W. Holzwarth, L.A. Girifalco, S. Rabii, Phys. Rev. B18, 5206 (1978)

Part III

Stability and Phonons

Lattice Distortions in Layered Compounds

C. Haas

Laboratory of Inorganic Chemistry
Materials Science Centre of the University
Nijenborgh 16, 9747 AG Groningen, The Netherlands

1. Introduction

Many inorganic compounds with highly polarizable anions crystallize in layer structures; examples are the metal di- and trihalides (CdI_2, $FeCl_2$, $CrCl_3$, etc.) and the transition metal dichalcogenides (MoS_2, $TiSe_2$, TaS_2, etc.). As a consequence of the anisotropic arrangement of the atoms in these compounds, the physical and chemical properties are unusual and interesting; characteristic are highly anisotropic electrical conductivity, dielectric constant and lattice vibrations, and easy intercalation of atoms and molecules in the Van der Waals gap between the layers.

Many of the layered compounds show lattice distortions at low temperature. In metallic crystals these distortions are usually discussed in terms of the charge density wave model, the distortion being caused by an instability of the conduction electrons [1].

In this paper we try to explain some of the properties of layered compounds with a simple ionic model. The stability of a layered structure for compounds with highly polarizable ions is caused by the polarization energy. The polarizable ion model is also used to explain the anisotropy of the dielectric constant and the lattice vibrations, and the occurrence of distorted layer structures. Finally the relation of these considerations based on the polarizable ion model with the charge density wave model is discussed.

2. Stability of Ionic Compounds with a Layer Structure

In this section we discuss the stability of ionic layer compounds MX_2 with the CdI_2 or the related $CdCl_2$ structure. These structures consist of strongly bound X-M-X sandwich layers (Fig.1). The bonding between the sandwiches is weak (Van der Waals bonding), and this causes easy cleavage and easy intercalation between the sandwich layers. Many of the metal dihalides and dichalcogenides crystallize in this type of layered structures; the fluorides and oxides have different non-layer structures (Table 1).

A classical model for calculating the cohesive energy of crystals is the ionic model, in which the crystal is assumed to consist of ions interacting with one another by long-range electrostatic interactions and short-range repulsive forces. One of the remarkable features of the ionic model is that it enables us to make a reasonable estimate for the cohesive energy not only for ionic crystals but also for more covalent crystals, such as ZnS and GaAs. It can be shown that the ionic model gives a good account of the binding energy but not of the charge distribution in the crystal [2].

Table 1 Crystal structures MX_2 listed in the order of decreasing radius R_M of the cation [5]. Layered structures are indicated by an asterisk

	R_M(A)	F	Cl	Br	I
Ba^{2+}	1.36	C1	C1,C23	C23	C23
Pb^{2+}	1.18	C1	C23	C23	C6*
Sr^{2+}	1.16	C1	C1	C53	-
Ca^{2+}	1.00	C1	C35	C35	C6*
Cd^{2+}	0.95	C1	C19*	C19*	C6*
Mn^{2+}	0.82	C4	C19*	C6*	C6*
V^{2+}	0.79	-	C6*	C6*	C6*
Fe^{2+}	0.77	C4	C19*	C19*	C6*
Zn^{2+}	0.75	C4	C19*	C19*	C6*
Co^{2+}	0.74	C4	C19*	C19*	C6*
Mg^{2+}	0.72	C4	C19*	C6*	C6*
Ni^{2+}	0.70	C4	C19*	C19*	C19*
	R_M(A)	O	S	Se	Te
Th^{4+}	1.00	C1	C23	C23	-
Zr^{4+}	0.72	C1	C6*	C6*	C6*
Hf^{4+}	0.71	C1	C6*	C6*	C6*
Sn^{4+}	0.69	C5	C6*	C6*	-
Pt^{4+}	0.63	C6*	C6*	C6*	C6*
Ti^{4+}	0.61	C4	C6*	C6*	C6*

In the ionic model the lattice energy is given by

$$E= -NAZ^2e^2/R + NBR^{-n} . \qquad (1)$$

The first term is the Madelung energy, A is the Madelung constant, Ze the charge of the anion, N is the number of cations in the crystal, and R is the characteristic (shortest) cation-anion distance. The second term represents the repulsion between the ions.

One would expect the most stable crystal structure to be the one with the largest Madelung constant, i.e. the structure in which equal charges are as far apart as possible, and with a symmetrical coordination of cations by anions and vice versa. The layer structures are certainly not the most favourable structures in this respect, and their stability must be explained in a different way.

The reason for the stability of the layered compounds is the high anion polarizability [3]. In the layered structures CdI_2 (C6) and $CdCl_2$ (C19) the anions have an asymmetric coordination by three cations, which induce electric dipoles in the anions (Fig.1). The polarization energy is given by

$$E_{pol} = -\frac{CNZ^2e^2\alpha}{R^4}\left(1 + \frac{D\alpha}{R^3}\right)^{-1} \qquad (2)$$

In this equation α is the anion polarizability, C and D are constants for lattice sums over fields of charges and dipoles; for the ideal CdI_2 structure C= 7.29 and D= 3.613 [4]. In equilibrium the total energy is a minimum; with the condition $\partial E/\partial R = 0$ it is possible to eliminate B. The result is

$$E = -\frac{NAZ^2e^2}{R}\left(1 - \frac{1}{n}\right) - \frac{NZ^2e^2}{R} C\beta(1 + D\beta)^{-1}\left[1 - \frac{4}{n} + \frac{3D\beta}{n(1 + D\beta)}\right] \tag{3}$$

with $\beta = \alpha/R^3$. We now compare the CdI_2 structure (C6) with the cubic fluorite structure (C1) for which the polarization energy vanishes. The Madelung con-

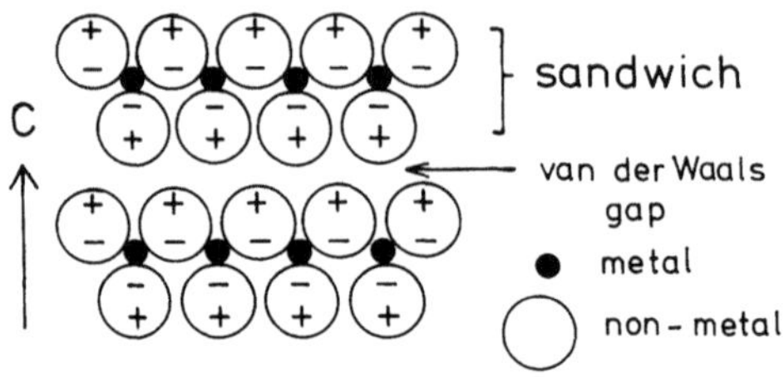

Fig.1 CdI_2-type layer structure

stants are A(C6)= 4.3819 and A(C1)= 5.0388. It is found that the CdI_2 structure is stable if $\beta > 0.2069$ (for n=9). This critical value is independent of Z but depends slightly on the value of n (for n=7 $\beta = 0.2360$, for n= 12 $\beta = 0.1857$). Thus we expect the CdI_2-type layer structure to be stable if the polarizability α is large, or if the cation-anion distance is small. This is precisely as is observed. In Table 1 we have listed the metal dihalides and dichalcogenides in order of decreasing R, and we find the layer structures C6 and C19 for small values of R. Moreover the stability region of the layer structures is larger for anions with a high polarizability. Layer structures are not found for the fluorides and oxides, due to the small polarizability. Thus these considerations based on a model with polarizable ions give a good explanation for the occurrence of layered structures for compounds with highly polarizable anions and small cations.

3. Anisotropy of Dielectric Constant and Lattice Vibrations

The static dielectric constant ε_0 of layered compounds is strongly anisotropic, with $\varepsilon_0(E\perp c) > \varepsilon_0(E/\!/c)$ (Table 2). This anisotropy is due to the contributions of the infrared-active lattice vibrations; the electronic contribution ε_∞ is nearly isotropic.

The reflectivity spectra of the layered compounds also show a pronounced anisotropy: the reflection maximum is larger and lies at a lower frequency for $E\perp c$ (E_u mode) than for $E/\!/c$ (A_{2u} mode), corresponding to a higher oscillator strength ($\varepsilon_0 - \varepsilon_\infty$) and a lower frequency for transverse optical phonons ω_{TO} for $E\perp c$ than for $E/\!/c$.

We discuss the infrared-active lattice vibrations in terms of a deformation dipole model [6] and we will show that the observed anisotropies are quite well accounted for by the presence of large dipoles at the anions. Each ion is represented by an effective charge ($+2Z^*e$ for the cations, $-Z^*e$ for the anions) and a polarizability (α_1 for the anions, α_2 for the cations). The normal modes (q=0) of the CdI_2 structure are shown in Fig.2. For each polarization, there is one infrared-active mode. The dielectric constant is given by

$$\varepsilon = \varepsilon_\infty + \frac{(\varepsilon_0 - \varepsilon_\infty)\omega_{TO}^2}{\omega_{TO}^2 - \omega^2 - i\gamma\omega} \, . \tag{4}$$

ω_{TO} is the frequency of the long-wavelength transverse optical mode with damping γ, the frequency of longitudinal modes ω_{LO} is given by $\omega_{LO}^2=\omega_{TO}^2\,(\varepsilon_0/\varepsilon_\infty)$. The static dielectric constant is $\varepsilon_0 = \varepsilon(\omega=0)$, ε_∞ is the electronic contribution.

Table 2 Parameters of infrared-active vibrations (ω_{TO}, ω_{LO} and γ in cm^{-1})

Material	polarization	ε_0	ε_∞	$\varepsilon_0-\varepsilon_\infty$	ω_{TO}	ω_{LO}	γ
PbI_2	E//c	9.3	5.9	3.4	96	121	20.2
PbI_2	E⊥c	26.4	6.1	20.3	52	108	4.7
CdI_2	E//c	5.4	4.3	1.1	136	152	16.3
CdI_2	E⊥c	12.9	4.6	8.3	79	132	16.6
NiI_2	E//c	6.2	5.8	0.4	178	184	6.0
NiI_2	E⊥c	8.0	6.4	1.6	156	176	5.6
SnS_2	E//c	6.2	5.7	0.5	340	356	12.2
SnS_2	E⊥c	17.8	7.6	10.2	205	314	5.9
$SnSe_2$	E//c	10.0	9.4	0.6	241	248	9.2
$SnSe_2$	E⊥c	23.2	10.7	12.5	144	204	5.9
β-GaS	E//c				318		9.5
β-GaS	E⊥c	11.5	9.0	2.5	294	333	1.5
2H-MoS_2	E//c	6.28	6.25	0.03	470	471	3.1
2H-MoS_2	E⊥c	15.4	15.2	0.2	384	386.5	1.0
2H-$MoSe_2$	E⊥c	17.8	17.1	0.7	286.0	291.8	4.0
2H-$MoTe_2$	E⊥c	22.3	20.3	2.0	234.5	245.8	3.8
TiS_2	E⊥c	80.1	13.7	66.4	175.1	423.4	20.8
$TiSe_2$	E⊥c	123.1	38.6	84.5	143.9	257.0	18.5
TaS_2	E⊥c	69.6	∼68.0	1.6	209.7	212.1	6.6
ZrS_2	E⊥c	34.5	9.23	25.27	181	350	16.3
HfS_2	E⊥c	23.1	6.20	16.9	166	318	26.6
$HfSe_2$	E⊥c	38.2	8.05	30.15	98	205	27.0
ε-GaSe	E//c	7.6	7.1	0.5	237.0	245.8	1.80
ε-GaSe	E⊥c	9.54	6.73	2.81	213.9	254.7	2.7

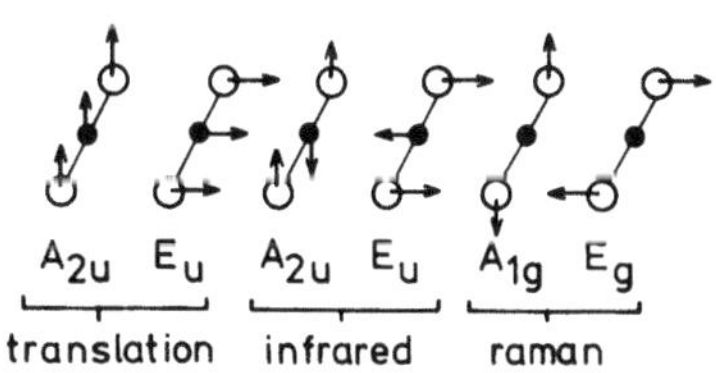

Fig.2 Normal modes (q=0) of CdI_2

The frequencies of the lattice vibrations are determined by short-range forces between neighbouring atoms, due to repulsion and covalent bonding, and by long-range Coulomb interactions between charges and dipoles. The local electric field E_j at ion j induces an electric dipole $\mu_j = \alpha_j E_j$. During the vibration, the interaction with neighbouring ions leads to a distortion of the charge clouds and to a redistribution of charges. In the deformation dipole model this effect is accounted for by the introduction of an effective charge Z* different from the static charge Z. The force on ion j is due to contributions of the electric field acting on the charge q_j and the field gradient acting on the dipole:

$$K_j = q_j E_j + \mu_j \nabla_j E_j \qquad (5)$$

The latter term vanishes in the harmonic approximation for cubic crystals; however in anisotropic crystals this term is quite important (although it is frequently neglected). We remark that both E_j and μ_j have static and oscillating parts. The calculation of the fields and dipoles involves lattice sums, which have been evaluated numerically for the ideal CdI_2 lattice [4].

Reflection spectra of a number of layered compounds have been analyzed with this theory [4]. It was found that a satisfactory explanation of the observed anisotropy could be given, using reasonable values for the effective charge and the polarizabilities (Table 3). The analysis shows that the force

Table 3 Ionic polarizabilities of anions (α_1) and cations (α_2), effective charge Z*, static dipole μ_0 at the anion, and force constants f, deduced from optical data

	α_1 [10^{-24}cm^3]	α_2 [10^{-24}cm^3]	Z*	μ^0 [10^{-18}esu]	$f(A_{2u})$ [10^{-5}dyn cm^{-1}]	$f(E_u)$ [10^{-5}dyn cm^{-1}]
PbI_2	8.4	2.0	0.68	3.8	0.29	0.42
CdI_2	5.6	2.4	0.58	3.0	0.36	0.43
NiI_2	5.6	2.1	0.30	1.5	0.42	0.46
SnS_2	3.8	2.9	0.59	2.5	1.34	1.25
$SnSe_2$	6.0	2.0	0.48	4.0	1.06	1.01

constants for the short-range forces are nearly isotropic, as expected. The observed anisotropy is mainly due to the electrostatic interaction; the various contributions are shown in Fig. 3. It is found that the presence of dipoles

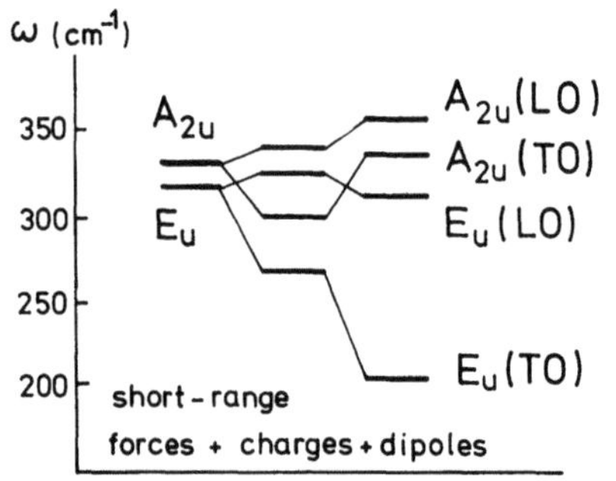

Fig.3 Influence of ionic charges and dipoles on the frequencies of optical phonons in SnS_2

at the anions causes a lowering of the frequency of vibrations of symmetry E_u and a larger oscillator strength for these vibrations (Fig.4). The vibrating cations induce oscillating dipoles at the anions; these dipoles are in phase with the vibrating cation charge for the E_u mode, but in antiphase for A_{2u}. The result is a large oscillator strength for the E_u mode and a small one for the A_{2u} mode.

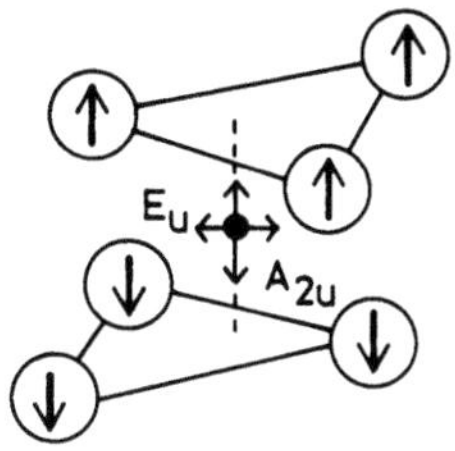

Fig.4 Anisotropy of the vibration of a cation in (undistorted) octahedron; due to the anion dipoles $\omega(E_u)<\omega(A_{2u})$

4. Lattice Distortions in Ionic Crystals

In the previous section we have shown that the presence of static dipoles oriented perpendicular to the layers results in a lowering of the frequency of vibrations involving displacements of ions in the layers. This softening of perpendicular lattice modes (E_u) can be considered as a precursor of lattice distortions which break the hexagonal symmetry of the layers. If the effect is strong, a spontaneous lattice distortion will occur.

We discuss the effect of a small lattice distortion on the total energy of the crystal. For convenience we replace the Madelung energy and Born repulsion by a symmetrical potential energy function which opposes the distortion. The distortions we consider correspond to an even more asymmetric coordination of the anions by the cation than already present in the layered structure. We consider a type of distortion which lowers the polarization energy. The total energy change is

$$\Delta E= p\,u^2 + q\,u^4 - s\,u^2 \qquad (6)$$

The first two terms represent the elastic energy (Madelung + Born), the last term is the change of the polarization energy due to the distortion u. We have carried out approximate calculations of the polarization energy of two simple distorted structures represented in Fig.5, with zigzag chains (MnP-type) and with triangles (1.t. NbS-type) of metal atoms. Distortions of this type have been observed in many compounds [7]. For simplicity we assume that the anion lattice remains unchanged, so that only displacements of cations are considered. Moreover only the influence of displacements of nearest-neighbour cations are taken into account.

In the distorted structures the coordination of the anion by the cations has changed; we show in Fig.6 the coordination of some typical anions in the distorted structures.

Calculations show that the polarization energies of MnP-type and 1.t.-NbS type distortions are equal (up to quadratic terms in the distortion amplitude u). The distorted structures are stable if $\alpha_1/R^6>p/9Z^2e^2$. Thus we expect for compounds with highly polarizable anions and small cations a further stabilization of the layer structure due to a distortion within the layers. This is observed indeed; many compounds with the highly polarizable Te^{2-} ions have distorted layer structures; examples are $MoTe_2$, WTe_2, $NbTe_2$, $TaTe_2$, VTe_2 and $ReTe_2$.

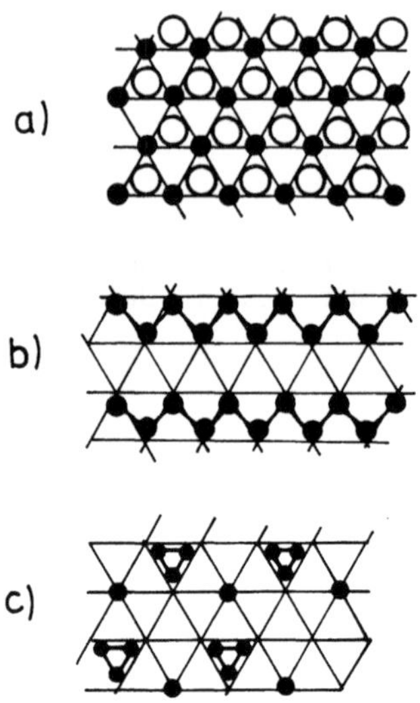

Fig.5 Distortions of a hexagonal layer of metal atoms embedded in a sandwich of anions. a) undistorted b) distortion $q= (0\frac{1}{2})$ (MnP-type) c) superposition of distortions $q_1= (\frac{1}{2}0)$, $q_2= (0\frac{1}{2})$ and $q_3= (\frac{1}{2}\frac{1}{2})$ (l.t. NbS-type). (the non-metal atoms are drawn only in the undistorted structure a)

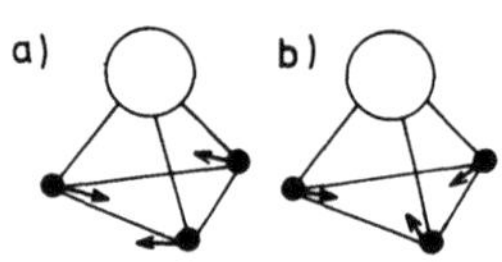

Fig.6 Distorted anion coordination in MnP-type (a) and l.t. NbS-type (b) distorted layers

In the polarizable ion model the crystal structure is a compromise between the Madelung energy favouring large cation-cation distances and the polarization energy favouring an asymmetric coordination of the anions. This principle can also be used to understand the crystal structures of other inorganic compounds [8,9]. The metal trihalides $FeCl_3$, $CrCl_3$, $CrBr_3$, BiI_3, etc. crystallize in layered structures derived from the CdI_2 or $CdCl_2$ structure by leaving one third of the metal sites unoccupied. In the partly occupied layers the metal atoms form an ordered array in which the repulsive energy between the cations is a minimum. In the intercalated materials A_xMX_2 (X= S, Se) the structure is usually such that the cations with the highest charge (M) occupy alternate layers, as in the CdI_2 structure. The remaining cations of lower charge A occupy sites in the Van der Waals gap. In this manner the anion coordination by cation charges is as asymmetric as is compatible with the composition of the crystal. Examples are Cr_3S_4 (fully occupied Cr^{3+} layers alternate with half occupied Cr^{2+} layers), Cr_2S_3 (fully occupied Cr^{3+} layers alternate with 1/3 occupied Cr^{3+} layers), Na_xTiS_2, Ag_xCrS_2, etc. (fully occupied Ti, Cr layers; Na^+ and Ag^+ in the Van der Waals gap).

Thus a simple model with polarizable ions can explain a considerable amount of structural data of inorganic solids. However, anion polarization is not the only mechanism which can drive a lattice distortion. $TiSe_2$ has at low temperature a distorted layer structure with a unit cell 2ax2ax2c, but with atomic displacements different from those of the l.t.-NbS-type distortion [10]. The atomic displacements in $TiSe_2$ are of a type which is not favoured by anion polarization. Recent calculations in which the distortion in $TiSe_2$ is attributed to a band Jahn-Teller effect (metal d - non metal p hybridisation) are in excellent agreement with experimental data [11]. The lattice deformations of TaS_2, $NbSe_2$, VSe_2, $TaSe_2$ are usually attributed to the charge density wave mechanism [1].

5. Lattice Distortions Caused by Metal-Metal Interactions

It is well-known that a one-dimensional metal with one electron per unit cell is unstable with respect to a Peierls distortion leading to a doubling of the unit cell (dimerization). However, the same distortion is obtained if there is an attractive interaction between the atoms such that the energy of a short, strong metal-metal bond and a long, weaker metal-metal bond is lower than the energy of two bonds of intermediate length (Fig.7). Thus the natural

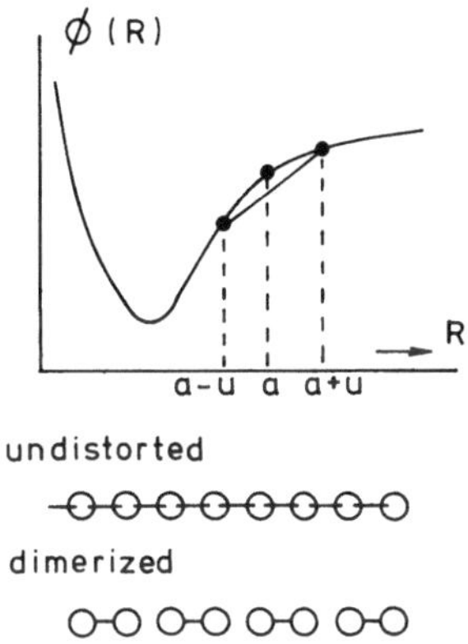

Fig.7 Lattice distortion of linear chain occurs for $\partial^2\phi/\partial R^2>0$. The average distance a is fixed by the presence of other atoms in the crystal

distortion of a linear chain with nearest neighbour interactions is a dimerization. If also interactions with more distant neighbours are taken into account, one obtains more complicated patterns of distortion, also incommensurate [12].

A similar situation can occur in the hexagonal layers, present in layered structures. For the sake of simplicity, let us consider a rigid hexagonal lattice of anions, with the metal atoms in the interstitial sites. Two types of interactions are taken into account: an isoptropic interaction ϕ_i which tries to keep the metal atom i in the center of the interstitial site, and an attractive interaction ϕ_{ij} between metal atoms i and j. Calculations show that for an attractive interaction between nearest-neighbour metal atoms two types of distorted structures are possible, a structure with zigzag-chains (MnP-type, observed in VS, NbS, $MoTe_2$, WTe_2) and a structure with triangles and isolated atoms (low temperature (l.t.) NbS-type, observed in NbS, Cu_xVS_2) [13] (Fig.5). If interactions with more distant metal atoms are included, one finds also more complicated types of distortion with clusters of metal atoms (Fig.8). In this case it is also possible to obtain incommensurate distortions.

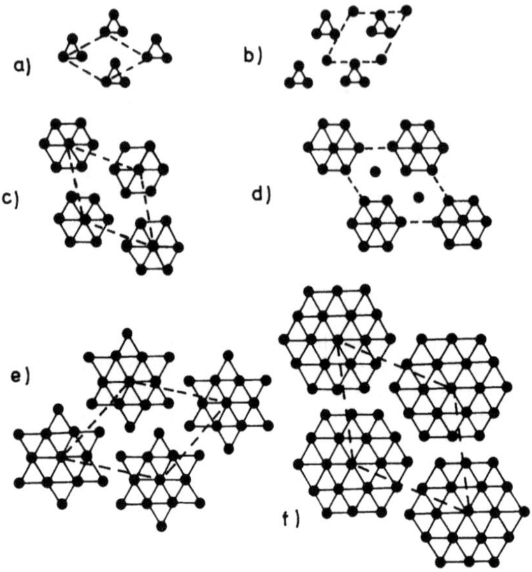

Fig.8 Clustering of metal atoms in hexagonal layers
a) q= (2/3 1/3)-FeS b) q= (½0)-1.t.NbS, Cu_xVS_2
c) q= (2/7 1/7)-VSe_2 d) q= (1/3 0)-VSe_2, 2H-$TaSe_2$, 2H-$NbSe_2$
e) q= (4/13 1/13)-1T-TaS_2, 1T-$TaSe_2$
f) q= (2/19 1/19)-$NbTe_2$, $TaTe_2$

6. Discussion

We have shown that lattice distortions may occur as a result of an attractive interaction between metal atoms in the crystal. However, we did not specify the nature of this interaction. A first possibility is a direct covalent metal-metal bonding due to overlap of metal orbitals. This is presumably the mechanism in several transition metal oxides (dimerization of V^{4+} in VO_2, etc.), but in the sulfides, selenides and tellurides the metal-metal distances are too large for appreciable direct overlap of metal orbitals. However, an effective metal-metal attraction is possible via the intermediate of other electrons, i.e. via a polarization mechanism. Both the charge density wave model and the model involving (anti)ferroelectric distortions due to anion polarization are of this type; in both cases it is essentially the polarization of the electrons which drives the distortion. The polarizability α of a system is given by

$$\alpha = \sum_i \frac{|\langle o|\mu|i\rangle|^2}{\varepsilon_i - \varepsilon_o}\,, \qquad (6)$$

where $\langle o|\mu|i\rangle$ is the dipole matrix element of an electronic excitation from the ground state ε_0 to an excited state ε_i. In insulators the electronic excitations are of an interband type across an energy gap ε_G. In this case all electrons of the highest occupied valence band (usually the anion p band) contribute to α. The total polarizability of the crystal can be approximated as a sum of ionic polarizabilities. For highly polarizable materials the gap ε_G is small and α is large.

For metals there is a large contribution to the electronic polarizability of electronic excitations near the Fermi energy; this contribution of a relatively small number of electron is large because the energy denominator $(\varepsilon_i - \varepsilon_0)$ is small. The interaction of metal atoms with conduction electrons produces a long-range oscillating effective interaction between metal atoms i and j, given by

$$\phi(r_{ij}) \sim \chi(q=2k_F) \cos 2k_F r_{ij} \; , \qquad (7)$$

where $\chi(q)$ is the generalized susceptibility (polarizability) of the (free) conduction electron gas, and k_F is the wave vector at the Fermi energy. The susceptibility $\chi(q)$ is strongly q dependent and may become large for $q= 2k_F$, due to nesting of electron and hole surfaces. In that case a spontaneous distortion with periodicity $q= 2k_F$ (charge density wave) will occur [1].

In actual metals the contribution of electrons near the Fermi surface (charge density wave mechanism) and the contribution of valence band electrons (ionic polarizability model) will operate simultaneously, and it may be difficult to decide which mechanism dominates. The contribution of the conduction electrons near the Fermi surface easily leads to complicated and in many cases incommensurate lattice distortions, due to the long-range oscillating metal-metal interactions involved. However even in some insulating ionic crystals, complicated and incommensurate lattice distortions have been reported (Rb_2ZnCl_2 [14], Na_2CO_3 [15]).

7. References

1 J.A. Wilson, F.J. DiSalvo and S. Mahajan, Adv. Phys. 24, 117 (1975).
2 W.C. Nieuwpoort and G. Blasse, J. Inorg. Nucl. Chem. 30, 1635 (1968).
3 Y.M. de Haan, in Molecular Dynamics and Structure of Solids, R.S. Carter and J.J. Rush Ed. (Nat. Bur. Standards, Washington, 1969), p. 233, and Y.M. de Haan, thesis, Delft, 1965.
4 H.J.L. van der Valk and C. Haas, phys. stat. sol. (b) 80, 321 (1977).
5 R.D. Shannon and C.T. Prewitt, Acta Cryst. B25, 925 (1969).
6 J.R. Hardy, in Dynamical Properties of Solids, G.K. Horton and A.A. Maradudin Ed. (North-Holland Publ. Co., Amsterdam, 1974), vol. 1, p. 157.
7 H.F. Franzen, C. Haas and F. Jellinek, Phys. Rev. B10, 1248 (1974).
8 A.E. van Arkel, Moleculen en Kristallen (Van Stockum & Zoon, Den Haag, 1961).
9 F. Jellinek, in Inorganic Sulphur Chemistry, G. Nickless Ed. (Elsevier Publ. Co., Amsterdam, 1968), p. 670.
10 F.J. DiSalvo, D.E. Moncton and J.V. Waszczak, Phys. Rev. B14, 4321 (1976).
11 Y. Yoshida and K. Motizuki, J. Phys. Soc. Japan 49, 898 (1980).
12 R. Phynn, Nature 281, 433 (1979).
13 C. Haas, Solid State Commun. 26, 709 (1978).
14 C.J. de Pater and C. van Dijk, Phys. Rev. B18, 1281 (1978).
15 C.M. Fortuin, Physica 86A, 224 (1977).

Experimental Studies of Phonon Dispersion and C-Axis Structure in Graphite Intercalation Compounds

P.C. Eklund and J. Giergiel

Department of Physics, University of Kentucky
Lexington, Kentucky 40506, USA and

P. Boolchand

Department of Physics, University of Cincinnati
Cincinnati, Ohio 45221, USA

I Introduction

In this paper we present new results, obtained from a variety of experiments, which probe the structure and lattice dynamics of donor- and acceptor-type graphite intercalation compounds (GIC's). Specifically, we wish to discuss conclusions deduced from x-ray diffraction, Mössbauer, inelastic neutron scattering and Raman scattering investigations of well-staged GIC's. The paper is organized as follows: in section II we discuss the experimental details, in section III we collect results pertaining to the chemistry, layer structure and c-axis dispersion of low-frequency phonon branches in $SbC\ell_5$:graphite; in section IV and V we present the results of Raman scattering studies; section IV - the stage-dependence of two-phonon scattering from high-frequency graphitic modes in $SbC\ell_5$:graphite (acceptor system) and Rb:graphite (donor system); section V- the observation and study of the temperature dependence of low frequency shear modes ($\omega \simeq 15$, 20 cm^{-1}) in stage 2 Rb:graphite.

II Experimental Details

The experimental investigations reported in this paper were performed on predominantly single stage material. The samples were prepared by reacting rectangular plates of highly oriented pyrolytic graphite (HOPG) with the respective atomic or molecular vapor in a closed pyrex tube by the two-temperature technique. Rubidium-graphite samples remained in sealed pyrex ampoules which were tipped off from the parent reaction tube, whereas the $SbC\ell_5$:graphite samples are air-stable [1] and therefore, were removed from the reaction tubes for subsequent experiments. The stage of intercalation was characterized by (00ℓ) x-ray diffraction scans (Rb-HOPG, $SbC\ell_5$-HOPG) as well as by sample dilation and weight-uptake ($SbC\ell_5$-HOPG). The various experimental techniques required samples of different dimensions. Listed here are typical values for the intercalated material: (a) x-ray and Raman studies $\simeq$5x5x0.4 mm (b) neutron scattering experiments $\simeq$10x10x10 mm (this volume was generated by stacking several plates $\simeq$10x10x1.5 mm). (c) Mössbauer spectroscopy $\simeq$10x10x0.3 mm.

The Mössbauer spectra were taken [2] using a conventional drive in conjunction with a He exchange gas dewar. All measurements were made using a $Ca^{121m}SnO_3$ source and intercalated graphite absorber. The source (37.2 KeV γ-rays) probes the Sb environments in the bulk of the sample. To obtain spectra as a function of sample thickness, we cleaved the original (as-grown) samples to obtain successively thinner absorbers. The samples were oriented with their c-axes parallel to the propagation direction of the gamma rays.

The Raman scattering studies were performed in the usual Brewster-angle, backscattering geometry using low laser powers. To further prevent laser-induced damage to the samples, the incident laser beam was focused to a 2 mm tall stripe using a cylindrical lens. The Raman spectrometer has been described elsewhere [3]. Studies of the temperature dependence of the low-frequency shear modes in stage 2 Rubidium:graphite were made with the sample ampoule mounted in a He exchange-gas cell located inside an optical dewar.

Integrated intensity measurements of the (00ℓ) Bragg reflections in stage 1, 2, 3 $SbC\ell_5$:graphite were made using a four-circle Enraf-Nonius-Delft diffractometer interfaced to a minicomputer. The data were taken using Mo K_α radiation. The c-axes of the samples were automatically placed in the scattering plane by computer control of the instrument using strong Bragg reflections near $\theta \approx 10°$. Integrated intensities for the (00ℓ) peaks used in the data analysis were first corrected for background contributions.

The neutron scattering experiments on stage 2 and 4 $SbC\ell_5$:graphite were performed on a triple-axis spectrometer located in the High Flux Isotope Reactor facility, Oak Ridge National Laboratory. These studies were made in collaboration with Dr. H.G. Smith. Elastic scattering measurements of the (00ℓ) peak profiles in the stage 2 and 4 compounds indicated that the entire scattering volume was predominantly single stage material.

III Chemistry, Layer Structure, and c-Axis Dispersion of LA and LO Phonons in $SbC\ell_5$:Graphite

The reactions of graphite with the group V pentahalides (AsF_5, $SbC\ell_5$, SbF_5) are known to produce lamellar compounds with high in-plane electrical conductivities. In the case of AsF_5:graphite, conductivities approximately one-half that of copper have been observed [4]. Only in the last few years has it become evident that the intercalate layers in these compounds contain large concentrations of molecular species other than the vapor phase pentahalide molecule participating in the reaction [4]. There now appears to be general agreement that the MX_6^- species (M = As, Sb and X = Cℓ, F) is present in all these group V pentahalide intercalation compounds. That the MX_6^- species is present in $SbC\ell_5$:graphite is one of the important results from our ^{121}Sb Mössbauer studies in these compounds [5,6]. The concentration of the MX_6^- sites is directly related to the high conductivities in these materials. There is considerably less agreement, however, about the potential presence (or absence) of the neutral species MX_3, MX_5. We briefly review the literature pertaining to these neutral species.

Bartlett and co-workers [7] have suggested that the intercalate layers in equilibrium AsF_5:graphite are comprised of AsF_6^- molecular ions and AsF_3 (neutrals) in a 2:1 ratio. However, Moran et al. [8] on the basis of XPS experiments find a significant amount of neutral AsF_5 in the intercalate layers within the skin depth (≈30 Å) of the XPS technique. ^{121}Sb Mossbauer spectra of Ballard and Birchall [9] on $SbC\ell_5$:graphite were interpreted by them to indicate the presence of $SbC\ell_5$ (not $SbC\ell_6^-$) and $SbC\ell_3$. More recent Mössbauer studies in the SbF_5:graphite system by Forsman et al. [10] report the observation of only an Sb^{5+} resonance in their spectra; they, furthermore, conclude that SbF_3 is probably formed at the surface of the compound and does not intercalate. They have assigned the Sb^{5+} resonance to the presence of both SbF_5 and SbF_6^- in the intercalate layer and infer a chemical formula of $C_{48}{\cdot}SbF_6^-{\cdot}6.3\ SbF_5$. However, in contrast to the results of Foresman et al., we do observe a strong Sb^{3+} resonance in SbF_5:HOPG samples [11].

It is clear that all the above mentioned results do not yet generate a consistent picture for a general chemistry in the group V pentahalide graphite compounds, and that more work is needed in this area. In what follows we restrict the discussion to our recent results in the $SbCl_5$:graphite system [5,6]. A more comprehensive study intended to unravel the general chemistry of MX_5:graphite compounds is underway and will be published elsewhere [11].

As can be seen from Fig. 1a, the Mössbauer spectra of stage 2 $SbCl_5$:HOPG show two dominant features: a narrow feature centered at v = 2.89(4)mm/s which we identify with Sb^{5+}, and a broader feature centered at v = -14.25(12) mm/s which we identify with Sb^{3+}. Each of these features was first analysed in terms of one Sb^{3+} and one Sb^{5+} chemical site using the usual 8 line quadrupole pattern characteristic of the $7/2^+ \rightarrow 5/2^+$ ^{121}Sb transition. The ratio of the nuclear quadrupole moments $Q(7/2^+)/Q(5/2^+)$ was fixed at a value of 1.34 [12].

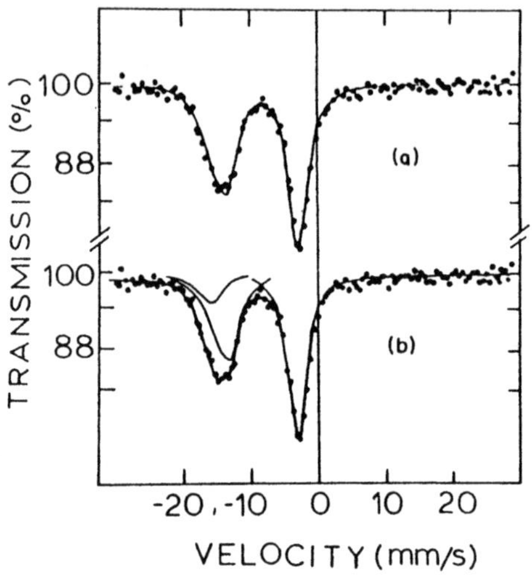

Fig. 1a,b Mössbauer spectrum (4.2K) of stage 2 $SbCl_5$:HOPG. Solid lines represent the results of lineshape analyses: Fig. 1a-two chemical site analysis, Fig. 1b-three chemical site analysis. See text.

The values for the isomer shift δ and quadrupole coupling e^2qQ for the Sb^{5+} chemical site (site 1) virtually coincide with the values reported for $SbCl_6^-$ in the molecular cyrstal $RbSbCl_6$ [13]. The isomer shift of neutral $SbCl_5$ as in solid $SbCl_5$ is about +0.5 mm/s relative to the δ of site 1. This difference represents ≈20% of the observed line width of Sb^{5+} resonance in our GIC samples. These results suggest that $SbCl_6^-$ molecular ion is the dominant Sb^{5+} species present in the intercalate layers, although the presence of some neutral $SbCl_5$ cannot be excluded from the data. This conclusion is further supported by our value of $e^2qQ \simeq 0$ for site 1, which is consistent with the octahedral symmetry of $SbCl_6^-$ (octahedral symmetry at the Sb site requires that the local electric field gradient eq be zero). Table I summarizes the ^{121}Sb isomer shift and quadrupole coupling constants relevant to this work. Entries in the table based on our spectra refer to 4.2K spectra, all other entries refer to spectra taken at 78K.

Our spectra of $SbCl_5$-graphite (HOPG) samples exhibit, in general, a small systematic misfit in the Sb^{3+} region of the spectrum when the data is analysed in terms of a single chemical site. This can be seen in Fig. 1a where the fitted curve (single Sb^{3+} site) shows a systematic misfit over the entire Sb^{3+} region and particularly at the peak. On re-analysing the Sb^{3+} resonance in terms of two chemical sites, significantly better fits to the observed lineshape resulted, as shown in Fig. 1b. More significantly, several spectra

Table I: ^{121}Sb isomer shift δ, quadrupole coupling e^2qQ in indicated absorbers. See text for site identification.

Absorber	Site	$\delta^{\dagger}$(mm/s)	e^2qQ^*(mm/s)
$SbC\ell_5$-HOPG	1	- 2.89(4)	$0^{\dagger\dagger}$
	2	-14.25(12)	+15.2(1.5)
	3	-16.06(40)	+11(5)
$SbC\ell_5$-powdered graphite	1	- 2.78(8)	$0^{\dagger\dagger}$
	2	-14.40(10)	+17.5(1.5)
$(SbC\ell_5+SbC\ell_3)$-HOPG	1	- 2.93(4)	$0^{\dagger\dagger}$
	2	-14.48(5)	+14.2(8)
$SbC\ell_3$	-	-14.19(7)	+18.03(1.18)
$SbC\ell_3$ (Ref. 14)	-	-14.16(20)	+12.2(1.7)
$SbC\ell_5$	-	- 3.41(5)	- 5.37(10)
$SbC\ell_5$ (Ref. 14)	-	- 3.56(3)	- 4.4(2)
$SbC\ell_6^-$ in $RbSbC\ell_6$ (Ref. 13)	-	- 2.81	0
$PyHSbC\ell_4$ (Ref. 15)	-	-16.56	9

†Isomer shifts δ quoted relative to $Ca^{121m}SnO_3$ at 4.2K.
*Anisotropy parameters for sites 1, 2, 3: η_1 = 0.01; η_2 = 0.18, η_3 = 0.18. These parameters were kept fixed in the lineshape analysis.
$^{\dagger\dagger}e^2qQ$ of $SbC\ell_6^-$ was kept fixed to zero, since on floating, it oscillated about zero with large uncertainties.

taken on samples with varying thickness, when analysed in terms of 2 sites for the Sb^{3+} region of the spectrum, all invariably refined to the same set of values for δ, e^2qQ for the two Sb^{3+} sites, which we refer to in Table I as sites 2 and 3. Values for the isomer shifts of sites 2 and 3 fall quite close to those reported for $SbC\ell_3$ (in the molecular solid) [14] and $SbC\ell_4^-$ (in $PyHSbC\ell_4$) [15], respectively. It is on this basis that we identify sites 2 and 3 with $SbC\ell_3$ and $SbC\ell_4^-$, respectively. The area ratio A_3/A_2 of the line associated with site 3 to that associated with site 2, when extrapolated to zero sample thickness, approaches a value 0.25 ± .05. Thus, a two site analysis of the Sb^{3+} resonance region of our GIC samples suggests that while Sb^{3+} is clearly present in the intercalate layers as $SbC\ell_3$, it is apparently also present in substantial amounts as $SbC\ell_4^-$. The separate contributions of the Sb^{3+} and Sb^{5+} species are also shown in the composite spectrum of Fig. 1b.

Sample thickness studies of the ratio of the area under Sb^{5+} resonance to the area under the Sb^{3+} resonance lead to a zero thickness value of 1.9 ± 0.1. This ratio, under the assumption of equal recoil free fractions for the two intercalant species $SbC\ell_6^-$ and $SbC\ell_3$, represents the concentration ratio of these species in the intercalate layers. If the oxidation reaction due to Bartlett et al. [7] in the AsF_5-graphite system were to describe the chemistry in $SbC\ell_5$-graphite, one would expect this concentration ratio to have the value 2. We should point out, however, that the small isomer shift of $SbC\ell_6^-$ relative to $SbC\ell_5$ hampers quantitative conclusions concerning the Sb^{5+} resonance. In fact, based on results of a recent two-site analysis (e.g. $SbC\ell_5$, $SbC\ell_6^-$) of the Sb^{5+} region of our Mössbauer spectra, we cannot rule out as high as a 30% contribution from neutral $SbC\ell_5$ [11].

In view of the potential co-existence of several distinct molecular species (e.g. MX_5, MX_3, MX_6) in the intercalate layers of the MX_5:graphite compounds, it is indeed suprising to find long-range, in-plane order in the intercalate layers. In-plane order at room temperature has been reported in both $SbC\ell_5$:

graphite [16,17,18] and SbF_5:graphite [19]. The in-plane arrangement of the various molecular species in the MX_5:graphite compounds remains a relatively unexplored, yet interesting, area.

We report here on the c-directional aspect of the molecular ordering in $SbC\ell_5$:graphite as determined by integrated intensity measurements of the (00ℓ) Bragg peaks in the stage 1, 2 and 3 compounds. We have been able to achieve a good fit to the x-ray data with a centrosymmetric model [20] for the intercalate layer. The constituent molecules are oriented relative to the graphitic c-axis to allow the chlorine atoms to contact the bounding carbon layers. This orientation gives rise to atomic sub-layers within the intercalate layer--two exterior chlorine layers and a central antimony layer (containing both Sb^{3+} and Sb^{5+} species), as indicated in Fig. 2. The c-axis coordinates, atomic densities and Debye-Waller factors for the various layers represent the parameters to be optimized in a non-linear, least-squares analysis of the data. The position of the Sb^{5+} layer was fixed at the center of the intercalate layer by geometric arguments. Details of our data analysis, similar to that of Leung et al. [20], will be published elsewhere [18,21]. In Table II we list the values of the parameters which minimized the chi square function for the set of calculated and experimentally determined (00ℓ) structure factors. Atomic densities have been converted to the more convenient compound stoichiometries. The data suggest that a general formula $C_{14n}SbC\ell_5$ is applicable to the coherantly scattering regions of these compounds, where n is the stage number. Of particular significance is that our analyses of the stage 1,

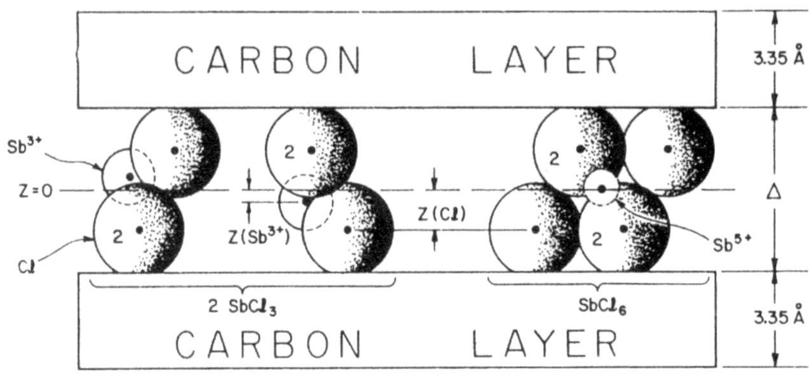

Fig.2. Geometry of the centrosymmetric intercalate layer sandwhich of $SbCl_5$: graphite. Layer coordinates in Table II are measured relative to the center of the sandwhich

Table II Layer Parameters for $SbC\ell_5$:Graphite Determined by (00ℓ) Integrated Intensity X-ray Data.

STAGE	LAYER COORDINATES (Å)				Sb^{5+}/Sb^{3+}	Compound Stoichiometry
	Sb+3	Chlorine	Carbon I	Carbon II		
1.	.050	1.405	4.700	-	1.92	$C_{14.0}SbC\ell_{5.0}$
2.	.049	1.400	4.654	-	1.94	$C_{28.2}SbC\ell_{5.0}$
3.	.051	1.410	4.670	8.001	1.94	$C_{42.5}SbC\ell_{5.0}$

2, and 3 x-ray data all lead to the same stage-independent set of intercalate layer parameters. The results of the x-ray analysis for the three stages can be visually compared in Fig. 3 where we show the c-axis charge density ρ(z) obtained from the Fourier transform of the experimental structure factors using the usual expression. Unfortunately these x-ray results are not able to distinguish molecular contributions to the Sb^{5+} layer from an $SbC\ell_5$ or an $SbC\ell_6^-$ molecule. However, the x-ray analysis is sensitive to the concentration ratio of the Sb^{5+} species to the Sb^{3+}. As seen in Table II, a value for $[Sb^{5+}/Sb^{3+}]$ of ≈1.93 was obtained from our x-ray data analysis. This result is in excellent agreement with a value of 1.9 ± 0.1 determined from the areas under the Sb^{3+} and Sb^{5+} Mössbauer resonances [6].

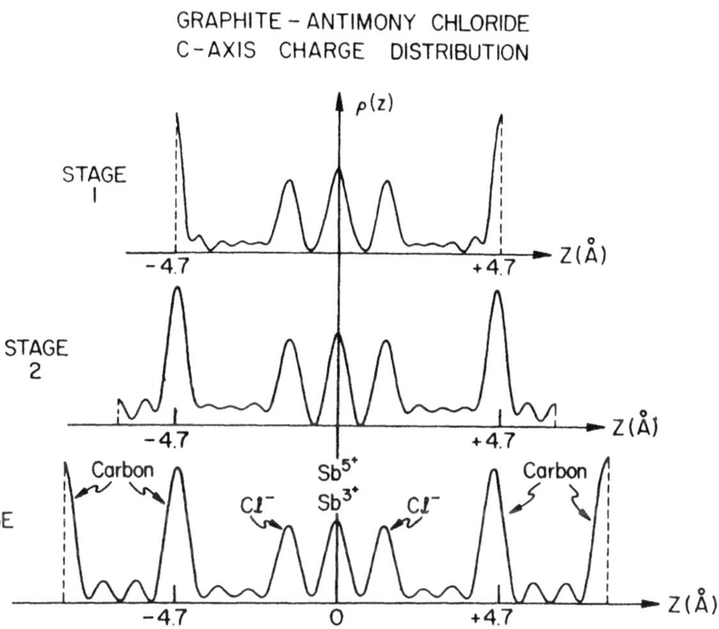

Fig.3. C-axis charge density of stage 1-3 $SbCl_5$: graphite obtained from the Fourier transform of the experimental x-ray structure factors. The central peak contains contributions from Sb^{3+} layers ($z = \pm 0.05$ Å) and the Sb^{5+} layer ($z = 0$)

We next turn to the results of inelastic neutron scattering experiments on stage 2 and stage 4 $SbC\ell_5$:graphite. We report here on the c-axis dispersion of the low frequency (ω<4THz) LO and LA phonon branches in these compounds. These branches are associated with rigid layer displacements parallel to the graphitic c-axis. The coordinates and atomic densities of the layers in the unit cell have been determined by our x-ray structure factor analysis discussed above. Previously, inelastic neutron scattering results in stage 2 $FeC\ell_3$:graphite [22] and low-stage alkali metal-graphite compounds [23,24] have been reported.

At this time we comment only on the qualitative aspects of our data. Quantitative efforts are underway which attempt to fit not only the neutron data on these low frequency LO and LA branches but also the high frequency (ω>1200 cm^{-1}) intralayer modes associated with the Γ, Σ, T and T' positions in the graphitic hexagonal Brillouin zone [25]. The frequency of these modes have been determined by one- and two-phonon Raman scattering experiments and infrared reflection spectroscopy [26].

The x-ray structure factor analysis of the stage 1, 2 and 3 compounds indicate three, or more precisely, five atomic sublaylers in the intercalate layer, if the small displacement of the two Sb^{3+} layers relative to the central Sb^{5+} layer is taken into account. In a stage 2(4) compound there are then 2(4) carbon, 2 chlorine and 3 antimony layers, respectively. The branch counting for the longitudonal modes is a one-dimensional exercise: 1 acoustic, and m-1 optic branches, where m is the number of layers in the unit cell. Thus for the stage 2(4) compound one predicts 6(8) longitudonal optic branches for a 5 sub-layer intercalate sandwhich, or 4(6) for a 3 sub-layer intercalate sandwhich (in which the Sb^{3+} and Sb^{5+} are both assigned to the central sub-layer). As can be seen in Fig. 4a only three branches, one LA, and two LO branches are observed. Thus most of the optic branches associated with internal optic modes of the atomic sub-layers in the intercalate layer are missing. This result is consistent with the recent neutron scattering studies of stage 2 $FeC\ell_3$:graphite by Axe et al. [24]. The highest LO branch we observe is tentatively identified with the center-of-mass mode where the intercalate sandwich beats against the carbon layers. In Fig. 4b we show the data for the stage 4 compound. Counting optic branches we again observe 2 or 4 less branches then the number predicted, depending on whether the intercalate sandwich is described in terms of 3 or 5 atomic sub-layers. As in the stage 2 data, the upper branch is tentatively indentified with the center-of-mass motion of the intercalate layer beating out of phase with the carbon layers. The gaps between branches at the zone center and zone edge seen in Fig.'s 4a and 4b are a measure of the zone-folding interactions in stage 2 and 4 $SbC\ell_5$:graphite.

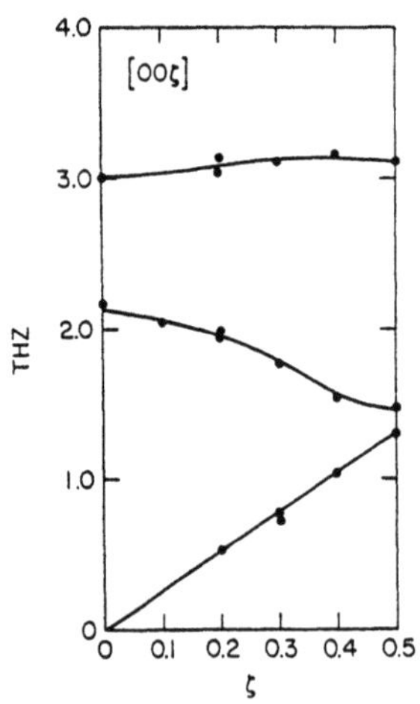

Fig. 4a C-axis dispersion of LA and LO phonons in stage 2 $SbC\ell_5$:graphite. The lines are guides for the eye

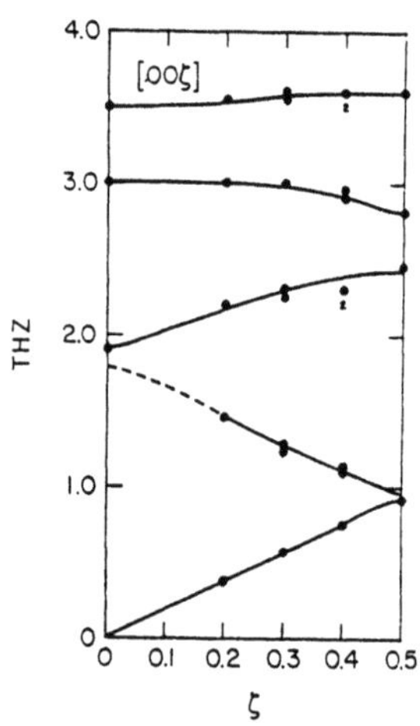

Fig. 4b C-axis dispersion of LA and LO phonons in stage 4 $SbC\ell_5$: graphite. The lines are guides for the eye.

IV Two-Phonon Raman Scattering

In the last several years considerable effort has been expended to document the stage and intercalate species dependence of the graphitic, high-frequency, intralayer-modes in acceptor- and donor-type GIC's [27-29]. These Γ-pt modes are observed in 1st order Raman and infrared spectra. Relatively little effort has been made to study the second-order Raman scattering processes in these compounds [26].

In the 1st-order Raman spectra of stage n>2 GIC's a doublet is observed at frequencies near 1582 cm^{-1} which is the value of the E_{2g} intralayer Γ-pt

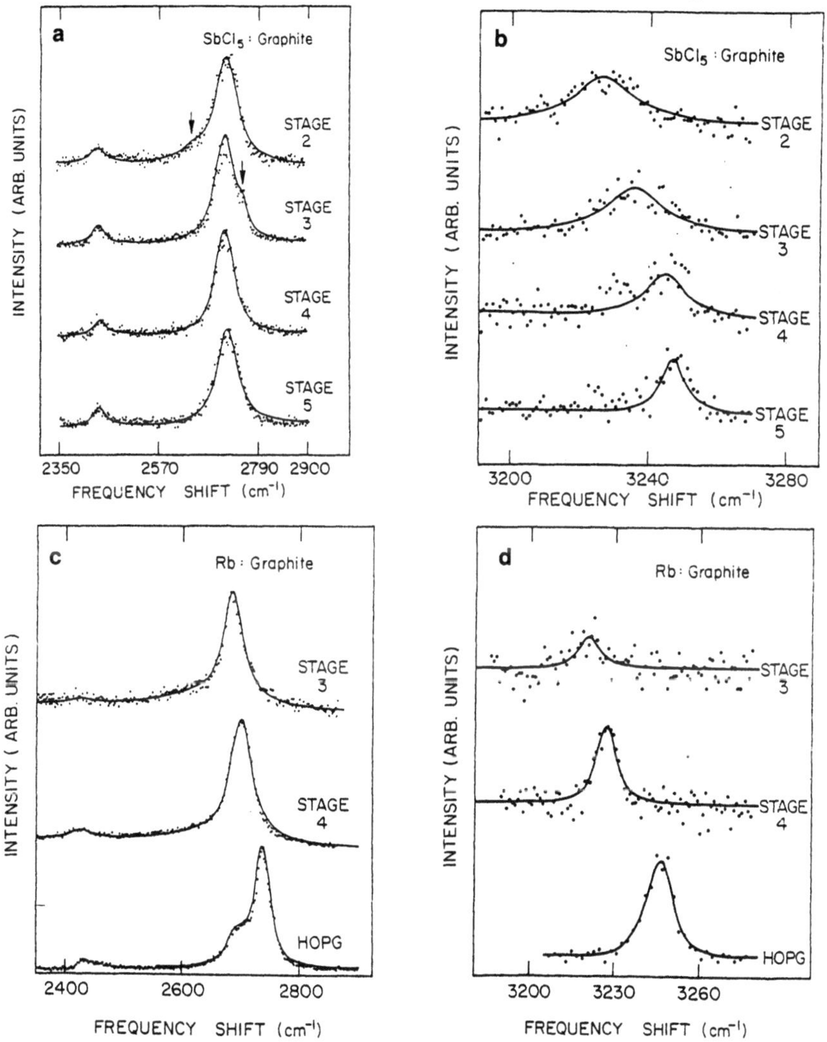

Fig. 5a,b,c,d Second-order Raman spectra of HOPG, $SbC\ell_5$:HOPG and Rb:HOPG. The solid lines represent the results of Lorentzian lineshape analyses.

mode in pristine graphite. The low and high frequency components of the doublet in the GIC are identified with the interior and bounding carbon layers, respectively. Stage 1 and 2 compounds exhibit a single line in this region of the spectrum because they contain only bounding layers. In the case of the stage 1 donor compounds the E_{2g} bounding layer line is asymetrically broadened due to an interference with a continuum scattering process [27-30].

The second-order features of pristine graphite are reasonably sharp and have been attributed to overtone scattering [31,32]. Thus the frequencies associated with high one-phonon density of states can be easily obtained from the Raman spectra by simply halving the frequency shifts. The second-order features are located at 2429 cm^{-1} (T'), 2737 cm^{-1}(T') and 3248 (T,Σ), where the symbols T', Σ and T refer to the zone edge (M-T'-K) and zone interior (Γ-Σ-M), (Γ-T-K) positions in the hexagonal Brillouin zone [31,32]. In the GIC's, second-order Raman spectra similiar to that of pristine graphite are observed [26]. However, in the stage 1 alkali metal (donor) compounds strong, high-frequency continuum scattering has so far prevented us from detecting the analogs of the second-order features seen in the higher stage compounds. In Fig. 5 a,b,c,d we show these second-order features for $SbC\ell_5$:graphite and Rb:graphite. The solid lines represent the results of Lorentzian lineshape analyses.

Contributions to the spectra of Fig. 5 are expected from both bounding and interior intralayer optic branches as well as from the high-frequency, LA branch. Because the second-order spectra are both uncharacteristically sharp and associated with overtone scattering, it is possible to extract characteristic phonon frequencies and plot their stage dependence without calcula-

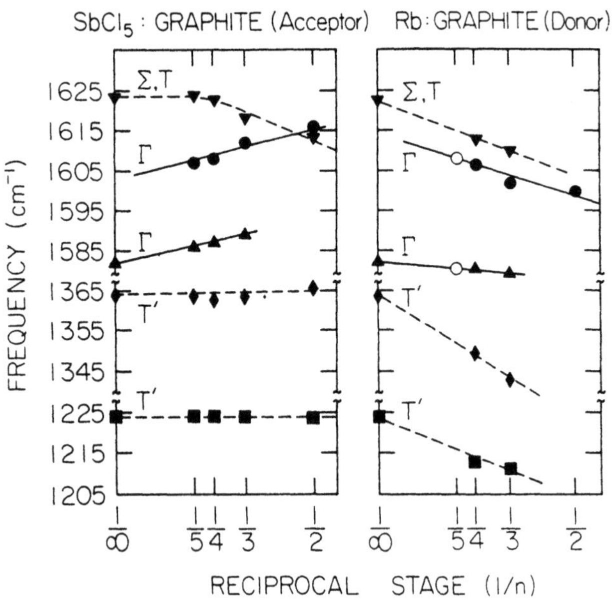

Fig.6. Characteristic intralayer mode frequencies associeated with high one-phonon density-of-states in $SbCl_5$: graphite and Rb : graphite plotted vs reciprocal stage (1/n). The open circles are from Ref.35

ting the two-phonon density of states. In Fig. 6 we plot the characteristic intralayer phonon frequencies obtained from our 2nd-order Raman spectra as a function of reciprocal stage (1/n) for Rb:graphite (right hand panel) and $SbCl_5$:graphite (left hand panel). As can be seen from these panels there exists fundamentally different, yet simple, stage dependences for the various mode frequencies in the donor and acceptor compounds. A quantitative explanation of these results awaits further lattice dynamics calculations.

V Further Raman Scattering Results: Low-frequency, Shear Modes in $C_{24}Rb$

We report the observation of a low-frequency doublet in the Raman spectrum of $C_{24}Rb$ which we attribute to Γ-pt modes involving rigid layer shearing motion of rubidium and carbon layers. The 300 K Raman (anti-stokes) spectrum is shown in Fig. 7a. In the inset to Fig. 7a we show the stokes- antistokes pair of features associated with these modes. The phonon shear modes tentatively assigned to the unresolved doublet in Fig.7a are associated with the relative, in-plane motion of the three layers in the stage 2 unit cell. In both of these Γ-pt modes the in-plane rubidium layer displacement and carbon layer displacement are out-of-phase; the carbon layer displacements in the two modes are in-phase and out-of-phase, respectively. Parameters for the unresolved doublet were determined by a non-linear, least-squares analysis. On the basis of this analysis we find the frequencies of two Lorentzian components to be $\omega_1 = 15 \pm 0.5\ cm^{-1}$ and $\omega_2 = 20 \pm 0.5\ cm^{-1}$.

The rigid-layer modes in stage 2 alkali metal:graphite compounds are of interest not only because they provide valuable low-frequency calibration points for current lattice dynamics calculations [32] but also these modes may probe the interesting phase transitions in the rubidium layers reported at ≃ 135 K and ≃ 95 K [33]. We have obtained the Raman spectra of these rigid layer modes as a function of temperature in the interval (80K<T<300K).

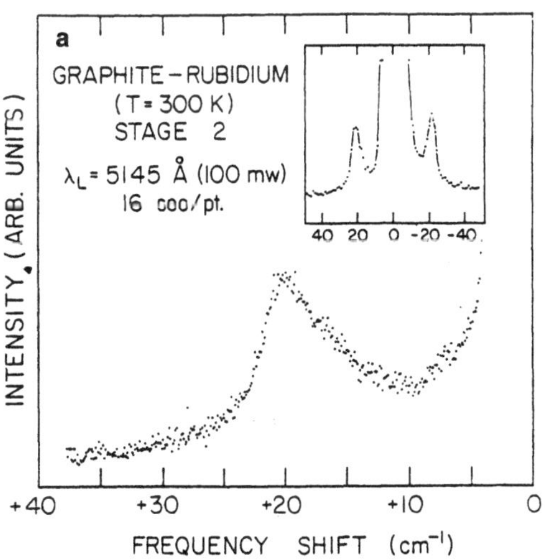

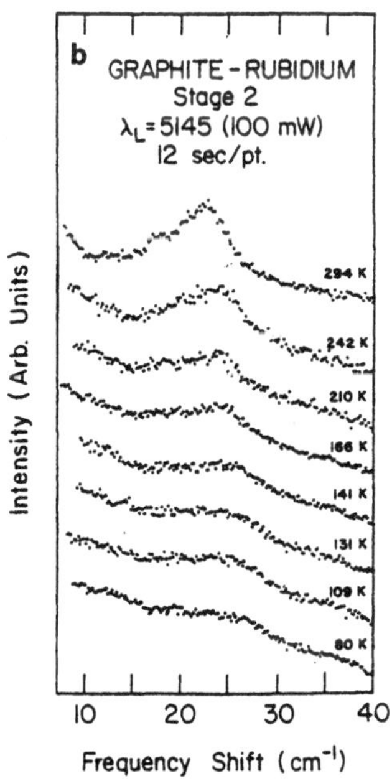

Fig.7a,b. Raman spectra of stage 2 Rb : graphite

These spectra are shown in Fig. 7b. We have not, as yet, finished the computer analysis of these spectra, so we must restrict ourselves here to qualitative remarks only. As can be seen in the figure, dramatic changes in the spectra near the phase transition temperatures (95 K, 135 K) are not readily apparent. The extraction of reliable information, important to the understanding of the phase transition, will require a careful computer analysis to extract frequencies and FWHM's. We note that the doublet shifts to higher frequencies with decreasing temperature, exhibiting approximately a 2-3 cm^{-1} increase in going from 300 → 80 K. This result is in agreement with unpublished neutron scattering results on low-frequency, in-plane modes in $C_{24}Rb$ [34]. Further quantitative work on the Raman spectra shown in Fig. 7b is underway and will be reported elsewhere.

Acknowledgements

We gratefully acknowledge helpful discussions with W. Kamitakahara, N. Wada, D.M. Wang, and G. Dresselhaus. We give special thanks to H.G. Smith and Oak Ridge National Laboratory for their expertise and hospitality during the collaboration on the inelastic neutron scattering experiments. The HOPG was kindly provided by A.W. Moore, Union Carbide. This research was supported from D.O.E. contract #DE-AS05-81ER10936.

References

1. V.R.K. Murthy, D.S. Smith and P.C. Eklund, Mat. Sci. Eng. 45, 77 (1980).
2. M. Blizzard, G. Langouche and P. Boolchand, J. Appl. Phys. 50, 2075 (1979).
3. J. Giergiel, P.C. Eklund and K.R. Subbaswamy, Solid State Commun., accepted for publication.
4. T.E. Thompson, E.M. McCarron and N. Bartlett, Syn. Met. 3, 251 (1981).
5. P. Boolchand, W.J. Bresser, D. McDaniel, K. Sisson, P.C. Eklund and V. Yeh, Extended Abstracts of the 15th Biennial Conference on Carbon, p. 119 (1981).
6. P. Boolchand, W.J. Bresser, D. McDaniel, K. Sisson, V. Yeh and P.C. Eklund, Solid State Commun., accepted for publication.
7. N. Bartlett, R.N. Biagioni, B.W. McQuillan, A.S. Robertson and A.C. Thompson, J. Che. Soc. Chem. Comm. 200, (1978).
8. M.J. Moran, J.E. Fischer, W.R. Salenek, J. Chem. Phys. 73, 629 (1980).
9. J.E. Ballard and T. Birchall, J. Chem. Soc. Dalton 1859 (1976).
10. W.C. Forsman, D.E. Carl and T. Birchall, Extended Abstracts of the 15th Biennial Conference on Carbon, p. 353 (1981).
11. P. Boolchand, W.J. Bresser, D. McDaniel, V. Yeh and P.C. Eklund, unpublished results.
12. L.H. Bowen in Mossbauer Effect Data Index, edited by J.G. Stevens and V.E. Stevens, Plenum, N.Y., p. 71 (1972).
13. T. Birchall, B. Dellavalle, E. Martineau and J.B. Milne, J. Chem. Soc. A 1855 (1971).
14. L.H. Bowen, J.G. Stevens and G.G. Long, J. Chem. Phys. 51, 2010 (1969).
15. J.D. Donaldson, A. Kjekshus, D.G. Nicholson and M.J. Tricker, Acta Chem. Sand 26, 3215 (1972).
16. R. Clarke and H. Homma, Bull. Am. Phys. Soc. 26 (3), 452 (1981).
17. J. Melin and A. Herold, C.R. Acad. Sc. 269, 877 (1969).
18. M.L. Saylors, M.H. Boca, D.S. Smith and P.C. Eklund, Extended Abstracts of the 15th Biennial Carbon Conference, p. 117 (1981).
19. F.L. Vogel, submitted for publication.
20. S.Y. Leung, C. Underhill, G. Dresselhaus, T. Krapchev, R. Ogilvie and M.S. Dresselhaus, Solid State Commun. 32, 635 (1979).

21. M.H. Boca, M.L. Saylors, D.S. Smith and P.C. Eklund, manuscript in preparation.
22. J.D. Axe, C.F. Majkrzak, L. Passel, S.K. Satija, G. Dresselhaus and H. Mazurek, Extended Abstracts of the 15th Biennial Carbon Conference, P. 52 (1981).
23. A. Magerl and H. Zabel, Phys. Rev. Letters 46, 444 (1981).
24. W.D. Ellenson, D. Semmingsen, D. Guerard, D.G. Onn and J.E. Fischer, Mater. Sci. Eng. 31, 137 (1977).
25. H.G. Smith, P.C. Eklund, V. Yeh, R. Al-Jishi and G. Dresselhaus, manuscript in preparation.
26. P.C. Eklund, D.S. Smith, V.R.K. Murthy and S.Y. Leung, Syn. Met. 2, 99 (1980).
27. S.A. Solin, Physica 99B, 443 (1980).
28. M.S. Dresselhaus and G. Dresselhaus in Physics and Chemistry of Layered Materials, ed. F. Levy, Vol. 6, Reidel, Dordrecht, Holland (1979).
29. M.S. Dresselhaus and G. Dresselhaus, Advances in Physics (in press).
30. P.C. Eklund and K.R. Subbaswamy, Phys. Rev. B20, 5157 (1979).
31. R.J. Nemanich and S.A. Solin, Phys. Rev. B 20, 392 (1979).
32. R. Al-Jishi, B.S. Elman and G. Dresselhaus, Extended Abstracts of the 15th Biennial Conference on Carbon, p. 34 (1981).
33. D.G. Onn, G.M.T. Foley and J.E. Fischer, Phys. Rev. B19, 6474 (1979).
34. W. Kamitakahara, private communication.
35. S.A. Solin, Mat. Sci. Eng. 31, 153 (1977).

Phonons in Graphite Intercalation Compounds Measured by Neutron Scattering

A. Magerl

Institut Laue Langevin, 156X
F-38042 Grenoble Cedex and

H. Zabel

Loomis Laboratory of Physics and Materials Research Laboratory
University of Illinois at Urbana-Champaign, Urbana, IL 61801, USA

1. Introduction

The entire phonon dispersion of a solid can be measured by neutron scattering under favorable conditions. In particular, triple axis spectroscopy allows the phonon energies to be determined for a chosen polarization at well defined wave vectors q throughout the Brillouin zone (BZ). For an unambiguous assignment of the excitations, single crystals have to be available. However, in the case of graphite a complete measurement of the phonon dispersion is not possible because of the limited size of (natural) single crystals, and because of the high energy (~ 200 meV) of some branches. Therefore, only a limited amount of neutron data have been collected with pyrolytic graphite (PG) [1,2]. Together with q = 0 modes from light scattering techniques [3], they have been used to calculate the whole phonon dispersion of graphite [2,4]. It now serves as a guide for the discussion of the phonon dispersion of graphite intercalation compounds (GIC).

In the following we focus on the [001]L phonon modes. In pure graphite as well as in GIC's these modes are characterized by the vibrations of rigid graphite and/or intercalate planes with both the amplitude and wave vector along the hexagonal c-axis. As there are no intraplane coordiantes involved the [001]L dispersion is equivalent to the dispersion of a linear chain with different mass points, representing graphite and intercalate planes, stacked along the hexagonal axis. Because of the one-dimensional character of these modes, they are insensitive to the random orientation of the planes in the PG material normal to the hexagonal axis. Therefore, a measurement of the [001]L dispersion is not obscured by using PG instead of single crystals.

The [001]L phonon branch of pristine graphite is well described by just one sinusoidal function, indicating that only nearest neighbor interactions between basal planes are relevant for these vibrations. Upon intercalation the symmetry, mass and interplanar interactions change. The aim of this paper is to describe the effect of these changes on the [001]L phonon branches of GIC's.

In the forthcoming discussion we will restrict ourselves to the phonon dispersion of alkali GIC's (AGIC). Since we can neglect the stacking sequence, the repeat period of a stage n AGIC is given by the distance between two intercalated layers. There are n+1 planes within the unit length[1]. Therefore

[1] The intercalated layer for AGIC's is represented by one rigid plane. The description can be easily modified to account for more complex systems like $FeCl_3$.

the [001]L phonon dispersion will consist of 1 acoustical branch and n optical branches. The q dependence and the splitting of the branches at the BZ center and edge will depend in detail on the masses and forces involved. A comparison of the phonon dispersion obtained from model calculations and experimentally determined phonon energies yields insight about the kind and the strength of the interplanar interactions.

The first [001]L phonon dispersion measurement on a GIC has been reported for C_8Rb [5]. Clear evidence for zone folding effects together with an opening of frequency gaps at the BZ center and edge have recently been given by the authors on $C_{36}K$ [6]. Measurments have also been performed at stage 1 C_6Li [7]. However, the branches could not be assigned unambiguously to the symmetry of [001]L modes. First information on the [001]L phonon dispersion of an acceptor compound, graphite - $FeCl_3$ has also been reported recently [8].

In the second part of this paper we will compare the phonon dispersion of $C_{36}K$ with several lattice dynamical model calculations. In addition, new experimental data on stage 1 and 2 K and Cs, and stage 2 Rb GIC's will be discussed[2]. A more detailed description of the phonon dispersions together with a description of sample preparation and experimental set-up will be given elsewhere.

2. Results and Lattice Dynamical Models

2.1 Born-von Kármán Models

Three out of four phonon branches have been presented in a recent publication on $C_{36}K$ [6]. The data are reproduced in Fig. 2a. The highest optical branch, which has a small inelastic structure factor is missing. A two-force-constant Born - von Kármán model (BvK) as shown in Fig.1a describes well the measured dispersion. The fit is given in Fig.2a by the full line. The model predicts a flat uppermost branch at 25.5 meV. However, in a later experiment this mode was found at considerably lower energies between 22.5 and 23.5 meV.

A fit of all four phonon branches with the same BvK model is shown in Fig.2b. The weight factor for each data point is chosen individually depending on the error bar. The discrepancy for the high energy branch is reduced. However, we observe significant discrepancies for the lower three branches. Forcing the uppermost branch to cross the data points (by changing the weight factors) results in a completely inadequate description for the other branches.

To improve the fit, several three-parameter BvK models were tested, as outlined in Fig.1b to 1d. In Fig.1b, an interaction between two alkali layers is added. Such a force constant, if it exists, could be responsible for the stability of pure stage compounds. In addition, a stage dependence of this interaction would be of considerable theoretical interest. Fig.1c shows a model with a second neighbor interaction between alkali and carbon planes. Finally in Fig.1d an interaction between the graphite planes adjacent to an alkali layer has been considered. With these models, it was hoped to decrease the energy of the uppermost branch without deteriorating the description of the lower three branches.

However, none of the proposed models led to a satisfactory description of the measured dispersion. The additional force constants represent second neighbor or even longer ranged interactions. Their contributions to the pho-

[2]These experiments have been performed at the NBS research reactor.

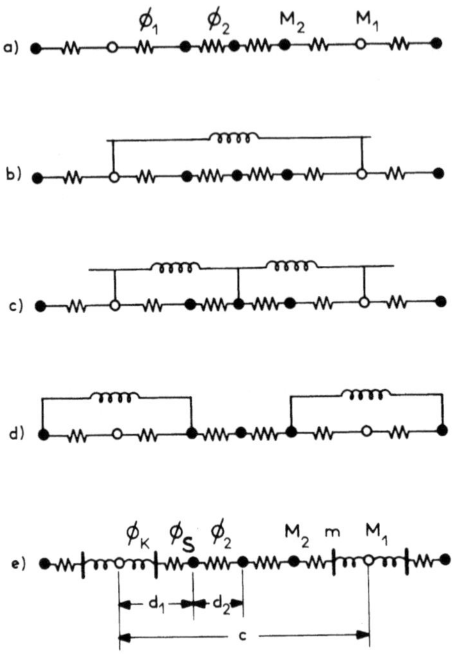

Fig.1 Linear chain models for a stage 3 GIC. For $C_{36}K$: d_1=2.68 Å, d_2=3.35 Å, c=12.06 Å, M_1=3.25 amu, M_2=12.00 amu, m=0. The various models are discussed in the text.

Fig.2 [001]L phonon dispersion for $C_{36}K$ fit by various models. For details refer to the text. Error bars about the size of the dots, otherwise indicated.

▽

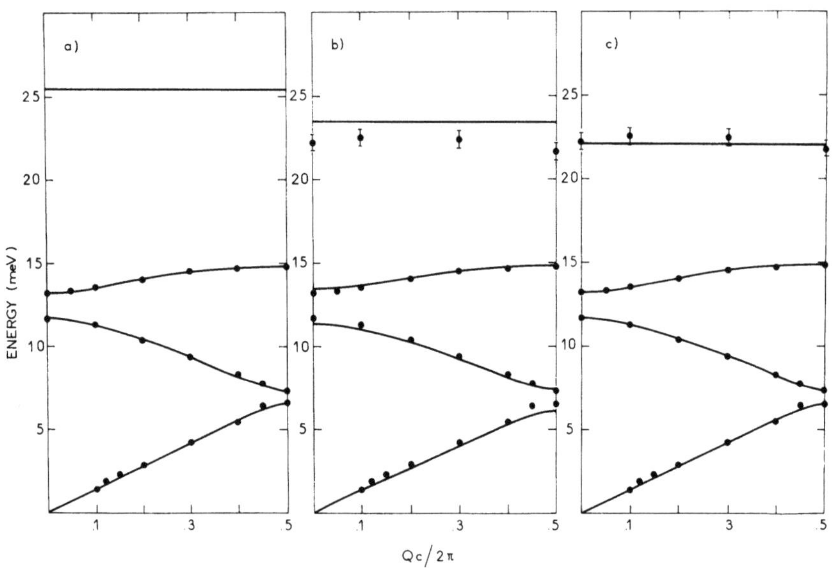

non branches vary more rapidly with wavevector than for a nearest neighbor interaction. If the additional force constants are large enough to be significant for the fit, the uppermost branch will in general show considerable dispersion. This is in disagreement with the experiment. By the same new force constants, additional dispersion is introduced also in the lower three branches. Consequently the good description as shown in Fig.2a is no longer possible.

In addition, we note that in some cases the correlation between the force constants became very high, demonstrating that irrelevant parameters have been added to the model.

A model that is able to fit the dispersion can be constructed in the following way: the uppermost branch is predominantly determined by the vibrations of the relatively light weighted K planes as compared to the C planes. An increase in the mass density by 25% will shift the highest branch down to about 23 meV. A reasonable description of the lower three branches can still be maintained in a two-force-constant BvK model by slightly adjusting the strength of the interactions. However the change of the mass density is outside the generally accepted concentration range for a stage 3 compounds.

We have also tried to fit the dispersions of stages 1 and 2 for K, Rb and Cs with BvK models from Fig.1. With the exception of stage 1 Rb, no model could reproduce the measurements reasonably. For example, for stage 1 K the calculated splitting at the BZ boundary was always too large, whereas for stage 1 Cs it was too small. In summary, we are led to the conclusion that a BvK model with a reasonable number of parameters cannot describe the [001]L dispersion for donor compounds.

2.2 Shell Model

Alternatively, we used a simple shell model to calcualte the dispersion. Fig.3 demonstrates the characteristic features of a monatomic linear chain. Each mass point is replaced by a heavy core and a shell with zero mass (adiabatic approximation). ϕ_K and ϕ_S are the force constants between the carbon core and the shell and between shell and the adjacent carbon, respectively. If ϕ_K is very large as compared to ϕ_S, the core-shell unit acts as a rigid entity and the dispersion is purely sinusoidal. However, for finite ϕ_K the phonon modes mainly near the BZ boundary are decreased, whereas the initial slope of the dispersion remains unchanged. This feature is distinctly different from a dispersion as calculated by a BvK model. Consequently a shell model is of particular interest for us; it seems feasible to keep the lower

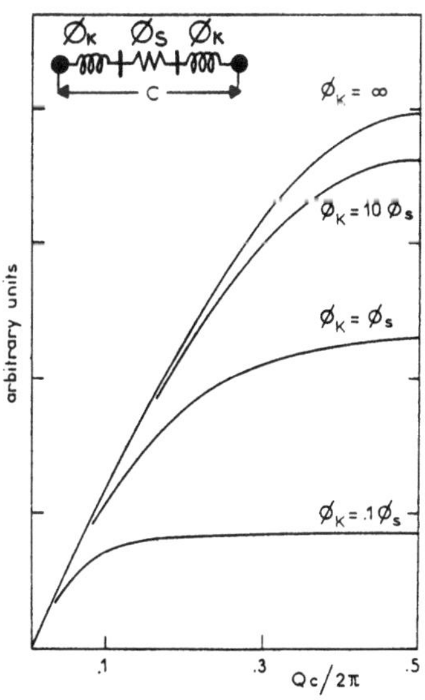

Fig.3 Phonon dispersion of a monatomic linear chain by a shell model. Various ratios of ϕ_K/ϕ_S are shown.

three branches of $C_{36}K$ almost the same (which have a close resemblance to the first part of the dispersion in Fig. 3), while decreasing the energy of the uppermost branch (which has a resemblance to the dispersion near the edge of the BZ in Fig.3).

A shell model for AGIC's seems justified because an increased electron density between the alkali planes and the binding graphite planes builds up as a consequence of a charge transfer between these planes [9]. We also note that shell models have been proven very successful for other systems like alkali halides [10].

Fig.1e represents a simple shell model for $C_{36}K$. In addition to ϕ_K and ϕ_S, there is also an interaction ϕ_2 between graphite planes. It is described by one spring only as is appropriate for pure graphite. This model has three interactions (2 interactions for a stage 1 compound). The same number of parameters was used above for the extended BvK models in Fig.1b to 1d.

Fig. 2c shows a fit of the shell model to the dispersion of $C_{36}K$. We note the good description of the measurements. In particular, we want to point out that all the mode splittings are well reproduced and that the high energy mode resembles a flat branch at 22.4 meV.

New data on stage 2 (K, Rb, Cs) and stage 1 compounds (K, Cs) together with stage 1 Rb from Ref. 5 are shown in Fig. 4 and 5, respectively. It is interesting to note the decrease of the uppermost branches when going from K to Rb to Cs compounds. This is mainly due to the various mass ratios between alkali and graphite planes. Also note the change in the dispersion of the uppermost branch for the various stages, most clearly observed for K compounds. Stage 3 essentially shows a flat branch. This is a characteristic feature of a localized mode.

The full lines in Fig. 4 and 5 represent fits by shell models, analogous to Fig.1e. We find that this model describes well the measured [001]L dispersions.

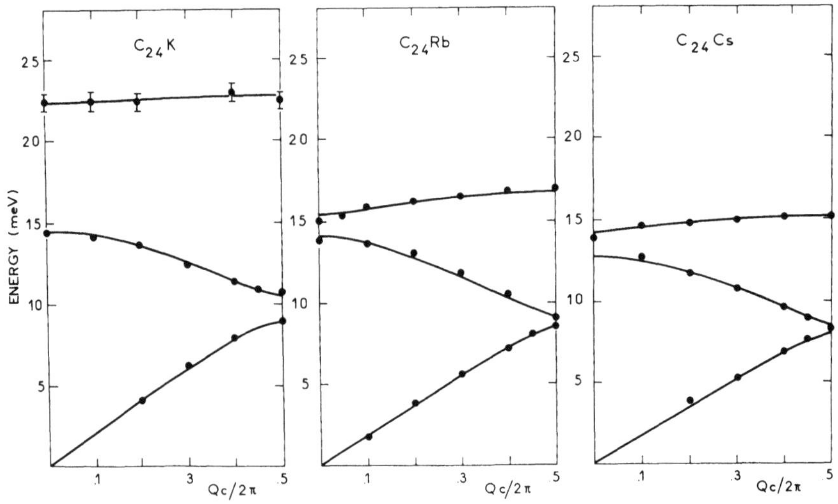

Fig. 4 [001]L phonon dispersion for stage 2 AGIC's. The solid lines represent fits by a shell model.

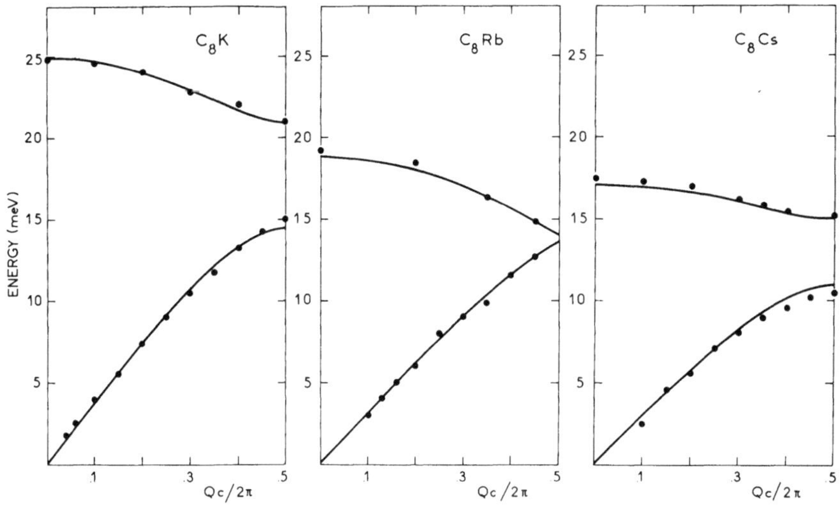

Fig. 5 [001]L phonon dispersion for stage 1 AGIC's. The solid lines represent fits by a shell model.

Table 1 Force constants in dyn/cm for PG and for AGIC's form a shell model. ϕ_K, ϕ_S and ϕ_2 represent the alkali core - electronic shell interaction, the bonding graphite layer - electronic shell interaction and the graphite - graphite interaction, respectively. See also Fig. 1e.
* For an easy comparison the results for the stage 1 compounds have been multiplied by 2/3. So the values as shown represent a C_{12}-alkali compound.

	ϕ_K	ϕ_S	ϕ_2
PG			2800
$C_{36}K$	22460 ± 2350	3510 ± 80	2885 ± 30
$C_{24}K$	21120 ± 300	3690 ± 65	3000 ± 50
$C_{24}Rb$	19870 ± 570	3350 ± 35	2950 ± 80
$C_{24}Cs$	15650 ± 1320	3420 ± 65	3065 ± 65
C_8K*	37770 ± 1150	3232 ± 10	
C_8Rb*	33020 ± 4700	3040 ± 50	
C_8Cs*	20940 ± 1700	3480 ± 50	

Force constants from the fits are summarized in Table 1. The interaction between two graphite planes in the stage 3 comound ϕ_2 = 2880 is slightly changed from the value in PG: ϕ_2 = 2800. However, for all stage 2 compounds we find ϕ_2 significantly increased by about 7%. This clearly indicates an influence of intercalation on the interatomic potentials beyond the nearest neighbor graphite planes. The shell-graphite force constants ϕ_S are considerably larger than ϕ_2. In addition, we find both different values for the stage 2 compounds and characteristic changes from stage 2 to stage 1. As usual for shell models ϕ_K is much larger than both ϕ_S and ϕ_2. Here we want to point out the markedly different values for the stage 2 compounds. Also very interesting is the dramatic change of ϕ_S from stage 2 to stage 1.

A more detailed discussion of these force constants with respect to the physical properties of AGIC's on an atomistic scale will be presented in a forthcoming paper.

3. Summary

The high energy branch of stage 3 K together with complete [001]L dispersions for stage 2 K, Rb and Cs and stage 1 K and Cs are presented for the first time. We have shown conclusively that a BvK formalism is inappropriate to model the [001]L dispersion of AGIC's (with the exception of stage 1 Rb). However, including an interaction with an electronic layer between alkali and bonding graphite planes results in a shell model which can consistently describe the measurements for all compounds.

Acknowledgements

We acknowledge stimulating discussions with J.M. Rowe and J.J. Rush. A.W. Moore donated HOPG material used in this study. This work was in part supported by the U.S. Department of Energy under Contract DE-1C02-76ER01198.

References

1 G. Dolling and B.N. Brockhouse, Phys. Rev. 128, 1120 (1962)

2 R. Nicklow, N. Wakabayashi and H.G. Smith, Phys. Rev. B 5, 4951 (1972)

3 R.J. Nemanich and S.A. Solin, Solid State Comm. 23, 417 (1977)

4 M. Maeda, Y. Kuramoto and C. Horie, J. Phys. Soc. Japan 47, 337 (1979)

5 W.D. Ellenson, D. Seemingsen, D. Guérard, D.G. Onn and J.E. Fischer, Mat. Sci. Eng. 31, 137 (1977)

6 A. Magerl and H. Zabel, Phys. Rev. Lett. 46, 444 (1981)

7 J. Rossat-Mignod and D. Fruchart, Synthetic Met. 2, 143 (1980)

8 See contribution of P.C. Eklund in this volume

9 See various contributions about electronic properties of GIC's in this volume

10 B.G. Dick Jr. and A.W. Overhauser, Phys. Rev. 112, 90 (1958)

The Raman Spectrum of EuC_6: A Clue to Understanding Light Scattering from Stage 1 Alkali Gic's

D.M. Hwang, Department of Physics, University of Illinois at Chicago Circle
Chicago, IL 60680, USA and

S.A. Solin, Department of Physics, Michigan State University
East Lansing, MI 48824, USA and

D. Guérard, Laboratoire de Chemie du Solide Mineral
Universite de Nancy, Nancy, France

ABSTRACT

The room temperature Raman spectrum of EuC_6 has been recorded and is found to exhibit a Fano-broadened band at 1500 ± 5 cm^{-1} and an additional feature at 1840 ± 10 cm^{-1}. The latter feature is attributed to crystallites of EuC_2. The Raman spectrum of EuC_6 will be compared with the spectra of LiC_6 and MC_8, where M = K, Rb and Cs. A model incorporating the static structure factor for phonon scattering and the momentum selection rules in the Raman process will be presented and shown to account for both disorder induced scattering and zone folding effects. We suggest that these effects are merely the extreme cases of a continuous range of phenomena which depend in detail on the degree of order in a given GIC sample.

1. Introduction

The Raman spectra of donor and acceptor GIC's have been heavily studied during the past several years [1,2]. The spectra of stage 3 and higher stage GIC's consist of a pair of lines at ∿ 1600 cm^{-1}, both of which are associated with intralayer modes of the graphite host lattice. The occurrence of this Raman doublet has been accounted for both qualitatively and quantitatively by the nearest layer model [3,4].

All stage 2 GIC's and all stage 1 acceptor GIC's studied to date exhibit a single relatively sharp Raman band which again occurs at ∿1580 cm^{-1}. In contrast, the stage 1 alkali GIC's MC_8 with M = K, Rb, Cs, yield a Raman spectrum that is qualitatively unique [5,6]. It consists of a continuous background extending from 0 to at least 2000 cm^{-1} and perhaps much higher Raman shifts. Superposed on that background is a triplet or doublet of lines at a shift of about 580 cm^{-1} and a broad resonance feature which is generally agreed to arise from a Fano interaction between the continuum background and the intralayer phonon mode of the host graphite material [1-6]. It is the continuum background and the above mentioned 580 cm^{-1} feature which have generated apparently conflicting explanations of the Raman spectra [1-6].

On the basis of recent calculations of the phonon dispersion curves of MC_8 GIC's [7] it is possible to associate the 580 cm^{-1} feature with zone folded M point phonons which are rendered first order Raman active by the enlarged inplane (2 x 2)R0° real space unit cell of the intercalated compounds. Thus, in the zone folding approach it is the fact that the stage 1 alkali GIC's possess a 3-dimensionally ordered commensurate structure which makes them unique and results in their unusual Raman spectra. The zone folding approach also yields a series of discrete Raman active phonon modes in the frequency range between 1275 and 1600 cm^{-1}. MIYAZAKI et al. [8] have shown that these phonons when coupled to an electronic continuum background can give rise to the apparent Fano lineshape observed in the Raman spectra of MC_8 compounds at $\sim$ 1500 cm^{-1}.

The zone folding approach cannot account for the continuum background referenced above. Moreover, CLARKE et al. [9] have shown that CsC_8 retains the stage 1 stacking sequence above 608°K but that the intralayer structure becomes liquid-like disordered. Yet CASWELL and SOLIN [6] have shown that the Raman spectra of this disordered CsC_8 is virtually indistinguishable from the corresponding spectra of the ordered CsC_8. They thus concluded that zone folding could not account for the uniqueness of the MC_8 Raman spectra and attributed its features to disorder induced scattering associated with c-axis stacking faults. The continuum background was shown to be single phonon in origin at least for $\omega \leq$ 600 cm^{-1} and was attributed to disorder induced scattering.

To date proponents of the zone folding explanation for the Raman spectra of MC_8 GIC's and proponents of the disorder induced scattering explanation have maintained that the two views are mutually exclusive. In addition, weaknesses of each view have been emphasized by its opponents. In fact, both views suffer serious weaknesses if taken independently. For instance, disorder induced scattering cannot easily explain the lack of additional Raman bands in higher stage alkali GIC's, while zone folding cannot explain the lack of such bands in the low temperature ordered phases of those same compounds [1-6].

It is the main purpose of this paper to show that in fact the zone folding and disorder induced scattering approaches are not mutually exclusive, but rather that they are extreme views of the same central phenomena. We were led to this conclusion while attempting to analyze the Raman spectra of EuC_6 and LiC_6. Thus, these two GIC's, together with CsC_8, will provide the focus of the following discussion in which an explanation for light scattering from stage 1 and higher GIC's is presented.

2. Experimental Results

Stage 1 europium intercalated graphite, EuC_6, used in this study was made from highly oriented pyrolytic graphite (HOPG) by the method of vapor phase intercalation. Details of sample preparation and characterization have been given elsewhere [10,11]. Room temperature back-scattered Raman spectra recorded from two different sample regions of the EuC_6 sample are shown in Fig. 1(a) and (b). Raman spectra obtained for CsC_8 and LiC_6, prepared from HOPG by vapor phase intercalation and recorded with the same light scattering configuration [11], are also presented in Fig. 1(c) and (d) for comparison.

Our CsC_8 spectrum is identical to that reported previously [6]. The main features are a very broad asymmetric Fano peak at 1500 ± 5 cm^{-1} and a triplet structure at 574, 594 and 620 ± 5 cm^{-1}. Our spectrum for LiC_6 is quite similar to the result of ZANINI et al. [12] and reveals only one relatively sharp peak at 1594 ± 5 cm^{-1}. We do not observe the strong frequency dependent background reported in the work of EKLUND et al. [13].

The spectra of EuC_6 exhibit a very broad asymmetric peak at 1500 ± 5 cm^{-1} which is similar to that [5,6] observed in MC_8 (M = K, Rb and Cs), but no sharp feature around 580 cm^{-1} is detectable. A sample dependent peak was found at 1833 ± 5 cm^{-1} and 1843 ± 5 cm^{-1} in Fig. 1(a) and (b), respectively,

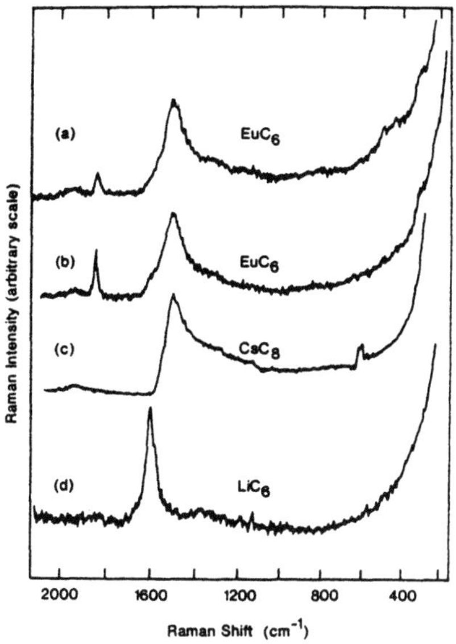

Fig. 1 The Raman spectra for (a) and (b) EuC_6, (c) CsC_8 and (d) LiC_6. The spectra were excited by the 5145 Å argon ion laser radiation at 20 mW, recorded at room temperature with back scattering geometry and with a spectral slit width of 15 cm^{-1}

with different intensity and linewidth. It was assigned [11] to the C=C stretching mode of europium carbide, EuC_2, an impurity formed on the sample surface during intercalation [10,14].

3. Raman Scattering in Partially Disordered Solids

We will now present an explanation of Raman scattering from partially disordered solids. This explanation is based on the momentum selection rules for light scattering and the static structure factor and embraces both the zone folding and the disorder induced scattering discussed above. The static structure factor $S(\vec{k})$ should have the same form for X-ray scattering, elastic neutron scattering, elastic electron scattering, as well as elastic phonon scattering, by the crystal lattice. $S(\vec{k})$ represents the relative probability of an incoming particle with momentum $\vec{k}_1$, exchanging a momentum $\vec{k}$ with the crystal lattice and becoming an outgoing particle with $\vec{k}_2 = \vec{k}_1 \pm \vec{k}$.

For perfect crystals, first order Raman scattering is allowed from zone center phonons only. However, in crystals with impurities or structural defects, the structure factor contains the Brag peaks of the host crystal, superimposed on a continuous background that results from the disordered system. Non-zone-center phonons of the host crystal can be elastically "scattered" into zone center phonons by this continuous structure factor. The resulting disorder induced Stokes Raman spectrum should be proportional to the phonon density of states of the host crystal convolved with the disorder induced structure factor multiplied by the Bose factor $n(\omega) + 1$ and the square of the appropriate Raman matrix element. The latter two factors vary slowly with frequency and will, therefore, not affect the peak positions of Raman bands.

In an ordinary crystal with dilute impurities or structural defects, the distribution of impurities or defects is random and uncorrelated. The resulting structure factor contains a flat background which represents a gas-like phase.

All the non-zone-center phonons of the host crystal have equal probability of being scattered into the zone center and yielding disordered induced Raman scattering which reflects the density of phonon states. If the impurities are so dense that they are spatially correlated, the phonon bands of the host crystal are heavily perturbed and this scheme no longer applies.

In GIC's, the density of the "impurities" is extremely high. However, due to the weak coupling between the graphite layer and the intercalated layer, the higher energy ($\omega \geq 100$ cm^{-1}) phonon dispersion curves of graphite are essentially preserved in GIC's. However, the graphite phonons can be scattered by the intercalated layer and exchange a momentum $\vec{k}$ with the intercalated lattice with a relative probability $S(\vec{k})$, the structure factor of the intercalated layer. Since the intercalated species are strongly spatial correlated, they result in a structure factor with essentially discrete peaks. Thus, non-zone-center graphite phonons with momenta corresponding to the peaks of $S(\vec{k})$ will have higher probability of being scattered into the zone center.

For GIC's with perfect commensurate ordered intercalated structures, $S(\vec{k})$ contains only δ-function like Bragg peaks. The above momentum selection scheme becomes the zone folding mechanism which has been used to explain the electronic [15-17] and phonon [1,4,5,12,13] properties of stage 1 donor GIC's.

The heavy alkali stage 1 GIC's, i.e., KC_8 RbC_8 and CsC_8, have a (2 x 2)R0° inplane structure. The graphite M point phonons will be Bragg scattered, or zone folded, to the zone center and become Raman active. The observed Raman shifts of $\sim$ 580 cm^{-1} agree with the calculated M point phonon energy [18]. Detailed phonon band structure calculations for KC_8 and RbC_8 confirm the above argument [7].

LiC_6 [19] and EuC_6 [10] have a ($\sqrt{3}$ x $\sqrt{3}$)R30° inplane structure. In this case the K points of the graphite zone will be folded back to zone center. To date such folded K point phonons, calculated to be at $\sim$ 430 and $\sim$ 1350 cm^{-1} [18] have not been observed in light scattering studies. This may be due to the small Raman matrix elements and/or small phonon density of states.

The zone folding scheme has been challenged as an interpretation of the 580 cm^{-1} Raman feature in GIC's, since the 580 cm^{-1} feature also appears in the high temperature liquid phase of CsC_8 [6]. Detailed temperature dependent studies [6] show that the 580 cm^{-1} peak only exhibits a small frequency discontinuity of $\sim$ 1.5 cm when CsC_8 undergoes at 608°K an order-disorder transition to a liquid-like, i. e. unregistered, phase [9]. Thus, this feature has been attributed to disordered induced scattering which would reflect the phonon density of states. If the 580 cm^{-1} feature is "disorder" induced, it is difficult to explain its absence in the spectra of LiC_6, EuC_6 and other GIC's with disordered structures. This discrepency can be understood by employing the structure factor momentum selection rule scheme discussed above.

The structure factors for ordered and disordered CsC_8 have been determined by X-ray scattering by CLARKE et al. [9]. The liquid-like structure factor exhibits a peak at q = 1.30 and 1.43 Å^{-1} for the excess and nonexcess samples, respectively, which correspond to 90% and 100% of the ΓM distance of the graphite Brillouin zone. The FWHM of these peaks are 20% of ΓM and its peak height is $\sim$ 5 times larger than the continuous background. Since the phonon bands near the M points are very flat [18] (which causes a sharp peak in the phonon density of states), only a small shift in the Raman frequency is expected when the Cs layer undergoes the order-disorder phase transition. The reported structure factor [9] was determined from samples prepared from HOPG and the angular information in the x-y plane is lost. We suggest that this liquid-like phase is also partially orientationally locked as in the case of low temperature stage 2 compounds [20], so that the phonons near the M points are enhanced in Raman scattering.

Using the structure factor momentum selection scheme, we can explain why this 580 cm^{-1} peak of the density of states of the graphite phonons does not show up in the Raman spectra of other disordered GIC's. Consider CsC_{24} as an example since its structure factor has been measured [20]. Although there are several low temperature phases for CsC_{24}, the first peak of the structure factor is always at $q = 1.16$ $Å^{-1}$ which is 78% of the ΓM distance. In this case M points are located near a deep minimum in the structure factor. Therefore, M point phonons contribute little to the disordered induced scattering.

The high frequency continuum in the Raman spectra of MC_8 and EuC_6 which results in the Fano peak at $\sim$ 1500 cm^{-1} has been attributed to multi-phonon scattering [1] and to electron-phonon coupling [8] . The very weak temperature dependence of this feature [5] rules out the multi-phonon process, while the electron-phonon coupling scheme would imply that the similarity of MC_8 and EuC_6 spectra are accidental, since the electronic structures and electron densities of MC_8 and EuC_6 are quite different. We suggest that this continuum may be due to disordered induced single phonon scattering. The phonon dispersion curves of pristing graphite [18] exhibits many critical points from 1275 to 1650 cm^{-1} contributed to by phonons of various momenta. Therefore, any crystal disorder will result in a broad continuum in this frequency region.

The weakness of this high frequency continuum of LiC_6 is associated with the fact that LiC_6 is a uniquely ordered GIC. It has been pointed out [11] that LiC_6 has a high stacking defect energy since it has the unusual stacking sequence of AαAαAα... . Therefore, the stacking sequence is highly ordered, which enhances the inplane order also. The weakness or absence of the high frequency continuum in the Raman spectra of acceptor GIC's may be due to the very weak coupling of the graphite phonons to the intercalated layer.

4. Concluding Remarks

Stage 1 donor type GIC's have the simplest 3 dimensional crystal structures among all intercalation compounds. The structural and the phonon properties for LiC_6, KC_8, RbC_8 and CsC_8 have been extensively studied. Nevertheless, the interpretation of their unusual Raman features has been a vexed problem for several years. We pointed out that the momentum selection rules in partially disordered solids are governed by the static structure factor. With this momentum selection scheme, we are able to explain why the 580 cm^{-1} graphite M-point phonons are observable in the Raman spectra for MC_8 (M = K, Rb and Cs) of both ordered and disordered phases but not in the spectra for LiC_6, EuC_6 and other disordered GIC's.

The origin of the high frequency continuum, which results in the Fano Peak at $\sim$ 1500 cm^{-1} in the Raman spectra for MC_8 and EuC_6, is still not conclusive. It may be due to the electronic scattering or the disorder induced single phonon scattering. Many new stage 1 donor type GIC's have been synthesized and their 3 dimensional crystal structures established [10,21]. It is important to study their Raman spectra in order to understand the nature of this continuum.

Acknowledgements

Thanks are due to Dr. A. W. Moore for providing the HOPG used in this study. This work is supported by the Atlantic Richfield Foundation Grant of Research Corporation 9422 and the Research Board of the University of Illinois at Chicago Circle (D. M. Hwang) and the NSF Grant DMR80-10486 (S. A. Solin). This work has also benefited from the NSF Grant DMR79-24007.

References

1. M. S. Dresselhaus, G. Dresselhaus, P. C. Eklund and D. D. L. Chung: Mater. Sci. Eng. 31, 141 (1977); M. S. Dresselhaus and G. Dresselhaus: in Physics and Chemistry of Materials with Layered Structures, edited by F. Levy (Dordrecht, Holland, 1978), Vol 5; and references therein.
2. S. A. Solin: Physics 99B, 443 (1980) and references therein.
3. R. J. Nemanich, S. A. Solin and D. Guérard: Phys. Rev. B 16, 2965 (1977).
4. S. Y. Leung, G. Dresselhaus and M. S. Dresselhaus: Syn. Metals 2, 89 (1980).
5. P. C. Eklund, G. Dresselhaus, M. S. Dresselhaus and J. E. Fischer: Phys. Rev. B 16, 3330 (1977); and references therein.
6. N. Caswell and S. A. Solin, Phys. Rev. B 20, 2551 (1979); S. A. Solin and N. Caswell: J. Raman Spectroscopy, in press.
7. C. Horie, M. Maeda, and Y. Kuramoto: Physica 99B, 430 (1980).
8. H. Miyazaki, T. Hatano, G. Kusunoki, T. Watanabe and C. Horie: Physica B, in press.
9. R. Clarke, N. Caswell and S. A. Solin: Phys. Rev. Lett. 42, 61 (1979).
10. M. Makrini, D. Guerard, P. Lagrange, and A. Herold: Physica 99B, 481 (1980).
11. D. M. Hwang and D. Guérard: to be published.
12. M. Zanini, L. Y. Ching, and J. E. Fischer: Phys. Rev. B 18, 2020 (1978).
13. P. C. Eklund, G. Dresselhaus, M. S. Dresselhaus, and J. E. Fischer: Phys. Rev. B 21, 4705 (1980).
14. R. E. Gebelt and H. A. Eick: Inorg. Chem. 3, 335 (1964).
15. T. Inoshita, K. Nakao and H. Kamimura: J. Phys. Soc., Jpn. 43, 1237 (1977); 45, 689 (1968).
16. D. M. Hwang, M. Utlaut, M. S. Isaacson and S. A. Solin: Phys. Rev. Lett. 43, 822 (1979); Physica 99B, 435 (1980).
17. D. M. Hwang, M. Utlaut and S. A. Solin: Syn. Metals 3, 81 (1981).
18. M. Maeda, Y. Kuramoto and C. Horie: J. Phys. Soc. Japan 47, 337 (1979).
19. D. Guérard and A. Hérold: Carbon 13, 337 (1975).
20. R. Clarke, N. Caswell, S. A. Solin, and P. Horn: Phys. Rev. Lett. 43, 2018 (1979).
21. D. Guérard and A. Hérold: C. R. Acad. Sc. Paris Serie C 280, 729 (1975).

Sound Velocities in Stage 1 and Stage 2 $FeCl_3$ Intercalated Graphite

D.M. Hwang and B.F. O'Donnell

Department of Physics, University of Illinois at Chicago Circle
Chicago, IL 60680, USA and

A.Y. Wu

Department of Physics, Purdue University
West Lafayette, Indiana 47907, USA

ABSTRACT

We report the first successful measurement of sound velocities in graphite intercalation compounds and provide the first experimental test of the clasical two force constant linear chain model for the elastic constants in intercalation compounds. Stage 1 and stage 2 $FeCl_3$ intercalated graphite were prepared from HOPG by the conventional two bulb technique and characterized by weight gain, thickness gain and X-ray diffraction. The longitudinal and transverse sound velocities along the c-axis were measured by the pulse echo overlap method with ultrasonic frequencies of 5, 11, and 30 MHz. The compressional elastic constants C_{33} were determined to be (1.58 ± 0.05), (3.1 ± 0.1) and (3.3 ± 0.1) x 10^{11} dyne/cm^2, for the stage 1 compound, the stage 2 compound , and HOPG, respectively. These data can not be fitted by the two force constant linear chain model, indicating that the forces between the intercalated layer and the graphite layer in the stage 1 compound are different from that in the stage 2 compound. This result can be interpreted by considering the stage dependence of the charge transfer and the charge distribution.

1. Introduction

Most recent research activities on graphite intercalation compounds (GIC) are focused on their electrical properties because of practical applications and on their structural phase transitions because of their importance for the physics in low dimensional systems. Nevertheless, at the present time, we know very little about the forces involved in the formation of the in-plane intercalation structure and the c-axis stacking sequence. Measurements of the elastic constants of GIC provide direct information about the interaction forces among the intercalated layers and graphite layers. This information is required in phonon band structure calculations and is important in establishing a better theory of the intercalation mechanism.

Compressional elastic constants C_{33} of GIC have been determined by neutron scattering for RbC_8 [1], LiC_6 [2], stage 2 $FeCl_3$-graphite [3], KC_{24}, KC_{36} and CsC_{24} [4], as well as by pressure dependent X-ray scattering for KC_8, KC_{24} [5], RbC_8 and CsC_8 [6]. These results show that C_{33} increases by 13% to 94% relative to pure graphite upon intercalation with alkali metals into stage 1 compounds.

Elastic constants of solids can also be accurately determined from sound velocities measured by ultrasonic techniques. Here, we report the first successful ultrasonic velocity measurement on GIC. The samples studied are stage

1 and stage 2 $FeCl_3$ intercalated graphite. They were chosen due to their stability in air and their inertness to many ultrasonic bonding agents. Longitudinal and transverse sound velocities were measured by the pulse echo overlap method [7] with ultrasonic waves propagating along the c-axis. The elastic constants were then derived and compared with the results of neutron scattering and X-ray scattering. We also test the validity of the classical two force constant linear chain model for the stage dependence of the elastic properties in GIC.

2. Experiments and Results

Homogeneous samples of a few mm in each dimension and free of internal cracks are required for ultrasonic velocity measurements. To our knowledge, the ultrasonic technique has been tried on GIC unsuccessfully by several groups. Therefore, details on the sample preparation are given here.

Highly oriented pyrolytic graphite (HOPG) of grade ZYB obtained from Union Carbide was cut by a wire saw into 6 x 6 mm squares with c-axis thickness from 1.6 mm to 6.5 mm. The c-faces were prepared for transducer bonding by carefully stripping with adhesive tape. Ultrasonic velocities were measured on these pristine samples to provide a reference standard and to select only those with decent ultrasonic properties for further intercalation.

The samples were intercalated with $FeCl_3$ by the conventional two-temperature technique [8,9]. However, due to the high vapor pressure of $FeCl_3$ at the intercalation temperature (1 atm at 310°C) and the large dilation factor along the c-axis upon intercalation (a factor of 2.8 for the stage 1 compound), extreme care has been taken in adjusting the temperature in order to avoid the formation of surface exfoliation and internal cracks.

HOPG samples were loaded into pyrex glass tubes with an excess quantity of sublimed anhydrous $FeCl_3$ in a nitrogen glove box and then sealed with 0.2 atm chlorine gas at room temperature.

For making the stage 2 sample, the sample temperature was first brought up to 350°C while the $FeCl_3$ end was kept under 100°C. Then the $FeCl_3$ temperature was slowly raised to 295°C over a period of 2 weeks and set there for another week to let the intercalation process be completed. We try to always keep the sample and the $FeCl_3$ vapor under quasi-equilibrium conditions and use the intercalation isotherm of $FeCl_3$ on natural graphite at 350°C of HOOLEY and BARTLETT [8] as a guideline for the temperature adjustment. The temperature increment rate was as slow as 2°C/day around the intercalation threshold temperature of 240°.

For making the stage 1 sample, the $FeCl_3$ temperature was further raised to 320°C. From 298°C to 305°C, the temperature increment rate was 1°C/day. After the $FeCl_3$ temperature reached 320°C, the sample temperature was then lowered to 325°C. It took 8 weeks to synthesize a stage 1 sample from HOPG.

The sample staging was established from measurements of 00ℓ X-ray spectra with a GE XRD-5 X-ray diffractometer using Mo $K\alpha$ radiation. The staging impurities were determined to be less than 5%. The thickness dilation confirmed the purity of staging. From weight gain, the chemical compositions were determined to be $FeCl_3C_{6.87}$ for the stage 1 sample and to be $FeCl_3C_{13.6}$ for the stage 2 sample.

Ultrasonic velocities were determined using the pulse-echo overlap method [7] with commercial electronic instrumentation [10]. X-cut and AC-cut quartz transducers were used to excite longitudinal and transverse waves, respectively. Ultrasonic frequencies of 5, 11, and 30 MHz were employed. Of various bonding agents we tried for coupling ultrasonic waves from the transducer to the sample, DC276-V9 resin was found to be the most suitable.

Figure 1 shows a typical ultrasonic echo train obtained from the as-grown stage 1 sample which is 4.5 mm thick. The echo separation was determined to be 3.520 ± 0.001 μs by the McSkimin tuning and detuning method [11]. The quality of the echo train indicates that the sample is free of internal defects on a macroscopic scale (the wavelength is 0.51 mm for the 5 MHz ultrasonic wave).

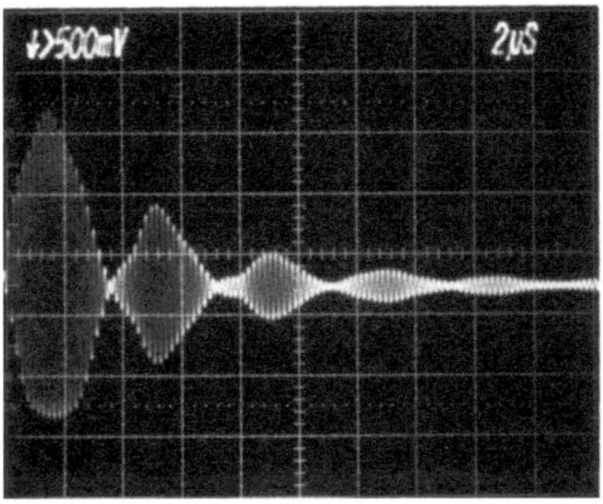

Fig. 1 The typical ultrasonic pulse echo train obtained from the stage 1 $FeCl_3$-graphite sample of 4.5 mm thickness with a 5 MHz longitudinal quartz transducer

After successful ultrasonic measurements, the as-grown stage 1 sample was cleaved into two pieces of thicknesses 2.5 and 2.0 mm, and the sound velocity measurements were repeated.

The stage 2 as-grown sample was 6.5 mm thick. However, a crack on one edge was observed with a microscope and it did not yield an acceptable ultrasonic signal. Therefore, it was cleaved along the crack. Successful measurements were obtained from the resulting samples of thicknesses 2.65 and 3.85 mm.

We found that the transducers were best bonded to the surfaces cleaved before intercalation which remained shiny after intercalation. The surfaces cleaved after intercalation yielded less satisfactory results. Each consecutive ultrasonic measurement may involve washing the sample in HCl acid and acetone, polishing the surface, and rebonding the transducer. The quality of the echo train is hard to reproduce. However, the measured sound velocities are always consistent with one another. This again confirms the homogeneity of our samples. Transverse sound velocities are more difficult to measure. So far we have not obtained a reliable transverse sound velocity from the stage 2 samples.

The elastic constants C_{33} and C_{44} were calculated from the measured longitudinal and transverse sound velocities, respectively, by the relation $C = \rho v^2$ [12]. The density ρ of the sample was determined from weight gain and thickness dilation, as well as from weight and volume measurements. The results are listed in Table 1. The uncertainty of the data mostly arises from the error in measuring the sample thickness with a micrometer.

Table 1 Elastic constants determined from ultrasonic velocity measurements

	ρ (gm/cm^3)	C_{33} (x 10^{11} dyne/cm^2)	C_{44}
Stage 1 $FeCl_3$-graphite ($FeCl_3C_{6.87}$)	2.39	1.58 ± 0.05	0.15 ± 0.02
Stage 2 $FeCl_3$-graphite ($FeCl_3C_{13.6}$)	2.36	3.1 ± 0.1	
HOPG	2.25	3.3 ± 0.1	0.021 ± 0.005
HOPG [12]		3.50 ± 0.15	0.028 ± 0.008
Fast-Neutron Irradiated HOPG [13]		3.50 ± 0.15	≥ 0.40

3. Discussions

3.1 Compressional Elastic Constants

The elastic constants C_{33} for HOPG determined by various methods and groups have been summarized in [14] and show a fair agreement with each other. Our result falls at the lower side of the uncertainty. This may reflect the different quality of HOPG used.

Upon intercalation with $FeCl_3$, C_{33} reduces to 48 ± 4% of that for HOPG for the stage 1 compound and to 94 ± 4% for the stage 2 compound. The stage 2 value is consistent with the recent neutron diffraction study [3]. For stage 1 alkali metal GIC, C_{33} increases by 94% for LiC_6 [2], 28% for KC_8 [5], 36% [1] or 13% [6] for RbC_8, and 75% for CsC_8 [6], relative to pure graphite. These results suggest that graphite becomes harder upon intercalation with donors and softer upon intercalation with acceptors. More detailed analysiled is required to understand this point.

It is interesting to examine the stage dependence of C_{33}. The classical two force constant linear chain model has been used to predict the stage dependence of the elastic constants in phonon band structure calculations [15]. In this model, the force constant between the intercalated layer and the adjacent graphite layer, as well as the force constant between two adjacent graphite layers, is assumed to be stage independent. Then the elastic constant of the stage n compound is expected to be

$$C_n = [(n - 1)d_\circ + d_1]/[(n - 1)d_\circ/C_\circ + d_1/C_1],$$

where $d_\circ$ is the distance between adjacent graphite layers, d_1 is the distance between two carbon layers adjacent to a given intercalated layer, $C_\circ$ and C_1 are the elastic constants of the pristine graphite and the stage 1 compound, respectively. The above equation applies to each component of the elastic tensor.

Our results on $FeCl_3$ compounds do not support the above model. Given the measured C_{33} for HOPG and stage 1 $FeCl_3$-graphite, the calculated C_{33} for stage 2 $FeCl_3$-graphite is (1.83 ± 0.05) x 10^{11} dyne/cm^2, 40% less than our observed value.

The deviation from the classical two force constant model in the stage 1 and stage 2 case is expected. In a stage 1 compound, each carbon layer is sandwiched between two intercalated layers, while in a stage 2 compound, each carbon layer is located between an intercalated layer and a carbon layer. If the same degree of charge transfer per unit intercalate were assumed, the charge density in carbon layers of the stage 1 compound would be twice of that of the stage 2 compound. Therefore, the force constant between the intercalated layer and the carbon layer in the stage 1 compound may be quite different from that in the stage 2 compound.

The elastic constants for the stage 1 and the ambient-pressure stage 2 heavy alkali GIC can not be used to test the two force constant model since their in-plane structures are different. Nevertheless, under high pressure, KC_{24} undergoes a phase transition and becomes a stage 3 compound which has the same in-plane structure as KC_8 [16]. The measured C_{33} for KC_8 and stage 3 KC_{24} are 28% and 71%, respectively, higher than that of HOPG [5]. This result can not be fitted by the two force constant model also.

3.2 Shear Elastic Constants

There is a wide discrepancy in C_{44} for HOPG determined by different groups. The value ranges from 0.013 to 0.45 x 10^{11} dyne/cm^2 [14]. It has been pointed out [14] that the presence of dislocations may appreciably reduce the value of C_{44} measured by ultrasonic techniques, while the measured C_{33} remain virtually unchanged. A rise in C_{44} toward the high value (0.45 x 10^{11} dyne/cm^2) has been observed after the dislocations in HOPG were pinned by fast neutron irradiation [13,17] or by boron doping of 0.2 atomic per cent [18].

C_{44} measured from our HOPG before intercalation is (0.021 ± 0.005) x 10^{11} dyne/cm^2. It becomes (0.15 ± 0.02) x 10^{11} dyne/cm^2 after intercalated into stage 1 $FeCl_3$-graphite. This indicated that intercalating $FeCl_3$ into HOPG results in a similar effect as doping with boron which pins the dislocations and increases the measured C_{44}. Nevertheless, stage 1 $FeCl_3$-graphite is still much softer than HOPG. Its C_{44} is about one third of that expected in a dislocation free HOPG.

4. Summary and Conclusion

Stage 1 and stage 2 $FeCl_3$ intercalated graphite compounds suitable for ultrasonic velocity measurements have been synthesized. The staging and the chemical formulae were determined by X-ray diffraction, thickness gain and weight gain. The homogeneity of the samples was further checked by successive velocity measurements after cleaving the samples into thinner pieces.

The longitudinal and transverse sound velocities were measured by the pulse echo overlap technique with ultrasonic waves propagating along the c-axis. The compressional elastic constants C_{33} for the stage 1 and the stage 2 compounds were determined to be 48 ± 4% and 94 ± 4%, respectively, of that for HOPG. These data can not be fitted by the classical two force constant linear chain model and indicate that the charge transfer plays an essential role in the coupling force constants between adjacent layers in GIC.

This work demonstrates the feasibility and usefulness of ultrasonic technique in studying the interlayer forces of GIC. The ultrasonic technique can detect a relative velocity change as small as 1 part per million. GIC have many interesting phase transitions which can be probed by this method. For example, C_{44} for GIC should strongly depend on the in-plane structure of the intercalated layer. Temperature dependent studies of the sound velocities in stage 2 alkali GIC will provide another probe in understanding the nature of their various in-plane structures.

Acknowledgements

We wish to thank R. J. Sladek and A. Edelstein for the use of their laboratories and A. W. Moore for providing the HOPG used in this study. Thanks are also due to G. Dresselhaus, N. Caswell and S. Sengupta for useful discussions. This research was supported by the Atlantic Richfield Foundation Grant of Research Coporation 9422 and the Research Board of the University of Illinois at Chicago Circle. A. Y. Wu was supported by NSF Grant DMR79-08538A1.

References

1. W.D. Ellenson, D. Semmingsen, D. Guerard, D.G. Onn and J.E. Fischer: Mater. Sci. Eng. 31, 137 (1977).
2. J. Rossat-Mignod, D. Fruchart, M.J. Moran, J.W. Milliken and J.E. Fischer: Syn. Metals 2, 143 (1980).
3. C.F. Majkrzak, J.D. Axe, L. Passell, S.K. Satija, G. Dresselhaus and H. Mazurek: Bull. Amer. Phys. Soc. 26, 452 (1981).
4. H. Zabel and A. Magerl: Bull. Amer. Phys. Soc. 26, 483 (1981).
5. N. Wada, R. Clarke and S.A. Solin: Solid State Comm. 35, 675 (1980).
6. N. Wada: Ph.D. Thesis, University of Chicago, to be published.
7. E.P. Papadakis: J. Acoust. Soc. Amer. 42, 1045 (1967).
8. J.G. Hooley and M. Bartlett: Carbon 5, 417 (1967).
9. N. Caswell and S.A. Solin: Solid State Comm. 27, 961 (1978).
10. D. Ng and R.J. Sladek: Phys. Rev. B11, 4017 (1975).
11. H.J. McSkimin: J. Acoust. Soc. Amer. 33, 12 (1961).
12. O.L. Blakslee, D.G. Prector, E.J. Seldin, G.B. Spence and T. Weng: J. Appl. Phys. 41, 3373 (1970).
13. E.J. Seldin and C.W. Nezbeda: J. Appl. Phys. 41, 3389 (1970).
14. J.E. Green, P. Bolsaities and I.L. Spain: J. Phys. Chem. Solids 34, 1927 (1973).
15. G. Dresselhaus: Bull. Amer. Phys. Soc. 26, 264 (1981).
16. R. Clarke, N. Wada and S.A. Solin: Phys. Rev. Lett. 44, 1616 (1980).
17. W.B. Gauster: Phil. Mag. 25, 687 (1972).
18. D.E. Soule and C.W. Nezbeda: J. Appl. Phys. 39, 5122 (1968).

Raman Measurements of Folded Interlayer Shearing Phonons in KC_{12n} ($n = 2$–6)

N. Wada,[a] M.V. Klein[a,b] and H. Zabel[b]

Coordinated Science Laboratory[a], Department of Physics[b] and
Materials Research Laboratory
University of Illinois at Urbana-Champaign, Urbana, IL 61801, USA

Introduction

Graphite intercalation compounds (GIC's) have been the subject of extensive research, not only because of practical applications [1],but because of fundamental reasons [2]. Their highly anisotropic structures, characterized by the strongly coplanar bonded graphite layers with relatively weak interlayer interactions, provide unique quasi-one and two dimensional systems to investigate. Along the c-axis, the intercalant layers usually form a well-defined periodic sequence, known as staging. A "pure" stage n compound consists of a single phase of one intercalent layer per every n graphite layers [3]. In alkali-metal GIC's (AGIC's), it is relatively easy to synthesize pure stage compounds up to stage 6 by controlling both alkali vapor pressure and sample temperature. In spite of numerous investigations, the origin of the long-range interlayer interactions responsible for these one-dimensional modulated systems is still not well understood. Between the graphite layers in GIC's, intercalant atoms or molecules form disordered or ordered layers, depending on stage and temperature. Recent x-ray [4,5,6], neutron [7] and electron microscope [8] experiments revealed interesting quasi-2D order-disorder phase transitions and dimensional crossover phenomena. High pressure x-ray [9,10] and Raman [10] experiments demonstrated staging phase transitions with formation of a 2 x 2 superlattice in high stage AGIC's ($n \geq 2$), It is of particular importance to study the dynamics of inter- and intralayer interactions in GIC's, since they play a crucial role in the staging mechanism and in-plane order-disorder phase transitions. In addition, information about the phonons should contribute to an understanding of their electronic structures, which result in such novel properties as superconductivity and extremely high conductivity [11].

Previously, there have been a number of Raman, IR [12,13], neutron [14,15, 16] and compressibility [17] experiments on the lattice dynamics of GIC's. Since the strong coplanar graphite structure persists upon intercalation, knowledge of the lattice dynamics of pristine graphite can be used to interpret the lattice dynamics of GIC's, and vice versa. Especially, the concept of Brilloin zone-folding, generated by the new periodicity of intercalated graphite, has been extensively applied to analyzing the phonon dispersion relations as well as electronic band structures [18]. Inelastic neutron diffraction measurements of AGIC's [14,15] clearly demonstrated zone-folding effects in the dispersion of c-axis longitudinal acoustic phonons. Raman measurements on stage 1, AGIC's gave some indication for the in-plane zone-folding effects [12,13,19].

In this paper, experimental observations of the folded shearing phonons will be presented for the first time. A simple nearest-neighbor linear chain model was used to extract the shearing force constants (SFC's) in these compounds. In our model we ignore the layer stacking sequence (ABAB ... for pristine graphite), since the perturbation by the stacking sequence on the transverse acoustic phonon branch should be small. In addition, we will show that the folded phonon energies depend on the specific sample used. This may indicate a variable layer stoichiometry of K layers in these compounds, although it is commonly believed that the stoichiometry of AGIC's is fixed, with the composition MC_{12n} (M = K, Rb, Cs and n = 2, 3 ...).

Experimental

Precautions during sample preparation were taken to ensure good sample surface conditions, since the penetration depth of visible light in these metallic compounds is less than 1000 Å, and since surface imperfections introduce unwanted stray laser light which covers real signals in the low frequency region. Even in pristine graphite, it is extremely difficult to detect the interlayer shearing Raman active E_{2g} mode at ~ 41 cm^{-1} [20].

Samples were prepared by the vapor transport method [21] from highly oriented pyrolytic graphite (HOPG). HOPG consists of microcrystals (~ 1 μ in size) whose c-axes lie in one direction (deviation is less than 1°) with the a-axes randomly oriented in the plane. The HOPG samples were cleaved by scotch tape to have an atomically smooth surface before intercalation. Each sample was characterized by (00ℓ) x-ray reflection measurements before taking Raman spectra.

Raman spectra were taken in the pseudo-Brewster angle configuration at room temperature with a 1 m focal length home-made double-grating monochromator equipped with concave holographic gratings [22], using a 5145 Å Ar^+ laser beam focused into a line image 50 μ wide by 1 mm long. Spectral resolution was 2 cm^{-1}. The power was limited to 300 mW to avoid local deterioration of the samples. Because of poor signal-to-background ratios and low real peak signals, each spectrum was taken with typical dwell times of 3 minutes per points. Sometimes spectra were added to yield better statistics.

Results and Discussions

Raman spectra obtained from KC_{12n} (n = 2 ~ 6) are shown in Fig.1. The spectrum (a) from a stage 2 sample exhibits an asymmetric band which consists of two Raman features at 19 and 23 cm^{-1}. The frequencies of these Raman peaks are surprisingly low. Since the mass per unit area for the K layers is much smaller than that of graphite, this observation should be attributed to small interlayer couplings in the compounds. (The "ideal" stoichiometry of the K layers is one potassium atom per twelve carbon atoms, which yields the mass ratio per unit area of a K layer to a C layer to be ~ 0.27.)

In fact, making use of our linear chain model, we find that $(\phi_1/\phi_G, \phi_2/\phi_G)$= (0.10, 0.58) reproduces the observed peaks well. Here, ϕ_1 and ϕ_2 designate the SFC's of K-C, C-C layers in KC_{24}, and ϕ_G that of C-C layers in pristine graphite (see Fig.2). The calculated phonon peaks are indicated with arrows in Fig.1. It is not surprising that the SFC of K-C layers was found to be quite small, since the K layers in KC_{24} are disordered at room temperature [6]. However, contrary to our expectations, the shearing interaction between the C-C layers was also "soft". Ordinarily, one would anticipate a much stronger interaction, because more electrons are available for the interaction due to charge transfer [2]. The reason for this anomolous C-C interlayer interaction is not well understood. Our speculative explanations are (1) the charged layers may coulombically repel each other and/or (2) the Pauli exclusion re-

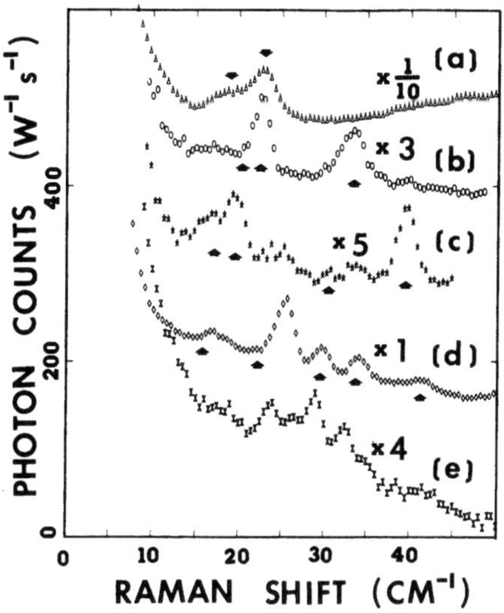

Fig.1 Raman spectra of KC_{12n} (n=2~6) taken in the depolarized configuration at room temperature. The spectra (a) ~ (e) were taken from stage 2~6 samples, respectively. The background leves are not absolute. The signal-to-background ratios were about 1/2 in (a), 1/4 in (b), 1/6 (c) and 1/10 in the other spectra. The arrows indicate the calculated values of the phonon frequencies (see text).

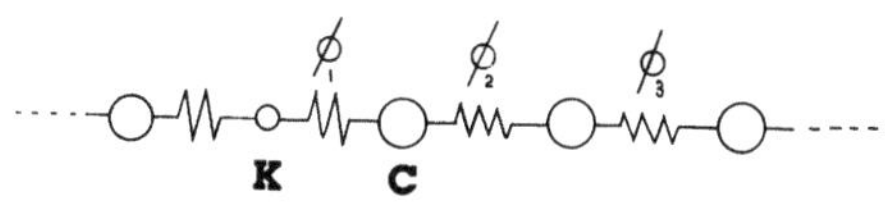

Fig.2 Schematic representation of the linear chain model.

pulsive force may become stronger, since the graphite π orbits are more filled in the compounds relative to graphite. These repulsive forces would result in smaller values for the interlayer overlap integrals of the π electron wavefunctions. We note that stage 2 KC_{24} was also found to be quite "soft" along the c-axis in the x-ray compressibility measurements [17].

The spectrum (b) in Fig.1 was obtained from a stage 3 sample. Two pronounced peaks were found, whereas three peaks are expected from zone-folding. For reference, the calculated peak positions using our model with the SFC's $(\phi_1/\phi_G, \phi_2/\phi_G) = (0.18, 0.84)$ are also indicated in the spectrum. We note that the highest frequency peak observed at 33.5 cm^{-1} is not sensitive to the K-C force constant ϕ_1. The motion of layers for that frequency is such that the interior C layer moves against the rest. Thus, we may conclude that the C-C SFC in stage 3 is also smaller than that of pure graphite. However, the value for ϕ_1 should not be taken too seriously, because of the missing third peak.

In stage 4 KC_{48}, three distinct peaks were observed at 16, 19 and 39 cm^{-1} in addition to some features at 24 and 33 cm^{-1}, as is shown in the spectrum (c) in Fig.1. The force constants $(\phi_1/\phi_G, \phi_2/\phi_G, \phi_3/\phi_G) = (0.14, 1.03, 1.13)$ reproduce well the observed three Raman peaks. The calculated values are indicated in the spectrum. The feature at ~ 24 cm^{-1} might originate from stage 5 regions on the surface of the sample.

Stage 5 KC_{60}, clearly exhibits five Raman peaks, as expected from zone-folding. The peak positions shown in the spectrum (d) of Fig.1 are at 17, 26, 30, 34 and 41 cm^{-1}. We attempted to find the SFC's which reproduce the observed phonon peaks, however we were not completely successful. Remember that we assumed only nearest-neighbor interactions and a constant areal mass for the K layers. For example, the force constants $(\phi_1/\phi_G, \phi_2/\phi_G, \phi_3/\phi_G) = (0.22, 1.07, 1.13)$ can reproduce the three highest frequency peaks well but give a deviation of 1 ~ 3 cm^{-1} in the two lowest peaks, as can be seen in the spectrum (d).

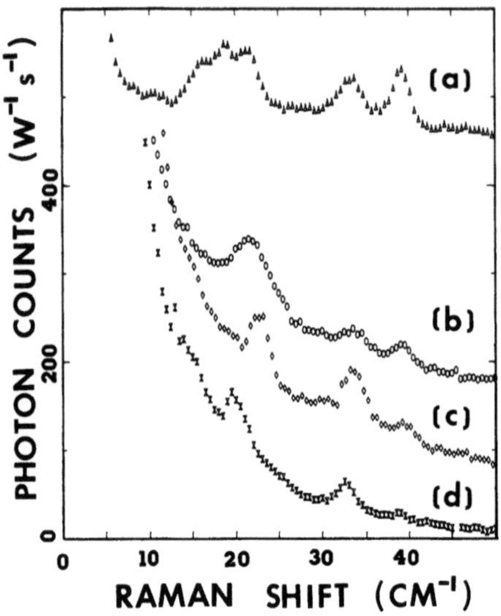

Fig.3 Raman spectra at four different samples which consists of a mixture of stage 3 and 4 regions. The background levels are not absolute

The final Raman spectrum (e) in Fig.1 was obtained from a stage 6 KC_{72} sample. Three peaks at 24, 29 and 33 cm^{-1} can be identified. Because of the poor quality of the spectrum, no attempt was made to extract to SFC's from the data.

Another purpose of this paper is to show evidence for a non-constant layer stoichiometry of K layers. Figure 3 shows Raman spectra of four different samples composed of a mixture of stage 3 and stage 4 regions. The peaks at 33 and 39 cm^{-1} originated from stage 3 and 4 regions, respectively (see Fig.1). By comparison of these spectra, we notice the following. First the spectrum (a) in Fig.3 may be reproduced by superimposing the pure stage 3 and 4 Raman spectra but the spectrum (b) cannot. Second, the Raman peak around 20 cm^{-1} shifts by as much as 3.3 cm^{-1} from sample to sample. Third, the bumps at 15 cm^{-1} in the spectra (c) and (d) imply another Raman peak which was not observed in either of the stage 3 or 4 spectra in Fig.1.

Even in mixed stage crystals, each domain of pure stage is known to be larger than 200 Å in size. Thus, it is unlikely that domain boundaries or defects may affect these folded phonons. We suggest that this sample dependence may originate from a variation in the stoichiometry of the disordered K layers. It may depend on the sample preparation procedures and probably varies less than a few percent. Recent x-ray experiments [10,23,24] also give some evidence for a non-constant stoichiometry of the intercalant layers.

In our Raman measurements of the folded phonons, no noticeable polarization dependence was found. This implies that these phonons are E_{2g}-like but not A_g-like. It should be mentioned, however, that the group theoretical analyses cannot be rigorously applied to this system because of the poorly defined crystalline structure in AGIC's ($n > 2$); (1) the intercalant layers are disordered and (2) there exist many faults in the c-axis stacking sequence in the compounds (the coherency of the stacking sequence extends over only ~ 40 Å in KC_{24} [4] and CsC_{24} [5]).

Summary

Systematic stage dependent Raman measurements have been conducted on potassium intercalated graphite. Folded phonons were observed and analyzed with a linear chain model. Unexpected softening of the SFC's of C-C layers were found in the stage 3 and 3 compounds. We suggest that Coulombic and/or Pauli principle repulsions between the charged layers may be responsible for the weak interlayer couplings in these intermetallic compounds. In the higher stage compounds (n = 4 and 5), the SFC's of C-C layers seem to be close to that of pure graphite. It should be pointed out that, if the interlayer repulsive interactions became stronger in the compounds relative to graphite, one would expect wider C-C layer separation. Unfortunately, no reliable measurements on the C-C interlayer distance in GIC's have been performed to date.

The SFC's of K-C layers in these compounds were found to be one order of magnitude smaller than in pristine graphite. The strength of the K-C interaction is consistent with the rough estimate from neutron scattering [16].

In general, since the low symmetry points in the phonon dispersion branches are folded back to the zone center in the presence of new periodicity, the Raman intensity for the folded-phonons are expected to be quite small. In fact, the only successful Raman observation of folded phonons has been made in the resonance experiments on the artificially-made semiconductor superlattice [25]. In GIC's, no confirming Raman measurements of folded phonons have been reported. It should be remarked that the large values of the (n + 1) factor, which the Stokes Raman intensity is proportional to, contributed to our successful observation of these very low frequency phonons. Finally, we report that, when low temperature Raman measurements were attempted in these compounds, the intensity of the Raman peaks diminished so rapidly with decreasing temperature that no detailed analyses were possible.

Acknowledgements

We thank W. A. Kamitakahara, D. M. Hwang and P. C. Eklund for valuable discussions. We also thank N. Caswell for useful information about preparing high stage samples, and A. W. Moore for providing the HOPG used in these experiments. This work was supported in part by NSF DMR-77-23999, DOE DE-AC02-76ER01198 and ONR N0014-79-C0424.

References

1. M. S. Whittingham, L. B. Ebert in Intercalated Layered Materials, edited by F. A. Lévy (Reidel, Dordrecht, 1979) p. 533.
2. J. E. Fischer, T. E. Thompson, Physics Today, 31, 36 (1978).
3. W. Metz, D. Holwein, Carbon 13, 87 (1975).
4. J. B. Hastings, W. D. Ellenson, J. E. Fischer, Phys. Rev. Lett. 42, 1552 (1979).
5. Roy Clarke, N. Caswell, S. A. Solin, P. M. Horn, Phys. Rev. Lett. 43, 2018 (1979).
6. H. Zabel, S. C. Moss, N. Caswell, S. A. Solin, Phys. Rev. Lett. 43, 2022 (1979).
7. S. Suzuki, H. Ikeda, H. Suematsu, Y. Endoh, H. Shiba, Proc. of the Yamada Conf. IV on Physics and Chemistry of Layered Materials (Sendai, 1980) Physica B in press.
8. N. Kambe, G. Dresselhaus, M. S. Dresselhaus, Phys. Rev. B8, 3491 (1980).
9. Roy Clarke, N. Wada, S. A. Solin, Phys. Rev. Lett. 44, 1616 (1980).
10. N. Wada, Phys. Rev. B in press.
11. J. E. Fischer, Physica 99B, 383 (1980).

12. S. A. Solin, Physica 99B, 443 (1980).
13. M. S. Dresselhaus, G. Dresselhaus in Intercalated Layered Materials, edited by F. A. Lévy (Reidel, Dordrecht, 1979) p. 423.
14. W. D. Ellenson, D. Semmingsen, D. Gruérard, D. G. Onn, J. E. Fischer, Mater. Sci. Eng. 31, 137 (1977).
15. A. Magerl, H. Zabel, Phys. Rev. Lett. 46, 444 (1981).
16. W. A. Kamitakahara, N. Wada, S. A. Solin, to be published.
17. N. Wada, Roy Clarke, S. A . Solin, Solid State Commun. 35, 675 (1980).
18. J. E. Fischer, in Intercalated Layered Materials, edited by F. A. Lévy (Reidel, Dordrecht, 1979) p. 481.
19. M. S. Dresselhaus, G. Dresselhaus, P. C. Eklund, D. D. L. Chung, Mat. Sci. and Eng. 31, 141 (1977).
20. R. J. Nemanich, G. Lucovsky, S. A. Solin, in Proc. of the Int. Conf. on Lattice Dynamics, edited by M. Balkanski (Flammarion, Paris, 1978) p. 619.
21. H. Hérold, Bull. Soc. Chim. Fr. 187, 999 (1955).
22. R. A. Field, Ph.D. Thesis, Univ. of Illinois at Urbana-Champaign (1977), unpublished.
23. Y. M. Jan, M. Mori, S. C. Moss, Bull. of Amer. Phys. Soc. 26. 452 (1981).
24. S. Y. Leung, C. Underhill, G. Dresselhaus, T. Krapchev, R. Ogilvie, M. S. Dresselhaus, Solid State Commun. 32, 635 (1979).
25. C. Colvard, R. Merlin, M. V. Klein, A. D. Gossard, Phys. Rev. Lett. 45, 298 (1980).

Part IV

Ordering and Phase Transitions

Structural Studies of Intercalants

J.B. Hastings

National Synchrotron Light Source, Brookhaven National Laboratory,
Upton, NY 11973, USA

Abstract

The structure of stage 2 potassium intercalated graphite, KC_{24}, is discussed in both the ordered and disordered phases. A one-dimensional model is used to illustrate the qualitative features of the KC_{24} diffraction patterns.

1. Introduction

The existence of graphite intercalation compounds has been known for almost fifty years. These systems have been of recent interst because of the high basal-plane conductivity and the highly two-dimensional nature of the electronic structure in some of these compounds [1]. Early x-ray structural studies on the stage 2 alkali metal compounds [2] MC_{24} (M = K,Rb, Cs) have shown the existence of an order-disorder transition. Following these studies a-axis resistivity measurements [3] showed not one, but two anomalies. The structural nature of these two anomalies has since been revealed by x-ray studies [4] for stage 2 potassium. Figure 1 shows a schematic phase diagram for KC_{24}, where above approximately 125K the potassium lattice is "disordered", between 125K and about 95K there is an ordered compound and at 95K there is a structural transition which involves the stacking of the potassium layers.

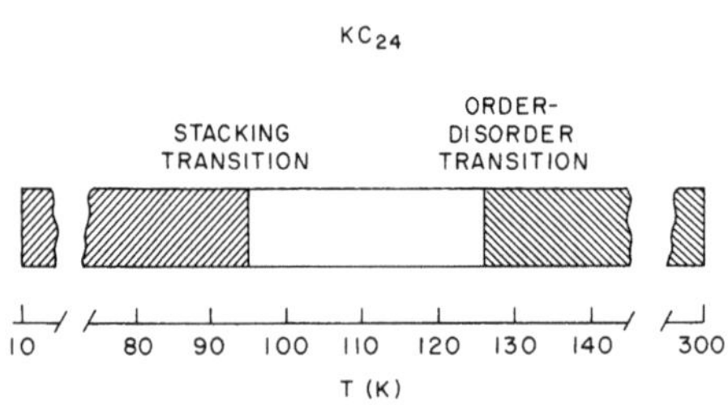

Fig.1 Phase diagram of KC_{24} as a function of temperature

The temperature evolution of the ordered phase at T = 125K was studied with x-ray scattering [4,5] and the order-disorder transition appeared continuous. Of particular interest was the observation of c-axis stacking faults in this intermediate phase [4] with ABC stacking (FCC like). BAK and DOMANY [7] studied the order-disorder transition in the presence of these stacking faults and found it to be a smeared first order transition which is consistent with the data [4,5].

The lower transition was shown to be structural in nature where the stacking went from ABCABC... to ABAB... [4,6] (an FCC to HCP like stacking). Thus, below this transition the c-axis stacking of the potassium no longer follows the graphite matrix and there is no evidence for stacking faults in the low temperature phase. Recent c-axis resitivity measurements [8], which should be sensitive only to the loss of a scattering mechanism at the lower transition, indeed show only one anomaly at the lower temperature.

The only remaining questions then are the in plane structure of the intermediate and low temperature phases. In the following sections these questions will be addressed albeit in a qualitative fashion and without any definitive conclusions. Clearly detailed crystallographic studies are needed to give the structural parameters of these interesting compounds.

2. X-ray scattering results and discussion

2.1 Experimental details

All the experimental data is on stage 2 potassium graphite. The sample is highly oriented pyrolytic graphite (HOPG) and intercalated by the standard two bulb method. The sample dimensions are typically 10 x 5 x 0.1 mm^3 and the c-axis is always in the scattering plane. After inercalation the sample was transferred to a beryllium can from its glass container in an inert atmosphere. The beryllium can reduces the background significantly. The can is attached to the cold finger of a closed cycle refrigerator that allows the temperature to be varied continuously from room temperature to about 10K with a stability of typ 0.05K.

The triple axis x-ray spectrometer uses a 1kW sealed Cu x-ray tube as a source. The beam is monochromated with a pyrolytic graphite monochromator, (00 2) reflection, set to reflect the CuK_α radiation giving an incident wavevector of 4.075$Å^{-1}$. An 00 2 graphite analyzing crystal is used on the third axis and in combination with the (002) monochromator gives a wavevector resolution, ΔQ, of typically .02$Å^{-1}$ at Q's of the order of 2$Å^{-1}$.

2.2 Disordered phase results

Scattering measurements have been made, referring to Fig. 1, in both the high temperature, disorderd phase, above 125K discussed below and the intermediate phase 95K < T < 125K discussed in the following section. As stated earlier the sample was intercalated from an HOPG sample and oriented with the c-axis in the scattering plane. Therefore scans perpendicular to the c-axis, $Q = (h,0)$, are powder scans with $h = |h\bar{i} + k\bar{j}|$ and thus the orientational dependence of the in plane scattered intensity of the potassium with respect to the graphite is lost.

A typical data set at $T = 130K$ is shown in Fig. 2. Plotted is the intensity as a function of h for $Q = (h,0)$. As noted previously [5] there is no variation of intensity with ℓ for fixed $h = h_0$, that is for $Q = (h_0,\ell)$. What is seen immediately is a fairly sharp feature at $Q \approx (1.2,0)$ and a broad feature at $Q \approx (2.40,0)$ typical of "liquid" like scattering. The sharp feature around $Q = (2.93,0)$ is the tail of the graphite (101) peak [4]. Between the features at $Q \approx (1.2,0)$ and (2.4,0) is another sharp feature at $Q \approx (1.7,0)$. These results are in contrast to the study of CsC_{24} [9] which shows a "liquid" like scattering and is well understood in terms of a close-packed 2 dimensional Cs liquid.

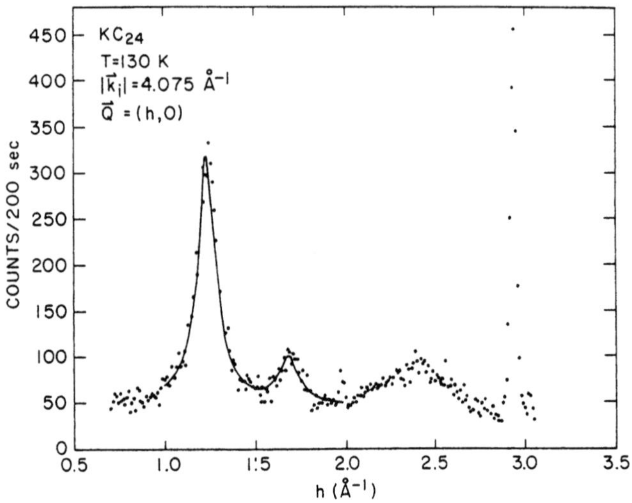

Fig.2 Scan along Q = (h,0) at 130 K in the disordered phase. The solid line is a fit of two Lorentzians.

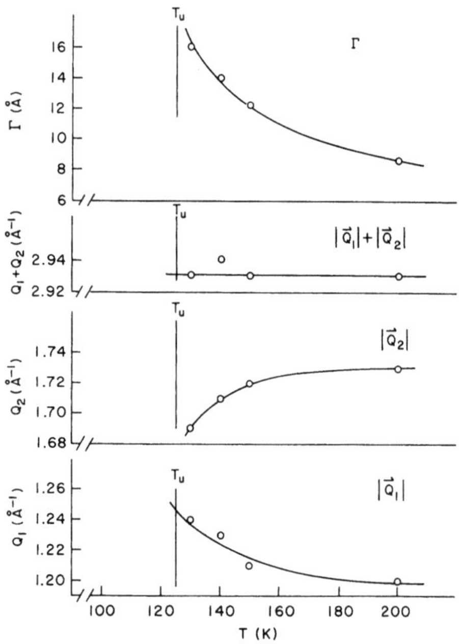

Fig.3 Parameters of the fits to scans along Q = (h,0) in the disordered phase. Γ is the width of the two Lorentzians, $\overline{Q}_1$ and $\overline{Q}_2$ the peak positions and $\overline{Q}_1+\overline{Q}_2$ the sum of the positions. The solid lines are guides to the eye.

In order to understand the differences between CsC_{24} and KC_{24} the two sharp features at $Q \approx (1.2,0)$ and $Q \approx (1.7,0)$ have been fit with two Lorentzians.

$$I(Q) = \frac{I_{01}}{1+\Gamma^2(Q-Q_1)^2} + \frac{I_{02}}{1+\Gamma^2(Q-Q_2)^2} + I_{Bkgd} \,. \tag{1}$$

The peak postions, intensities, and a single width were free parameters. The solid line in Fig. 2 is a fit and a summary of parameters as a function of temperature is given in Fig. 3. As the order-disorder transition is approached from above the in plane correlations grow as can be seen in the increase of Γ. The potassium-potassium distance decreases with decreasing temperature as well, noted by the variation of $|\bar{Q}_1|$. However, the decrease of $|\bar{Q}_1|$ is always matched by an increase of $|\bar{Q}_2|$ such that the sum $|\bar{Q}_1|+|\bar{Q}_2|$ is independent of temperature and equal to to the graphite (100) wavevector.

Thus unlike the Cs system where the disordered alkali metal is constrained to lie between the graphite layers but is otherwise independent of the matrix the K system is already "affected" by the in plane graphite structure far above its ordering temperature.

2.3 Intemediate phase results

As the temperature is lowered through the order-disorder transition into the intermediate phase the peaks sharpen and develop three dimensional character in the direction perpendicular to the layers [4,5]. An earlier study [4] has discussed the presence of stacking faults in both the graphite matrix and the intercalant lattice in this phase. A typical scan along the $\ell = 0$ layer, Q = (h,0) at 105K is shown in Fig. 4. In contrast to the broad features above 125K the peaks are resolution limited in the plane. The solid lines in the figure are merely guides to the eye. The relatively sharp feature at h ≈ 1.7 remains and in fact has grown relatively in intensity. The positions of the features of this scattering pattern seem to be well understood and will be addressed in the following discussion.

2.4 Discussion

The understanding of diffraction data involves both position and intensity. First is to understand the positions of the maxima. The peak positions shown in Fig. 4 as well as other experiments on KC_{24} [4-6] can be understood geometrically as coming from two lattices, the graphite matrix and the inter-

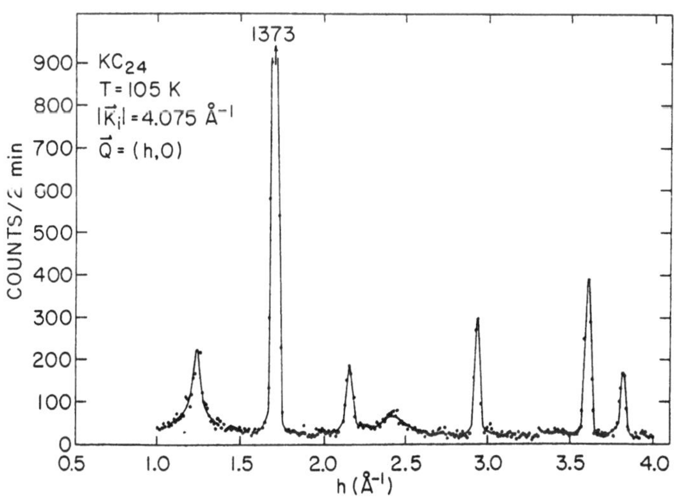

Fig.4 Scan along Q = (h,0) at 105K in the intermediate phase. The solid lines are guides to the eye.

calant as well as sums and differences of the wavevectors of these two lattices. This is eveidence for a modulated structure and following OVERHAUSER [10] directly both the positions and relative intensities can be calculated. The interaction betwe n the intercalant and the graphite will cause displacements of the atomic positions

$$\bar{U}(\bar{L}) = \bar{A} \sin (\bar{q}.\bar{L} + \phi)$$

where $\bar{U}$ is the displacement, $\bar{L}$ is the vector to the ℓth atom, $\bar{q}$ is a wave vector describing the modulation, $\bar{A}$ is the modulation amplitude and ϕ a phase factor. Assuming that one of the two lattices is undistorted then the density of the distorted lattice can be written as

$$\rho'(\bar{r}) = \sum_{\bar{L}} \rho(\bar{L})\, \delta\, (\bar{r} - \bar{L} - \bar{A} \sin (\bar{q} . \bar{L}))$$

neglecting the phase. The elastic scattering amplitude is just the Fourier transform of the scattering density

$$F(\bar{Q}) = \int \rho'(\bar{r})\, e^{i\bar{Q} . \bar{r}}\, d\bar{r}$$

which is, substituting from above

$$F(\bar{Q}) = \sum_{\bar{L}} f_{\bar{L}} (\bar{Q})\, e^{iQ . (\bar{L} + \bar{A} \sin \bar{q} . \bar{L})}$$

where $f_{\bar{L}}(\bar{Q})$ is the atomic scattering factor for the ℓth atom. Assuming that $f_{\bar{L}}(\bar{Q}) = f(\bar{Q})$ independent of $\bar{L}$ and using the Jacobi-Anger generating function for Bessel functions gives

$$F(\bar{Q}) = f(\bar{Q}) \sum_{\bar{L},n} J_n (\bar{Q}\cdot\bar{A})\, e^{i(\bar{Q} + n\bar{q}) \cdot \bar{L}} .$$

$F(\bar{Q})$ is only non-zero when $\bar{Q} \pm n\bar{q} = \bar{G}$ so the diffraction maxima of the modulated lattice occur when $\bar{Q} = \bar{G} \pm n\bar{q}$ where $\bar{G}$ is a reciprocal lattice vector of the "modulated" lattice, and $\bar{q}$ is the wavevector of the modulation. For $n = 0$ the peaks are the Bragg peaks of the modulated lattice, for $\bar{G} = 0$ there are peaks of the undistorted lattice and $n \neq 0$, $\bar{G} \neq 0$ are the sums and differences of the two lattices.

The relative intensities of the main Bragg peaks of the modulated lattice ($n = 0$) and the first order satellites ($n = \pm 1$) can be calculated. Neglecting thermal motion and phase fluctuations

$$I \propto F^2(\bar{Q}).$$

For $n = 0$, $\bar{Q} = \bar{G}$ and $F \propto J_0(\bar{Q}.\bar{A})$ and for $n = \pm 1$, $\bar{Q} = \bar{G} \pm \bar{q}$ and $F \propto J_1(\bar{Q}.\bar{A})$. Generally, the amplitude of the modulation is small so that the Bessel functions are in the limit of small arguments, X, where

$$J_n (x) \approx X^{2n} .$$

However, the magnitude of the modulation peak, the feature at $Q = (1.7,0)$ in

Fig. 4, indicates that the modulation is not small. Referring to Fig. 5 where $J_0^2(X)$ and $J_1^2(X)$ are plotted versus their arguments, one can see that in fact around x ≈ 2.2 there is a zero in $J_0^2(X)$ so that some primary Bragg peaks could have very small intensities.

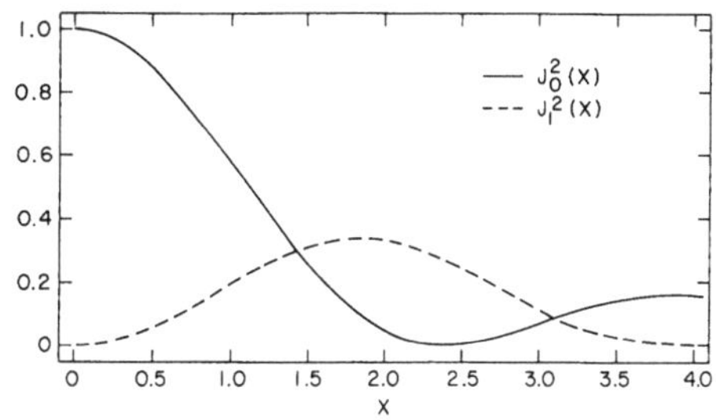

Fig.5 The square of the zeroth and first order Bessel functions versus their arguments

Using this formulation for the intensities, one can look qualitatively at the intermediate phase intensities in some simple limits, assuming the modulation is purely longitudinal, either rigid graphite with modulated potassium or the reverse. Roughly, then, for rigid graphite the ratio of the potassium (110) to the first order satellite, I_{SAT}, at Q ≈ (1.71,0) gives $|\overline{A}_L|$ ≈ 1.35 ± 0.2 Å for $\overline{U}(\overline{L})$ ≈ $\overline{A}_L \sin(\overline{q} \cdot \overline{L} + \phi)$ with $\overline{q}$ equal to the graphite (100) wavevector. Using this value of $|\overline{A}_L|$ predicts a ratio of I(100) to I_{SAT}. of 2:3 and measured is roughly 1:2 but the existence of the faulted structure makes it difficult to estimate the integrated intensity of the (100) potassium peak.

With the assumption of rigid potassium and modulated graphite and q the potassium (100) wavevector the ratio of the graphite (110) intensity to the satellite gives $|\overline{A}_L|$ ≈ 0.18 ± .05. Using this value however and comparing the graphite (104) to the satellite predicts a ratio of 9.6 compared with the measured value of 3.6. Furthermore, with a rigid potassium lattice the fall off in intensity of the potassium peaks I(100)/I(110) ~ 4.5 cannot be explained. However, with the modulation of the potassium lattice and referring to Fig. 5, with $|\overline{A}_L|$ ≈ 1.35, the (110) potassium peak falls near a zero in $J_0^2(X)$ while the (100) peak does not.

Finally, with this argument in hand it is interesting to ask, could the potassium ions sit "over" the centers of graphite hexagons, be registered, and still give a diffraction pattern that appears incommensurate with the graphite matrix. There are several observations that support this argument. First, in the early work of PARRY and co-workers [1], in order to explain the intensity ratios of the graphite peaks in the disordered phase a contribution of 1/6 of a potassium atom over each carbon hexagon (2 carbon atoms per hexagon and 12 potassium atoms per layer) was required. Second, the extended x-ray absorption fine structure measurements of CASWELL et. al. [11] are consistent with the potassium ions centered over the carbon hexagons. Lastly, the value of $|\overline{A}_L|$ for the modulated potassium calculated roughly above is just the difference between the average near neighbor distance in stage 2 and stage 1. The stage 1 compound KC_8 is a commensurate 2 x 2 structure and thus the stage 2 compound may be made up of "islands" of stage 1, perhaps as small as two or three atoms.

These ideas can be addressed for a simple one-dimensional model to see the qualitative features of the diffraction pattern from registered, disordered lattices. The model [12] consists of an infinite chain with atoms placed at

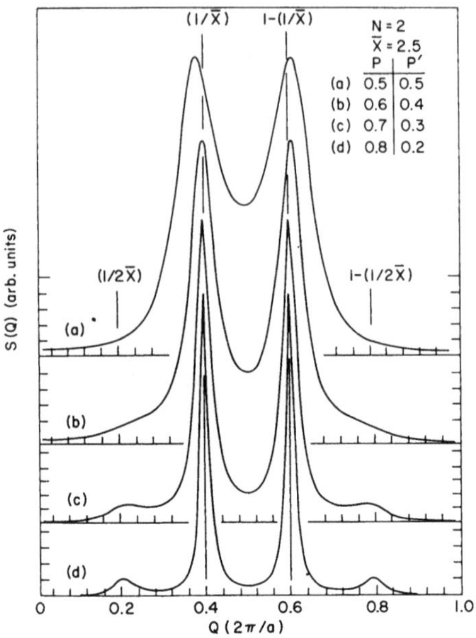

Fig.6 S(Q) versus Q for one-dimensional chains going from random occupation, panel (a) to almost ordered, panel (d). The average separeation between atoms on the chain is 2.5a

two spacings na and (n + 1)a, where 'a' is the near neighbor spacing along the chain. The atoms are then 'placed' on the chain by the flip of one of two coins. After choosing the first spacing, do by the flip of a coin, if $d_0 = (n + 1)a$ then one uses a 'coin' with the probability of d_1 being (n+1)a equal to p' and na being q' = 1 -p'. For d_0 = na a second 'coin' is used with probability, p, for $d_1 = (n + 1)a$ and q = 1 - p for d_1 = na. With these rules then, by varying p and p' one can go from a completely random lattice, p = p' = .5 to complete order, spacing n, n + 1, n, n + 1..., with p' = 0, p = 1. GRIMM, AXE, and KROHNKE [12] derived an analytic form for the structure factor, S(Q), and for n = 2, and several values of p and p' S(Q) is plotted as a function of Q in Fig. 6. X is just the average spacing, (n + (n + 1))/2. If the system were completely ordered there would be sharp, Bragg peaks at Q = 0.2, 0.4, 0.6, 0.8 and so on, however even in the random system, there remain two sharp features one at Q roughly (1/X) and one at (1 - 1/X). These peaks could be interpreted easily as an "incommensurate" peak and a modulation satellite and are reminiscent of the features seen in the disordered phase of KC_{24}, Fig. 2. As the system orders p→1, p'→o these features become sharp but still slightly in-commensurate, as can be seen in the progression from (a) to (d) in Fig. 6. This, too, is similiar to the behavior of KC_{24} in the intermediate phase.

The features of a two dimensional system using the observed 'average' lattice for the potassium and then forcing the potassium atoms to sit over the centers of the nearest graphite hexagons has been calculated [13] by computer and yields the ovserved peak positions for KC_{24} in the intermediate phase. This can be considered as an extreme of a modulation.

3. Summary

The phase diagram of KC_{24} is now reasonably understood but only a qualitative picture is available for the in-plane structure. This system may be an example of a registered disordered system or it could be simply incommensurate. Some of these issues could be addressed by calculation and others by detailed line shape measurement in the intermediate phase or a complete structure analysis of a single crystal. In any case there are still things to be done.

Acknowledgments

The author wishes to thank J.D. Axe, C.R. Fincher, Jr. and J.E. Fischer for discussions and ongoing collaborations. This work was supported by the U.S. Department of Energy Contract No. DE-AC02-76CH00016.

References

1 G.M.T. Foley, C. Zeller, E.R. Falardeau and F. L. Vogel, Solid State Commun. 24, 371 (1977); F. Batallan, J. Bok, I. Rosenman and J. Melin, Phys. Rev. Lett. 41, 330 (1978).

2 G.S. Parry, D.E. Nixon, D.M. Lester and B.C. Levene, J. Phys. C2, 2156 (1969); G.S. Parry and D.E. Nixon, Nature (London) 216, 909 (1967).

3 David G. Onn, G.M.T. Foley and J.E. Fischer, Mater. Sci. Eng. 31, 271 (1977).

4 J.B. Hastings, W.D. Ellenson and J.E. Fischer, Phys. Rev. Lett. 42, 1552 (1979).

5 H. Zabel, S.C. Moss, N. Caswell and S.A. Solin, Phys. Rev. Lett. 43, 2022 (1979).

6 H. Zabel, Ordering in Two Dimensions, edited by S.K. Sinha (North-Holland, Amsterdam, 1981).

7 Per Bak and Eytan Domany, Phys. Rev. B23, 1320 (1981).

8 Phan and Fuerst, unpublished.

9 R. Clarke, N. Caswell, S.A. Solin and P.M. Horn, Phys. Rev. Lett. 43, 2018 (1979).

10 A. Overhauser, Phys. Rev. B3, 3173 (1971).

11 N. Caswell, S.A. Solin, T.M. Hayer and S.J. Hunter, Physica 99B, 463 (1980).

12 H. Grimm, J.D. Axe and G. Krohnke, BNL 29083 (1981).

13 S.B. DiCenzo, unpublished.

X-Ray Diffraction from Graphite Intercalation Compounds Under Hydrostatic Pressure

S.A. Solin

Department of Physics, Michigan State University
East Lansing, MI 48824, USA

Abstract

X-ray diffraction studies of pressure-induced intralayer and interlayer structural phase transitions in alkali GIC's are discussed with particular emphasis on KC_{24} and CsC_{36}. It is shown that such transitions involve the well-defined layer stoichiometries MC_8 and MC_{12} which occur in ambient pressure compounds. Stage 1 MC_8 compounds exhibit no structural transitions at pressures up to 13 kbar but show a Hook's law behavior from which their compressibilities have been determined.

1. Introduction

Much of the knowledge gained in condensed matter physics during the past several decades resulted from experimental studies in which the response of the physical properties of a solid to various external fields, e.g. electric, magnetic, was examined. In particular, the application of a stress field resulting from exposure to hydrostatic pressure is one of the most popular perturbative techniques in condensed matter physics.[1,2] This popularity results in part because hydrostatic pressure can produce dramatic changes in interatomic distances as well as atomic configurations (e.g. structural phase transitions).[3] In turn, relatively large changes in interatomic distance and configuration invariably produce dramatic, readily observable effects on the physical properties of solid systems.

For the above stated reasons, CLARKE, WADA and SOLIN[4](CWS) decided to study the effect of hydrostatic pressure on the structural properties of graphite intercalated with potassium, which is one of the prototypical graphite intercalation compounds (GIC's).[5] The technique of choice for initial studies of the perturbative effects of pressure is, of course, x-ray diffraction, since it provides the most direct information on atomic configuration and pressure-induced structural alterations.[6] However, other less direct probes such as light scattering[7] and electrical resistivity also reflect pressure-induced changes. Accordingly, it is my purpose in this paper to discuss pressure effects on the structural properties of GIC's.

Section 2 contains a brief discussion of experimental methods. In Section 3, I will review briefly the seminal experiment of CWS[4] which exhibited for the first time the remarkable effect of pressure on interlayer and intralayer ordering in GIC's. In Section 4, I will highlight some of the most recent results on pressure-dependent structural alterations of alkali GIC's of various stages with particular emphasis on CsC_{36}. Finally, in Section 5 I will discuss quantitative results of the x-ray derived compressibilities of stage 1 alkali GIC's. Recent transport measurements of the pressure dependence of the resistivity of some GIC's will also be addressed. These measurements have provided evidence for new pressure-induced structural phase transitions in GIC's, transitions which have not yet been explored using the x-ray probe. Sections 6 and 7 of this paper contain, respectively, concluding remarks and acknowledgments.

2. Experimental Methods

Unless otherwise noted, all GIC samples referred to in this paper were prepared using graphite single crystal flakes and the standard two-bulb technique of HEROLD.[8] The details of the x-ray apparatus,[9] measurement techniques,[9] and high pressure cell[10] for x-ray and light scattering have been given elsewhere.

3. The First X-ray Pressure Studies - Potassium GIC's

The effect of hydrostatic pressure on the intra- and interlayer structure of stage 2 KC_{24} is exhibited in the (00ℓ) diffraction profiles of Figs. 1a-e. The profile in Fig. 1a which was recorded at ambient pressure is that of a "pure"[11] stage 2 potassium GIC. In particular, the width of each of the observed reflections is instrument limited, all of the observed reflections can be indexed to the stage 2 stacking sequence and they exhibit relative intensities which are consistent with the (00ℓ) structure factor based on a stage 2 stacking sequence. The insert in the lower right-hand corner of Fig. 1 indicates (using the "classical" model of staging)[12] that the intercalate layers (dashed lines) are planar disordered at ambient temperature and pressure. This planar disorder is confirmed by the precession photograph of Fig. 2a, which shows no evidence of potassium atom in-plane order (compare with Figs. 2b and 2c to be discussed forthwith).

As hydrostatic pressure is applied to KC_{24} it will be noted from Fig. 1b-d that the set of reflections corresponding to pure stage 2 KC_{24} persist, but lose intensity relative to a new set of reflections which can be indexed to a "pure" stage 3 stacking sequence. It is important to note that at elevated pressures the two sets of reflections exhibit intraset relative intensities consistent with the structure factors for the stage 2 and stage 3 stacking sequences, respectively. However, careful measurements reveal that the (10ℓ) reflections at intermediate pressures (1 bar < P < 7 kbar) are not instrument limited. Thus the data of Fig. 1a provide conclusive evidence that in the pressure range of 1 bar < P $\lesssim$ 7 kbar the KC_{24} sample consists of coexisting stage 2 and stage 3 regions of finite c-axis extent.[4] At pressures $\gtrsim$7 kbar the sample evolves to a pure stage 3 stacking sequence with no evidence for the stage 2 sequence which was observed at lower pressures.

The stage 2 - stage 3 pressure-induced transition discussed above was observed to be completely reversible as it was possible to repeatedly cycle between the two stacking sequences. This reversibility indicates that the transition occurred at constant KC_{24} bulk composition. Otherwise, any potassium which extruded from the sample would have coalesced in the pressure transmitting fluid (specially treated paraffin oil) and would not have reintercalated upon reducing the pressure. Thus at about 7 kbar the sample exhibits a "pure" stage 3 stacking sequence with a bulk stoichiometry of

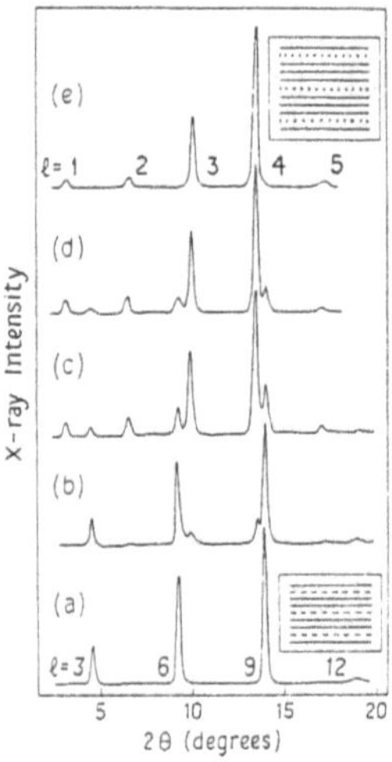

Fig.1 Pressure dependence of the room temperature (00ℓ) x-ray diffraction profiles of KC_{24} prepared from HOPG. (a) ambient pressure, (b) 2.5 kbar, (c) 4.0 kbar, (d) 5.5 kbar, (e) 6.5 kbar. Insets (a) stage 2 stacking (intercalant planes indicated by broken lines), (b) stage 3 stacking (ordered intercalant layers indicated by dotted lines).

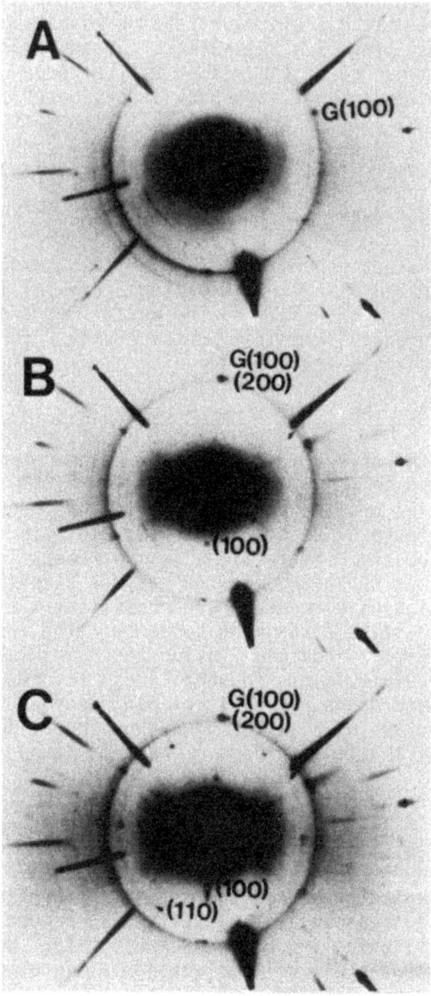

Fig.2 X-ray oscillation photographs of KC_{24} single crystals at A) 1 kbar, B) 4 kbar, and C) 11 kbar. The intense streaks are x-ray reflections from the diamond anvils in the pressure cell and the diffraction ring is from the Be-Cu gasket of the pressure cell. Reflections G(100) contain contributions from both host and potassium atoms. Other indices indicate potassium superlattice reflections.

KC_{24}. This implies a layer stoichiometry of KC_8 which in turn suggests an in-plane layer structure identical to the (2 x 2)R0° structure of the stage 1 compound KC_8. Indeed this densified layer structure of KC_{24} is perfectly consistent with the x-ray precession photographs shown in Figs. 2b and 2c. Notice the interior set of spots labeled (200) and (100) in that figure can be indexed on the (2 x 2)R0° structure. It is to be emphasized that that structure is the simplest and most plausible one with which to index the patterns of Fig. 2. Conceivably other, more complicated structures might also account for the observed patterns. Such other structures cannot be ruled out because the high pressure techniques used to record the patterns of Fig. 2 involve limited angular range over which Bragg diffraction spots can be recorded.

4. Other Alkali GIC's Under Pressure: CsC_{36}

In this section, I shall focus on the pressure dependence of the x-ray diffraction from cesium GIC's. Consider stage 3 cesium graphite which has the ambient pressure bulk composition CsC_{36}. The pressure dependence of the (00ℓ) x-ray diffraction profiles of that compound is shown in Figs. 3a-d.[13]

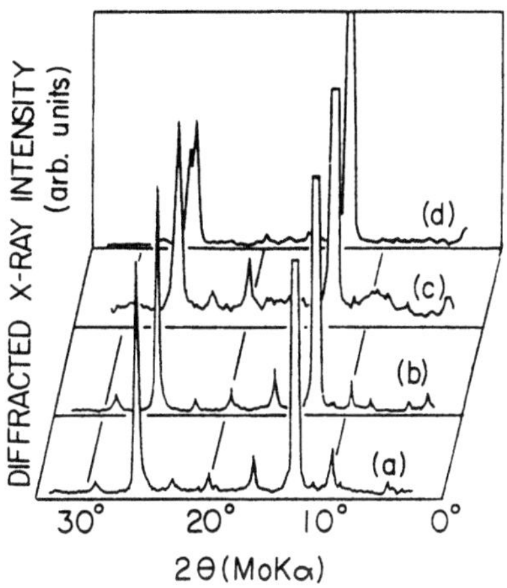

Fig.3 (00ℓ) x-ray profiles of CsC_{36} taken at (a) 1.6, (b) 6.3, (c) 7.3, and (d) 10.3 kbar.

One immediately notices that the pressure effects for this compound are much less dramatic than those associated with KC_{24}. Nevertheless, the qualitative behavior of these two compounds under pressure is similar. First, note again that at ambient pressure and temperature the (00ℓ) reflections of CsC_{36} indicate a "pure" stage 3 compound. As the pressure is increased some of the x-ray reflections (e.g. $\ell = 8$, $2\theta \cong 26°$) appear to split with the split components exchanging intensity with increased pressure. The splittings of the x-ray reflections from CsC_{36} are illustrated more graphically with the expanded 2θ scales of Fig. 4. Notice in particular that the reflection at $2\theta \cong 26°$ clearly splits with the line at higher angle gaining intensity with increasing pressure. A similar effect occurs for the reflection at $2\theta \cong 13°$, but the splitting is not nearly as well resolved.

The reflections of Fig. 3a-d can be indexed onto a superposition of stage 3 and stage 4 stacking sequences in which the set of reflections attributed to the latter gain intensity with respect to those attributed to the former. Also of interest is the fact that there is a slight lattice contraction, ~1.7% at 10 kbar, of the stage 3 regions. The set of reflections which grows with increasing pressure is attributed to stage 4 regions of the sample because the "package thickness" deduced from those reflections, 16.08 ± .13Å, is identical to within experimental error to that expected for a stage 4 cesium graphite compound at ambient conditions.[14] Here the "package thickness" is defined as the minimum repeat distance that characterizes a given stage GIC, i.e. the distance between nearest intercalate layers measured along the c-axis. In contrast the c-axis repeat distance is either equal to or an integral multiple of the package thickness since it is a measure of the c-axis stacking sequence itself.

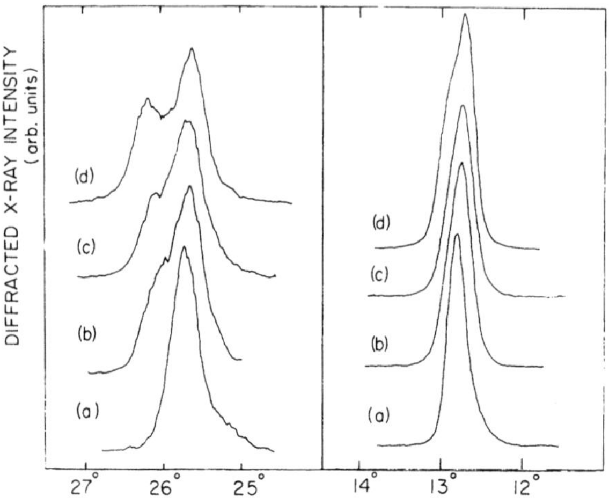

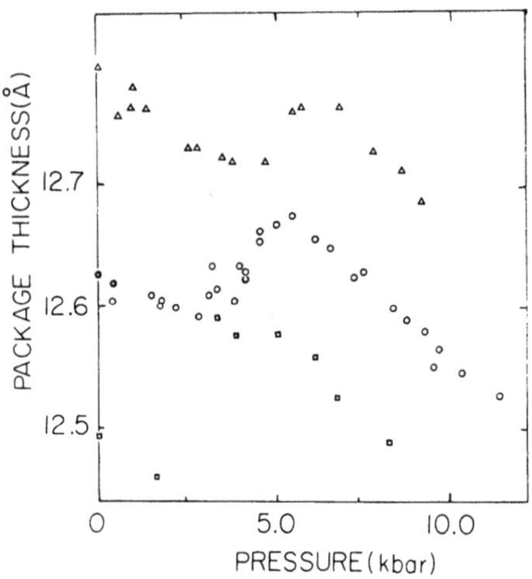

Fig.4 Expansion of the 2θ ≈ 13° diffraction peaks in CsC_{36} at (a) 0.0, (b) 5.0, (c) 6.5, and (d) 10.0 kbar.

Fig.5 Calculated package thickness versus pressure for three different CsC_{36} samples, designated as ▲, o, and ◘.

In Fig. 5, the pressure dependence of the package thickness of CsC_{36} is measured for three different samples. This figure clearly illustrates several key points. First, note that the package thickness of CsC_{36} varies by as much as 2% from sample to sample, a variation that is well outside the x-ray accuracy of $\lesssim 0.4\%$. Second, there is a transition associated with an increase in package thickness which is sample dependent and occurs at ≈ 5 kbar in CsC_{36}.

As in the case of KC_{24}, precession x-ray photographs yield an in-plane Cs spot pattern at elevated pressure which can be indexed on the (2 x 2)R0° structure of stage 1 alkali GIC's. However, there may also be present in the sample, regions containing disordered layers of Cs with the layer stoichiometry of CsC_{12} usually associated with MC_{12n} alkali GIC's, n ≥ 2. Such regions could be associated with the persistence of either stage 3 (00ℓ) reflections or stage 4 (00ℓ) reflections in the patterns of Fig. 4. For KC_{24} the pure stage 2 stacking sequence evolves to a pure stage 3 sequence which thus must be identified with the (2 x 2)R0° KC_8 layer structure. The transition can be described as

$$1\ KC_{2x12} \Rightarrow 1\ KC_{3x8}\ , \qquad (1)$$

where the symbol C_{nxs} implies both the stage, n, and the layer stoichiometry, s. For CsC_{36}, the sample does not acquire a single pure stage stacking sequence at the pressures applied. Both stages 3 and 4 coexist at all applied pressures. Therefore, if the sample contains layer stoichiometries other than C_{nx12} and C_{nx8}, it is not possible to definitely assign the (2 x 2)R0° in-plane structure to either stage. However, if we assume that only layer stoichiometries C_{nx12} and C_{nx8} are obtained in the sample, then the only viable transition scheme would be

$$1\ CsC_{3x12} \rightarrow 1/2\ C_{3x8} + 1/2\ CsC_{4x12}\ . \qquad (2)$$

Indeed that scheme accounts for the coexistence of stages 3 and 4 in CsC_{36} at all observed pressures. Within the intensity accuracy of the x-ray measurements of Fig. 5 it is not possible to experimentally confirm (2) above. However, it is possible in principle to determine the in-plane number density of intercalate atoms from careful structure factor measurements of the (00ℓ) reflections[15,16] and thus to experimentally verify (2).

As in the case of KC_{24}, the pressure-induced staging change in CsC_{36} occurs at constant stoichiometry as evidenced by the reversibility of the pressure-dependent x-ray patterns of Fig. 5 and by (2). In contrast to KC_{24} in which pressure caused an intercalate densification to (2 x 2)R0° in the higher stage (e.g. stage 3) of the observed mixed stage pair, the densification for CsC_{36} apparently occurs in the lower stage (also stage 3) of the mixed pair (see (2)). In either case, the microscopic mechanism which drives the pressure-induced interlayer and intralayer transitions in alkali GIC's appears to select out those layer stoichiometries, MC_{nx12} and MC_{nx8}, which naturally occur at ambient conditions. Note that up to 10 kbar the x-ray pressure data for CsC_{36} provides no evidence for layer stoichiometries such as C_{nx6} which are more dense than C_{nx8}.

5. Stage 1 Alkali GIC's Under Pressure

X-ray diffraction from stage 1 KC_8 has been studied at pressures up to 13 kbar.[17] However, in those studies no evidence was found for either an interlayer or intralayer structural phase transition. This observation is consistent with the assumption that for potassium the (2 x 2)R0° in-plane superlattice is the most dense structure consistent with the ionic size of potassium in KC_8.[5] However, FUERST et al.[18] have observed a discontinuity at ≈15 kbar in the pressure dependence of the c-axis resistivity of KC_8 which they attribute to a transition from a (2 x 2)R0° phase to a (2 x 1) or (1 x 1) densified intralayer phase. Since diffraction studies at pressures in excess of 13 kbar have not been carried out on stage 1 MC_8 GIC's, M = K, Rb, Cs, these superdense structures remain a hypothesis of FUERST et al. However, such structures are difficult to envision at the modest pressure of 15kbar. These structures are especially unlikely since they involve a potassium intralayer nearest–neighbor distance of 2.46Å. Such a small metal-metal distance has not been detected in any GIC to date under ambient conditions, not even in GIC's prepared from lithium, which has a much smaller ionic radius than potassium.

An alternative explanation for the observations of FUERST et al.[18] is that the KC_8 compound undergoes a first-order c-axis stacking transition from the AαAβAγAδ ambient temperature arrangement[19] to one with a shorter repeat distance along the c-axis. Here, A represents carbon layers (A-A-A-A implies a configuration in which all carbon bonds are eclipsed when viewed along the c-axis) and α, β, γ, δ refer to (2 x 2)R0° order of metal atoms arranged over the centers of carbon hexagons when viewed along the c-axis. Since the electrical anisotropy between the basal plane and c-axis resistivities of KC_8 is considerably less than that of graphite, there is clearly non-negligible electron wavefunction overlap along the c-axis in KC_8. In this case one could anticipate that a c-axis stacking transformation could markedly effect the c-axis resistivity of KC_8 even in the absence of an intralayer transition in the potassium layers.

Like KC_8, RbC_8 and CsC_8 exhibit no inter- or intralayer structural phase changes at pressures up to 13 kbar. However, by measuring the dependence of the package thickness, d, on pressure, P, it is possible to obtain a quantitative measure of the c-axis compressibility k_d of MC_8 GIC's, M = K, Rb, Cs. The compressibility is determined from the relation

$$k_d = \frac{1}{d}\left(\frac{\partial d}{\partial P}\right)_{P=0} \qquad (4)$$

if one assumes that the in-plane contractions of the GIC under pressure are negligible compared to the c-axis contractions.[17,10]

In Fig. 6 is shown a plot of the pressure dependence of the package thickness of CsC_8. The linearity exhibited in Fig. 6 clearly indicates that for P ≤ 13 kbar, CsC_8 obeys a Hook's law. Similar data have been obtained for RbC_8 and KC_8 as well as for pristine graphite in both the single crystal and highly oriented pyrolytic graphite form. The values of k_d determined from such data are tabulated in Table 1 as are the corresponding ambient

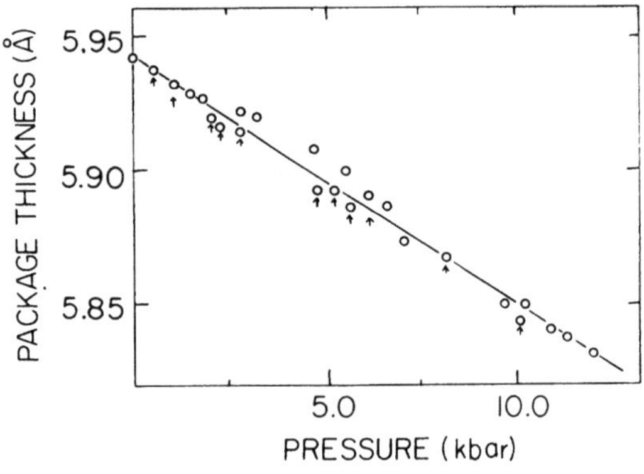

Fig.6 Package thickness of CsC_8 as a function of hydrostatic pressure. The circles without an arrow show the first-run data points, which were taken consecutively up to the highest pressure. The arrows indicate the second-run data points from the same sample. The straight line was obtained from the least-squares analysis of the data points.

Table 1 The compressibility k_d and the zero pressure package thickness d_0 of graphite and MC_8 GIC's

	k_d ($10^{-12}cm^2/dyn$)	d_0 (Å)
Graphite	2.73±0.09	3.353±0.003
KC_8	2.13±0.09	5.355±0.004
RbC_8	2.42±0.12	5.657±0.007
CsC_8	1.56±0.05	5.943±0.004

pressure package thicknesses. Note in particular that the value of k_d determined for pristine graphite using (3) is in excellent agreement with the value determined from ultrasonic measurements.[20] This fact supports the approximation in which basal plane contractions are ignored in computing k_d. Moreover, since the heavy alkali GIC's KC_8, RbC_8 and CsC_8 are "harder", i.e. less compressible, than pristine graphite, such an approximation should be equally valid for those compounds.

6. Concluding Remarks

With the discovery of pressure-induced phase transitions in GIC's several new directions of research are opened. One involves "redoing" all the old measurements on GIC's but with pressure applied. This activity might yield interesting new results and may indeed be worth pursuing. Another research direction involves continued careful study of selected model GIC's under pressure. It is particularly important to extend the x-ray studies to higher pressures, preferably in excess of 100 kbar. This task is straightforward, since x-ray diffraction from samples at 100 kbar pressure can be routinely recorded using a diamond anvil pressure cell.

7. Acknowledgments

I am grateful to my collaborators, N. WADA and R. CLARKE, for useful discussions and for their major contributions to the work reviewed in this paper. I also thank N. WADA for permission to include Fig. 6 in this paper prior to its publication in his Ph.D. thesis. This research was supported by the USARO under Contract No. DAAG20-80-K-0003.

8. References

1. P.W. Bridgman, The Physics of High Pressure, Bell, London (1952).
2. C.W. Christoe and H.G. Drickamer, Rev. Sci. Inst. 40, 169 (1969) and references therein.
3. P.W. Bridgman, Endeavor 10, 68 (1951).
4. R. Clarke, N. Wada and S.A. Solin, Phys. Rev. Lett. 44, 1616 (1980).
5. S.A. Solin, Advances in Chemical Physics, in press.
6. H.P. Klug and L. Alexander, X-Ray Diffraction Procedures for Polycrystalline and Amorphous Materials, Wiley, New York (1974), 2nd Ed.
7. S.A. Solin, Physica 99B, 443 (1979).
8. A. Herold, Bull. Soc. Chim. Fr. 187, 999 (1955).

9. R. Clarke, N. Caswell and S.A. Solin, Phys. Rev. Lett. 42, 61 (1979).
10. N. Wada, Phys. Rev., in press.
11. See references 5 and 7 and references therein.
12. N. Caswell, S.A. Solin, T.M. Hayes and S.J. Hunter, Physica 99B, 463 (1980).
13. N. Wada and S.A. Solin, Physica, in press.
14. W. Rudorff, Adv. in Inorg. Chem. Radiochem. 1, 223 (1959).
15. C. Underhill, T. Krapcher and M.S. Dresselhaus, in Proceedings of the International Conference on Intercalation Compounds of Graphite, ed. by. F.L. Vogel and A. Herold, Synthetic Metals, in press.
16. B.E. Warren, X-Ray Diffraction, (Welsely, Reading, Mass., 1969).
17. N. Wada, R. Clarke and S.A. Solin, Solid State Commun. 35, 675 (1980).
18. C.E. Fuerst, D. Moses, A. Devenstein, A.J. Heeger and J.E. Fischer, Bull. Am. Phys. Soc. 26, 451 (1981).
19. A.R. Ubbelohde and F.A. Lewis, Graphite and Its Crystal Compounds, Clarendon, Oxford (1960).
20. O.L. Blaklee, D.G. Proctor, E.J. Seldin, G.B. Spence and T. Wang, J. Appl. Phys. 41, 3373 (1970).

On the Structure of the First Stage Binary Compound of Metals Intercalated into Graphite

Daniel Guérard and Philippe Lagrange

Laboratoire de Chimie du Solide Minéral (LA 158) Service de Chimie Minérale Appliquée - CO 140 - F-54037 Nancy Cedex

1. Introduction

The general formula of the first stage graphitides of metals is MC_x. Up to now, one knows only two values for x ; x = 6 (Li, alkaline - earth metals, lanthanides) ; x = 8 (heavy alkali metals).

A compact arrangement of the metal atoms in epitaxy on the carbon matrix would involve a formula MC_2. Then, there are three or four possible sites for the metal atoms in each layer, according to the chemical composition.

The purpose of our structural study is essentially to determine how many sites are actually occupied and what is the exact stacking of the metallic layers along the $\vec{c}$ axis. In this work, we will study only the ideal structures with a regular stacking. In particular, we purposly ignore the possible double occupation of one site, for instance $\alpha\beta\beta\gamma\delta\alpha$... On the other hand, we never have a completely random stacking : there is an in-plane substructure comparable to graphite.

2. Possible Unit Cells

The possible unit cells corresponding to the regular occupation of one, two three (or four) sites in the MC_6 (and MC_8) compounds are described in the following table :

Table 1 Possible sites for the MC_6 and MC_8 compounds

	Number of occupied sites	stacking	System	Space Group	c	Ref.
MC_6	1	αα...	Hex	P6/mmm	Ic	[1]
	2	αβα...	Hex	$P6_3/mmc$	2Ic	[2]
	3	αβγα...	Rh.	$R\bar{3}m$	3Ic	
MC_8	1	αα...	Hex	P6/mmm	Ic	
	2	αβα...	Orth	Fmmm	2Ic	
	3	αβγα...	Hex	$P6_222$ or $P6_422$	3Ic	[3]
	4	αβγδα...	Orth	Fddd	4Ic	[4,5]

- There are 3 systems never formed : the type αβγα... for MC_6 and the types αα... and αβα... for MC_8.
- There is no theory able to predict what kind of stacking corresponds to a given compound.
- Through the example of the study of the 1st stage graphitide of potassium, we will point out what kind of disorder is involved by the presence and occupation of several possible sites for each metal atom.

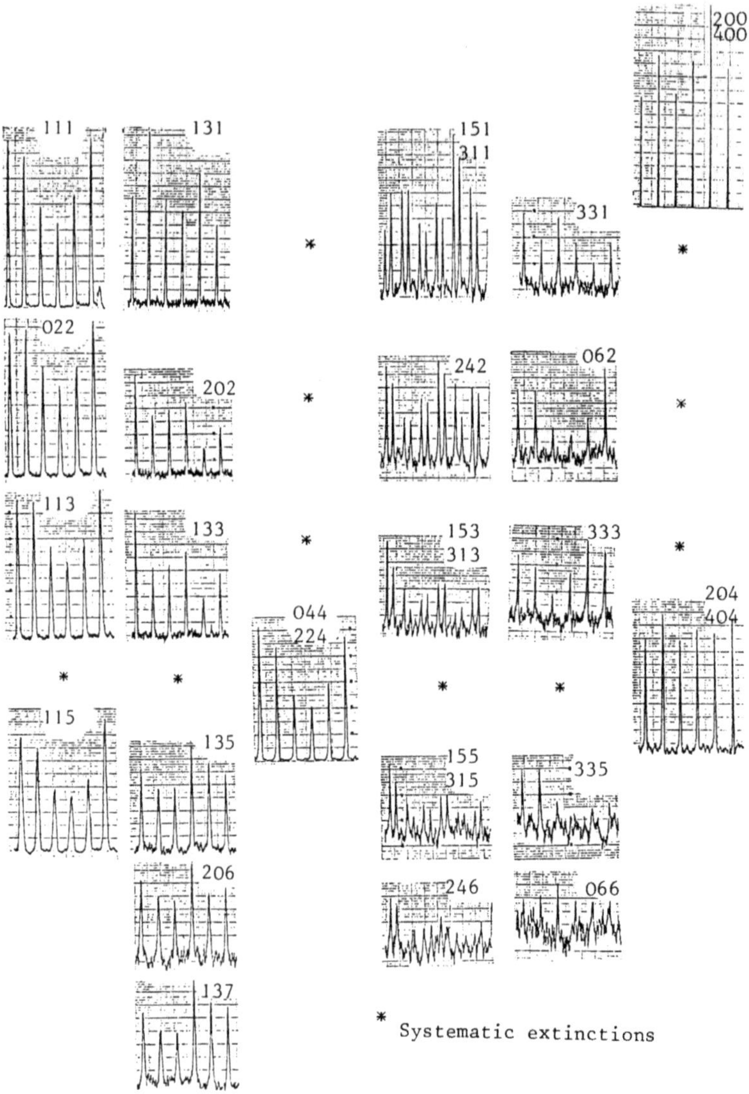

Fig.1 hk0 and hkl reflexions observed by rotation of the crystal around its $\vec{c}$ axis

3. First Stage Graphitide of Potassium

3. 1. Previous Observations

From single crystal photographs, one knows that the c parameter is equal to four times the identity period along the $\vec{c}$ axis [6].

This unit cell has to be orthorhombic, according to the table 1. If one performs laue photographs on a single crystal, one can easily see an hexagonal symmetry, found also in the other intercalation compounds of graphite.

Studying the kh0 and hkl reflexions given by the double rotation method [7], one can easily determine the multiplicity factor of each reflexion. The values are 6, 12 or 24, as shown in Fig. 1, are characteristic of an hexagonal unit cell, not to an orthorhombic one.

We will try now to reconciliate the different observations.

3. 2. Interpretation of the Results

If one starts from an ideal plane of graphite, one can suppose that the 1st interlayer spacing is filled by the metal atoms in the site α(open circles).

The filling of the second layer of metal atoms (open triangles) can be either at the β, γ or δ sites, which divides the plane in three domains.

For the 3rd layer (points), the first domain can be filled in γ or δ, the second in β or δ, the last one in β or γ. This divides each domain in two parts.

The last layer is obtained by the occupation of the last possible site (filled triangles).

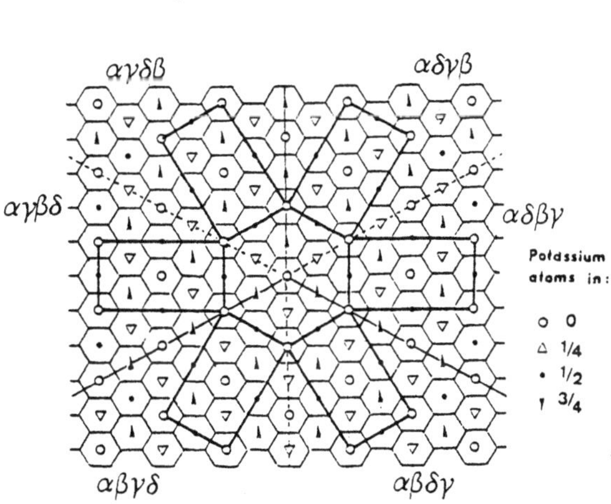

Fig.2 The six crystallographic domains of KC_8

Now, one ideal single crystal is divided into 6 domains, on each of which one can draw an orthorhombic unit cell. In fact, these 6 unit cells are two and two deductible by a translation vector $\frac{1}{4}, \frac{1}{4}, \frac{1}{4}$ and the first stage graphitide of potassium can be described by three orthorhombic Fddd unit cells ($\alpha\beta\gamma\delta$, $\alpha\gamma\delta\beta$ and $\alpha\delta\beta\gamma$). Perfectly superimposable, they are not equivalent when placed on the same graphite plane since they are deducible one from the other by rotations of 120 and 240°.

Fig. 3 shows the superimposition of some hk0 and hkl reflexions given by these three orthorhombic unit cells, which leads to an apparent hexagonal symmetry.

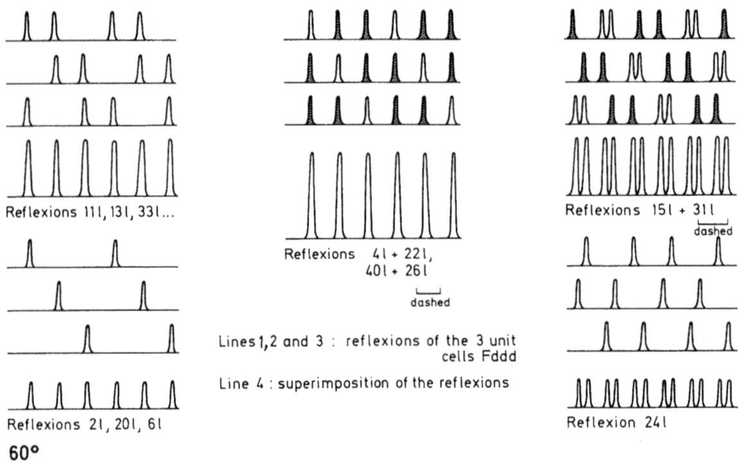

Fig.3 Superimposition of the three orthorhombic unit cells reflexions which leads to an hexagonal symmetry

4. Generalization of this Study

In our model, we started from a first unique site α for the whole metallic layer. In fact, while the α site is occupied in one part of the crystal, one other site can be occupied in an other part of the graphite plane. This fact means that the synthesis itself involves the formation of different domains : the intercalation compounds are polysynthetic ones.

This phenomenon can be generalized to every intercalation compound which includes separate atoms or molecules in epitaxy on the carbon matrix. One can think of compounds of graphite with ICl, $SbCl_5$..., but not for compounds with liquid molecules like HNO_3 or macromolecules like $FeCl_3$.

On the other hand, the 1st stage LiC_6, where only one site is occupied, is generally described as $\alpha\alpha$...,but there can be also $\beta\beta$ or $\gamma\gamma$ stacking in the same compound.

This kind of multiple stackings cannot be detected by X-rays but there is probably an influence by such defects on some physical properties (e.g. electrical conductivity) : the in-plane $\alpha\alpha$ and $\alpha\beta$ distances are different. However, this kind of defect, located at the frontier zone of domains does not constitute the main disorder, as it is discussed in an other paper [8].

5. Other First Stage Graphitides

5. 1. MC_6 Compounds (M = Li, Alkaline-earth metals, lanthanides)

The unit cells were determined from •the relative intensities of hk0 reflexions and

•the systematic extinctions

In all these cases, the hexagonal space group was $P6_3/mmc$ with C_2 $2I_c$.

5. 2. RbC_8 and CsC_8

The 1st stage rubidium graphitides are identical to KC_8 and the orthothombic unit cell belongs also to the F_{ddd} space group.

The 1st stage graphitide of cesium has an hexagonal unit cell with c = $3I_c$, and the space group are either $P6_222$ (right hand screw axis) or $P6_422$ (left hand screw axis).

In that last case, we could have optically active domains. Unfortunatly, the size of the domains is too small (4 to 500 Å) to be easily detected.

6. Conclusion

According to the occupation of the possible sites, the hexagonal structure of graphite remains or disappears. However, the intercalation process itself leads to polysynthetic crystals which create an apparent hexagonal symmetry.

1. D. Guérard and A. Hérold : Carbon 13.4 [1975] 337
2. D. Guérard and A. Hérold : C.R. Acad. Sc. Paris, série C, 279 [1974] 455/6 ibid 280 [1975] 729/30 ; ibid 281 [1975] 929/31
3. D. Guérard, P. Lagrange, M. El Makrini and A. Hérold : Carbon 16.4 [1978] 285/90
4. P. Lagrange, D. Guérard and A. Hérold : Ann. Chim. Fr. 3.2 [1978] 143/59
5. D. Guérard, P. Lagrange, M. El Makrini and A. Hérold : C.R. Acad. Sc. Paris, série C 287 [1978] 179/82
6. W. Rüdorff and E. Schultz : Angew. Chem. 66 [1954] 305
7. D. Guérard, M. Lelaurain and A. Aubry : Bull. Soc. Fr. Minéral. et Cristallogr. 98 [1975] 43/8
8. F. Rousseaux, A. Plançon, D. Tchoubar, D. Guérard and P. Lagrange : these Proceedings

Ordering in Intercalation Graphite Compounds by Means of X-Ray Diffraction

F. Rousseaux, A. Plançon and D. Tchoubar

Laboratoire de Cristallographie, Faculté des Sciences d' Orléans
F-45045 Orleans La Source and

D. Guérard and P. Lagrange

Laboratoire de Chimie Minérale Appliquée, Faculté de Nancy 1
CO 140 - F-54037 Nancy Cedex

1. Introduction

In recent years, some "ideal" structures of intercalation graphite compounds have been well established [1] [2] [3]. Starting from these works, we report in the present paper the results of a quantitative study performed on the first-stage graphitide of potassium KC_8 which reveals the nature and the probability of structural defects. This study implies the comparison between experimental intensities and intensities computed on the basis of models of disorder.

As these compounds have lamellar structures, the disorder can be described both by stacking faults in the "ideal" stacking of the layers and by intralayer defects.

2. X-Ray Diffraction Experiments

It is well known that, as soon as a sample has a disordered structure, scattered intensity appears not only at the (hkl) Bragg positions but also between them. It is therefore necessary to perform, with accuracy, scans along (hk) rods as a function of l, which is not, so far, interger but takes continuous values (Fig.3).

The measurements were performed on a 2-axis spectrometer using the $MoK\alpha_1$ radiation. The samples were made from highly oriented pyrolitic graphite HOPG. They were cut as parallelepipeds of 2x8x0.3 mm^3 in order to be sealed under controlled atmosphere in Lindman glass. During the measurements the $\vec{c}$-axis, perpendicular to the layer's plane, remained in the scattering plane.

As the HOPG behaves like a highly oriented powder, the reciprocal diffraction patterns consist in cylinders whose axes are parallel to the $\vec{c}$-axis direction. Consequently only a small fraction of this intensity is intercepted by the spectrometer. Moreover, it must be noticed that because of the mosaïc spread of the $\vec{c}$-axis of the samples (which broadens with the intercalation process) the experimental measure, at a given value of l, is not I(l) but corresponds to a partial integration. The conditions of this integration vary with the value of 1 and the thickness of the cylinder. Consequently, the experimental profiles recorded are systematically distorted (see Fig.2). This defect is quite common for all the intercalation graphite compounds and must be taken into account in the computed curves in order to avoid crude errors when comparing experimental and theoretical curves. The details of this correction will be published in another paper; see also [4].

3. Models and results for KC_8

To understand the interlayer stacking order we must first consider the superlattice unit cell for a single layer. In the case of KC_8 a single "layer" will be constituted both by a carbon layer and by a potassium layer whose atoms are situated at a distance Z = 5.35 Å/2 above the carbon plane. The unit cell of this entity is orthorhombic with $|\vec{a}|$ = 4.96 Å and $|\vec{b}|$ = 8.59 Å (Fig.1). It contains 16 carbon atoms for 2 potassium atoms. Throughout our discussion we will compare the data to this orthorhombic reciprocal lattice.

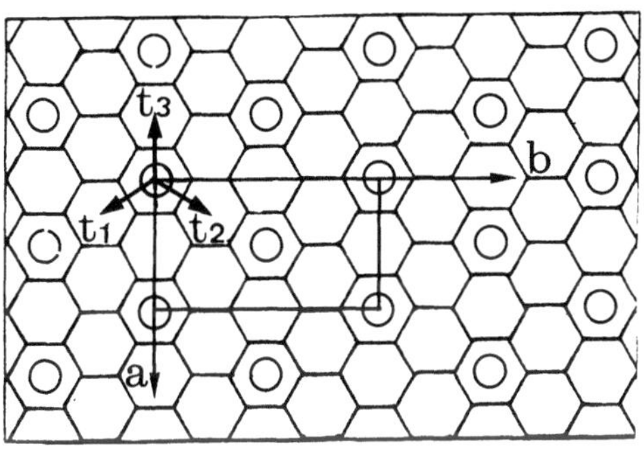

Fig.1 Unit cell of a "layer" constituted by one carbon plane and one potassium plane as in KC_8.

The description of the ideal stacking sequence in KC_8 can be done with only two of the three following translation vectors :

$$\vec{t}_1 = \frac{\vec{a}}{4} - \frac{\vec{b}}{4} + d_{001}\,\vec{c}$$

$$\vec{t}_2 = \frac{\vec{a}}{4} + \frac{\vec{b}}{4} + d_{001}\,\vec{c} \qquad \text{with } d_{001} = 5.35\ \text{Å}$$

$$\vec{t}_3 = -\frac{\vec{a}}{2} + d_{001}\,\vec{c}$$

For exemple, the αβγδ sequence is obtained as follows. Let us call "layer 0" the original layer, the next one, "layer 1", will be shifted by the translation t_1 relative to "layer 0". Then, the second one, "layer 2", will be shifted by the translation $\vec{t}_2$ relative to "layer 1". Then the third one, "layer 3", will be shifted by the vector $-\vec{t}_1$ relative to "layer 2" and the fourth one, "layer 4", will be shifted by $-\vec{t}_2$ relative to "layer 3". One can easily show that $\vec{t}_1$ is equivalent to $-\vec{t}_1$ and $\vec{t}_2$ is equivalent to $-\vec{t}_2$. The αβγδ sequence can be therefore described by only one kind of "layer" (carbon plane + potassium plane) and the successive repetition of the two translations vectors $\vec{t_1}$ and $\vec{t_2}$ ($\vec{t_1}\vec{t_2}\vec{t_1}\vec{t_2}\vec{t_1}$...) In the same way the others families αδβγ αγδβ... can be described either by the sequence $\vec{t_1}\vec{t_3}\vec{t_1}\vec{t_3}\vec{t_1}$... or by $\vec{t_2}\vec{t_3}\vec{t_2}\vec{t_3}\vec{t_2}$... A macroscopic sample of KC_8 has, naturally, domains of the three stacking types.

With this simplified description it is possible to imagine different kinds of stacking faults in only one of the three sequences defined above. If we choose, for example, the $\vec{t_1}\vec{t_2}\vec{t_1}\vec{t_2}\vec{t_1}$... sequence, a fault may be either a a repetition of translation as $\vec{t_1}\vec{t_1}$, $\vec{t_2}\vec{t_2}$, $\vec{t_3}\vec{t_3}$ and / or $\vec{t_1}\vec{t_3}$, $\vec{t_2}\vec{t_3}$ or a translation which is random with respect to the layer lattice parameters. We used a mathematical formalism developped by A. Plançon [5] in order to calculate the effect of this kind of defects and their probability. This formalism implies two -nearest- layer interactions. This part of the work will be developed in another paper.

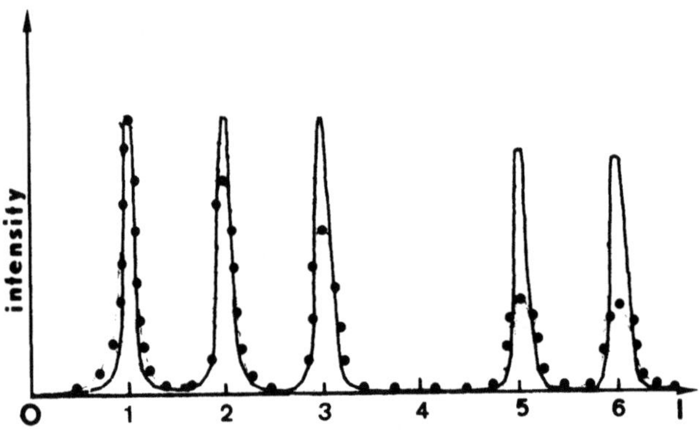

Fig.2 Solid line : Calculated scattering profile of (02,11) rod for a perfectly oriented powder without stacking faults
Black dots : Calculated scattering profile of (02, 11) rod when taking into account a mosaïc spread of the $\vec{c}$-axis of 2°.

Among the models of disorder elaborated by combination of the different types of stacking faults, only one model was convenient. Figure 2 shows the calculated scattering profile for the (02, 11) rod when the sample is considered as a perfectly oriented powder without stacking faults (solid line). On the same figure the black dots show the effect of a mosaïc spread of the $\vec{c}$-axis of 2° when all the others parameters are unchanged. One can notice the great decrease of the intensity I(l) as l increases.

Figure 3 illustrates the experimental scattering profile (02,11) (solid line) performed at room temperature. The discrepancy between figure 2 and figure 3 gives the evidence of the presence of defects in the compound. The peaks (hkl) are broadened and an anomaly can be noticed for the maxima (022), (026). They are lower than expected.

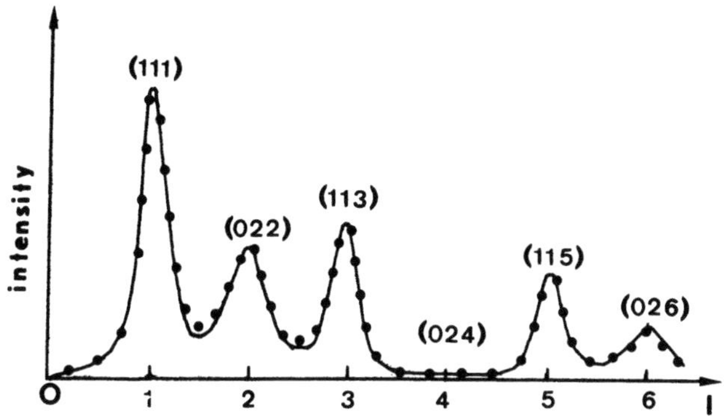

Fig.3 Black dots : Experimental scan of (02, 11) performed with a very good counting statistic (10 000 photons every step).

Solid line : Best fit obtained for a model of disorder which takes in account only the stacking fault $\vec{t_3}$ with a probability = 0.1.

The model which reproduces the anomaly of the (022, (026) reflections implies that there is only one kind of stacking fault in KC_8. The fault consists of dinding the translation $\vec{t_3}$ after $\vec{t_1}$ or $\vec{t_2}$ in an ideal sequence $\vec{t_1}$, $\vec{t_2}$, $\vec{t_1}$, $\vec{t_2}$... This stacking fault is distributed in a random manner. The best fit (black dots on figure 3) was obtained for a probability 0.1. This result means that the average number of layers in a perfect stacking $\vec{t_1}\vec{t_2}$ is layers. In others words, the average correlation length L (perpendicular to the layer) is about 50 Å. This length is small with regards to those we have measured in a HOPG precursor ($L_\perp$ > 1000 Å)

Scans were also performed on the (20, 13) rod. This profile shows the same intensity and anomaly for the (202), (206) reflections. A good fit was also obtained for the same model of disorder with the probability 0.1.

Other comments can be made :

- The nature of the stacking fault $\vec{t_3}$ seems very logical, when one remembers that the three sequences types ($\vec{t_1}\vec{t_2}\vec{t_1}$... $\vec{t_2}\vec{t_3}\vec{t_2}$... $\vec{t_3}\vec{t_1}\vec{t_3}$...) coexist with the same probability in the macroscopic sample.
- There are no distortions or displacements of atoms since the ideal structure factor of the "layer"(one carbon plane + one potassium plane) is able to reproduce the data.
- There are not any random translations.

Moreover, the in-layer correlation length $L_{//}$ is about 400 Å ± 20 Å, while in HOPG samples it can reach 1 μ [6]. So, we can conclude that the intercalation process has noticeably decreased the length of the perfect stacking domains and their planar dimensions.

4. Conclusion

This work shows that it is possible to obtain a consistent picture of defects in intercalation graphite compounds. This kind of study will now be extended to others samples, for example, poorer stages.

We also have studied a first stage graphite - ferric chloride. In that compound, the stacking disorder is quite different from KC_8, it almost only consists in random translations. Details of this study will be published in another paper.

Acknowledgments : We acknowledge Dr. A.W. Moore for providing HOPG samples.

References

1. P. Lagrange, D. Guérard, A. Hérold : Ann. Chim. Fr. 3.2 [1978] 143/59
2. D. Guérard, P. Lagrange, M. El Makrini and A. Hérold : Carbon 16.4 [1978] 285/90
3. P. Lagrange, M. El Makrini, D. Guérard and A. Hérold : Carbon 18 [1980] 257/64
4. A. Plançon : J. Appl. Cryst. 13 [1980] 524/28
5. A. Plançon : Act. Cryst. [1981] in press
6. A. W. Moore : Chem. and Phys. of Carbon. 17 [1980]

Liquids in a Periodic Potential

A. Bunde and W. Dieterich
Fakultät für Physik, Universität Konstanz
D-7750 Konstanz, Fed. Rep. of Germany

1. Introduction

A liquid subject to an external periodic potential represents a model system for submonolayer films adsorbed on surfaces: intercalation compounds and superionic conductors. The characteristic and common feature of these systems is that they consist of two species of particles: one species is fixed around certain equilibrium sites and forms a regular lattice, whereas the other species is mobile and moves in a fluid-like fashion through this lattice. In some systems the concentration of the mobile particles can be varied to a certain extent, and effects of commensurability between the fluid particles and the underlying lattice can be investigated.
In a previous work [1,2] we have studied commensurability effects in a one-dimensional model system with soft-core interactions. The present work also includes systems with hard-core interactions and it is our purpose here to compare the results for both kinds of interactions as a function of the average density of the mobile particles.
The actual computations are based on the inhomogeneous version of the Percus-Yevick (PY) theory [3] in connection with the first equation of the Yvon-Born-Green (YBG) hierarchy (see e.g. [4]).

2. Theory

Our model is defined in terms of the following total potential energy:

$$V(x_1,\dots,x_N) = \sum_k v^{(1)}(x_k) + \sum_{k<\ell} v^{(2)}(x_k - x_\ell)\,, \tag{1}$$

where

$$v^{(1)}(x) = -(V_0/2)\cos(2\pi x/a) \tag{2}$$

is the periodic potential simulating the influence of the periodic medium and

$$v^{(2)}(x) = V_0 \begin{cases} (b_0/x)^n - (1/3)^n & \text{for } |x|<3b_0 \\ 0 & |x|\geq 3b_0 \end{cases} \tag{3}$$

represents the pair potential. We take the values n = 2 and n = 6, which correspond to a soft-core- and a hard-core-like repulsion, respectively.

In the following we consider thermal energies k_BT which are of the order of the average interaction energy per particle $\overline{V}^{(2)}$. X-ray scattering performed in the high temperature region of the intercalation compounds C_{12n} Cs (n=1, 2,3,4) [5,6] and on Ag^+-type superionic conductors [7,8] have shown that the radial distribution function g(r) of the mobile particles is similar to that of a liquid where $\overline{V}^{(2)} \sim k_BT$, approaching unity after a few oscillations. In this temperature regime the full anharmonicity of the pair potential must be taken into account, in contrast to the case $\overline{V}^{(2)} \gg k_BT$, where the system can be mapped on the Frenkel-Kontorova model.
In order to obtain detailed information on the density distribution $\rho(x)$ and the pair correlation $g(x,x')$, we use the inhomogeneous version of the PY theory [3]

$$g(x,x')-1 = c(x,x') + \int_{-\infty}^{+\infty} dx'' \; c(x,x'')\rho(x'')[g(x'',x')-1], \tag{4}$$

$$c(x,x') = g(x,x')[1 - \exp(\beta V^{(2)}(x - x'))] \tag{5}$$

together with the first equation of the YBG hierarchy [4]

$$\frac{\partial}{\partial x} \log \rho(x) + \beta V^{(1)}(x) = -\beta \int_{-\infty}^{+\infty} dx' \; \frac{\partial V^{(2)}(x - x')}{\partial x} \rho(x')g(x,x') \tag{6}$$

where $\beta \equiv 1/k_BT$. The method of solution of this nonlinear coupled system of equations has been described before [1,2]. When $g(x,x')$ and $\rho(x)$ are known, the two-particle density $\rho^{(2)}(x,x') \equiv \rho(x)\rho(x') \cdot g(x,x')$ can be calculated and the structure factor $S(q) \equiv \sum_{i,j} \langle \exp(iq(x_i - x_j))\rangle$ is obtained according to

$$S(q) = (Na/\rho_o) \sum_k \delta_{q,k} \, \rho_k^2 + \tilde{S}(q) \tag{7}$$

with

$$\tilde{S}(q) = 1 + (1/N)\int dx \int dx' \exp(iq(x - x'))\rho(x)[g(x,x')-1]\rho(x') \; . \tag{8}$$

Here, N is the number of mobile particles, ρ_o is their average density, and k denotes the reciprocal lattice vectors defined by the periodic medium ($k = 0, \pm q_o, \pm 2q_o, \ldots$, with $q_o = 2\pi/a$).
According to (7), S(q) consists of a sum of Bragg peaks at the reciprocal lattice vectors plus a liquid-like contribution $\tilde{S}(q)$. From the intensity of the Bragg peaks, conclusions about $\rho(x)$ can be drawn. $\tilde{S}(q)$ can be conveniently discussed in terms of the structure factor $S_o(q)$ and the pair correlation $g_o(x-x')$ of the corresponding homogeneous liquid [2]. Expression (8) can be rewritten as

$$\tilde{S}(q) = S_o(q) + \sum_{k \neq o} (\rho_k/\rho_o)^2[S_o(q + k)-1] + \Delta\tilde{S}(q), \tag{9}$$

where

$$\Delta\tilde{S}(q) \equiv (1/N)\int dx \int dx' \exp[iq(x-x')]\rho(x)[g(x,x') - g_o(x-x')]\rho(x') \tag{10}$$

is a measure of the deviation of $g(x,x')$ from the pair correlation of the homogeneous liquid. If the approximation $g(x,x') \simeq g_0(x-x')$ holds we have $\Delta\tilde{S} \simeq 0$ and the structure factor is determined solely by the structure factor of the homogeneous liquid and its modulated side peaks at $q \pm q_0$, $q \pm 2q_0$ etc., the intensities of which are proportional to $(\rho_1/\rho_0)^2$, $(\rho_2/\rho_0)^2, \ldots$. Recent numerical analysis [2] of (4) - (6) has shown that $\Delta\tilde{S}(q)$ may be small compared with $\tilde{S}(q)$, even in cases where $g(x,x')$ and $g_0(x-x')$ differ rather strongly from each other.

The numerical results for $\rho(x)$ will be discussed in terms of the effective single particle potential

$$U(x) = -k_B T \log \rho(x) . \tag{11}$$

Its barrier height U_0 plays the role of the activation energy entering the mobility. This can be shown by regarding the diffusive particles as heavily damped Brownian particles whose mobility can approximately be expressed in terms of the density $\rho(x)$ [9].

3. Results and Discussion

We have solved (4) - (6) along the lines described previously [1,2]. The results depend on the commensurability between two characteristic lengths in the system. A commensurate situation occurs if the position y_0 of the main peak in the pair correlation function $g_0(y)$ of the corresponding homogeneous system coincides with the lattice constant a. Then the pair forces tend to lock the particles inside the potential well, and the barrier height U_0 of the effective potential $U(x)$ is enhanced.

There are two limits where a commensurate situation can be identified by geometrical considerations only. In the Frenkel-Kontorova limit (see section 2) U_0 assumes maxima as $1/\rho_0 = a$ (or multiples of a)[10]. For an ideal hard-core potential ($n = \infty$) with diameter b_0 this condition is replaced by $b_0 = a$. Consequently, in our hard-core-like case $n = 6$ with $b_0 \simeq a$ we expect the barrier height U_0 to increase with increasing ρ_0. The case $b_0 = 2/3\,a$, however, shows a non-monotonous behaviour of U_0 as a function ρ_0, as can be

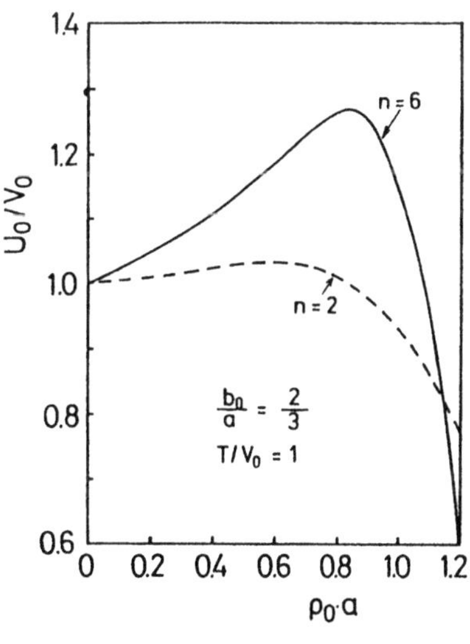

Fig.1 Effective barrier height U_0 as a function of the average density ρ_0 for a hard-core-like (n=6) and a soft-core (n=2) pair-potential.

seen from Fig. 1 . It turns out that the maximum in U_0 can be related to a maximum in $g_0(y=a)$ near $\rho_0 a \simeq 0.85$. On the other hand, for the soft-core potential (n=2) $g_0(y)$ and y_0 depend strongly both on density and temperature. With increasing ρ_0 the position of the main peak in g_0 passes through a near $\rho_0 a \simeq 0.65$, which leads to the weak maximum in the dashed curve of Fig. 1 . This effect becomes more pronounced at lower temperatures [2].

We have also calculated the pair correlations g(0,y) and g(a/2, a/2 + y), which, apart from a factor $\rho(y)$, give the probability density of finding a particle at distance y from a particle at the bottom and the top of the potential well, respectively. For cases where $U_0>V_0$, the structure in g(a/2, a/2 + y) is more pronounced than in g(0,y). In incommensurate cases, where $U_0<V_0$, the two functions do not differ very much from $g_0(y)$.

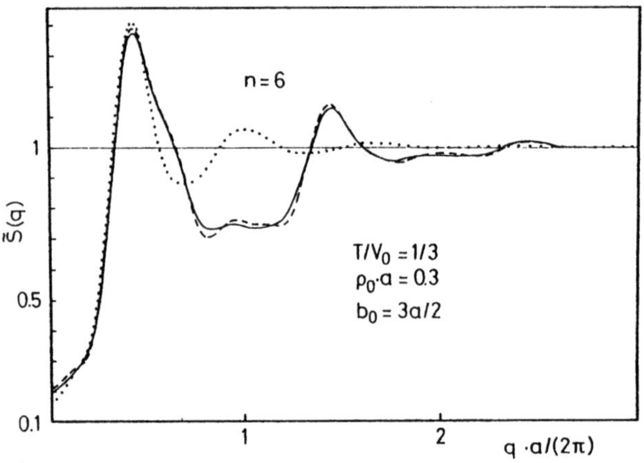

Fig.2 Full curve: numerical result for the static structure factor $\tilde{S}(q)$; dashed curve: neglection of $\Delta\tilde{S}(q)$; dotted curve: $S_0(q)$.

An example for the static structure factor is shown in Fig.2. For the parameters chosen it turned out that g(x,x') and $g_0(x-x')$ differ only by about 10 percent. Consequently, according to (10) $\Delta\tilde{S}(q)$ is small and the exact result (full curve) is well produced by a superposition of functions $S_0(q \pm nq_0)$ (dashed curve), as discussed in section 2. $S_0(q)$ is shown by the dotted curve. Besides the main peak in S(q) near $\bar{q} \simeq 0.4(2\pi/a)$ additional peaks at $q=q_0 \pm \bar{q}$, and, less pronounced, at $q = 2q_0 + \bar{q}$ are observed. Furthermore, there appears a weak maximum corresponding to the second peak in $S_0(q)$.

The general features described do not depend in an essential way on the parameters of the pair potential. In the majority of the cases studied, $\Delta\tilde{S}(q)$ turned out to be a small correction only so that S(q) can be described in terms of $S_0(q \pm nq_0)$.

References

1. A. Bunde, W. Dieterich: Solid State Commun. 37, 229 (1981)
2. A. Bunde: Submitted to Z. Physik B
3. J.K. Percus in: The Equilibrium Theory of Classical Fluids, H.L. Frisch and J.L. Lebowitz (eds.), New York: Benjamin (1964)

4. J.P. Hansen, I.R. Mc Donald in: Theory of Simple Liquids, London: Academic Press (1976)
5. R. Clarke, N. Caswell, S.A. Solin, P.M. Horn: Phys. Rev. Lett. 43 , 2018 (1979)
6. H. Zabel, S.C. Moss, N. Caswell, S.A. Solin: Phys. Rev. Lett. 43, 2022 (1979)
7. Y. Tsuchiya, S. Tamaki, Y. Waseda, J.M. Toguri: J. Physics C 11, 651 (1978)
8. Y. Tsuchiya, S. Tamaki, Y. Waseda: J. Physics C 12, 5361 (1979)
9. A. R. Bishop, W. Dieterich, I. Peschel: Z. Physik B 33, 187 (1979)
10. T. Geisel: Phys. Rev. B 20, 4294 (1979)

Effect of Relaxation on some Thermodynamic Properties of Lattice Gas Models with Applications to $AgCrS_2$ and Hollandite

P. Brüesch, T. Hibma, L. Pietronero and S. Strässler
Brown Boveri Research Center
CH-5405 Baden-Dättwil, Switzerland

1. Introduction

Recently, various aspects of interacting particles moving in a periodic substrate potential have been investigated. Many papers have considered the commensurate-incommensurate transition within various restricted assumptions. (Restriction to T=0 [1-3], continuum approximation [3-5], low temperatures [6-7]). Studies of the order-disorder regime have been restricted mainly to one dimensional models [8-10]. Here we want to discuss some aspects of this problem for one and two dimensional models. In order to discuss the order-disorder regime, we have chosen to relate the original Hamiltonian to an effective Cluster Hamiltonian. Then, standard methods can be applied. In the following we show how this can be achieved for two specific examples.

2. A one dimensional model

We consider first a one dimensional system of harmonically interacting particles of mass m in a periodic potential (with periodicity a). Its energy is written as (periodic boundary conditions)

$$H = \sum_{\ell=1}^{N} V(x_\ell) + \frac{1}{2} A \sum_{\ell=1}^{N} (x_{\ell+1} - x_\ell - b)^2 \quad . \tag{2.1}$$

Here x_ℓ denotes the position of the ℓ-th ion. A is the force constant. The substrate potential is taken as a piecewise parabolic potential:

$$V(x) = \begin{cases} V(x+a) \\ \frac{1}{2} m\omega_o^2 x^2 & |x| < \frac{1}{2} a \, , \end{cases} \tag{2.2}$$

b is the equilibrium distance in the absence of a substrate potential. ω_o is the vibrational frequency. The density of particles is given by

$$\rho = a/b = N/N_s \, , \tag{2.3}$$

where N_S is the number of pots defined by the substrate potential and N is the number of particles. The hard-core aspect of the particles can be taken into account approximately by allowing one particle per pot only. A particular configuration can then be specified by

$$\alpha = \{p_\ell \quad , \quad \ell = 1,N\} \quad , \tag{2.4}$$

where

$$p_\ell = 0,1,2,\ldots \tag{2.5}$$

denotes the number of empty pots which follow the ℓth ion.

The energy is now split up into a configurational part H_C and a displacement contribution H_P. H_C and H_P are defined through the condition

$$\frac{\partial H}{\partial x_{\ell,\beta}} = 0, \quad \ell = 1, \quad N \quad , \quad \beta = x,y,z \; . \tag{2.6}$$

It has been shown [7,8] that the configurational part can be written as

$$H_C = J \sum_{\ell<\ell} e^{-|\ell-\ell'|/\lambda} p_\ell p_{\ell'} \tag{2.7}$$

where

$$J = A\, a^2/S \quad , \tag{2.8}$$

$$S = (1 + \frac{4A}{m\omega_o^2}) \tag{2.9}$$

and

$$\lambda^{-1} = \log \frac{S+1}{S-1} \quad . \tag{2.10}$$

The displacement part becomes

$$H_P = \frac{1}{2} m\omega_o^2 \sum_{\ell=1}^{N} u_\ell^2 + A \sum_{\ell}^{N} (u_{\ell+1} - u_\ell) \quad , \tag{2.11}$$

where

$$u = x_\ell - x_\ell^{(\alpha)} \quad . \tag{2.12}$$

Here $x_\ell^{(\alpha)}$ are the metastable equilibrium position given by

$$x_\ell^{(\alpha)} = a \frac{\gamma}{1+\gamma} \sum_{\ell=1}^{N} \gamma^n \left[p_{\ell+n} - p_{\ell-n-1}\right] \tag{2.13}$$

where

$$\gamma = \frac{S-1}{S+1} \quad . \tag{2.14}$$

One should note that a metastable configuration α is defined only if

$$|x_\ell^o| \leq \frac{1}{2} a \qquad \ell = 1,N \quad . \tag{2.15}$$

3. A two dimensional model, relation to cluster Hamiltonian

We consider a hexagonal lattice with lattice constant a whose site vectors are $\vec{a}_i$. Site occupation numbers are denoted by $s_i=0,1$. The energy of the system is in analogy to (2.1) given by

$$H = \sum_{i=1}^{N} V(|\vec{r}_i-\vec{a}_i|)s_i + \sum_{(ij)} J(|\vec{r}_j-\vec{r}_i|)s_i\, s_j \quad . \tag{3.1}$$

Here J(r) represents a two body interaction between the particles i,j. $\vec{r}_i$ refers to the position of the particle in pot i (no double occupancy is allowed). V represents the periodic sublattice potential which in analogy to (2.2) we choose in the form

$$V(u) = \frac{1}{2}m\,\omega_o^2\, u^2 \qquad \vec{u} \in \text{cell} \quad . \tag{3.2}$$

In contrast to the one dimensional model, the configurational part of H can no longer be represented in a closed form. If the interaction J is not too strong, an expansion is possible:

$$H_c = \sum_{(ij)} J_{ij}\, s_i s_j + \sum_{(ij\ell)} J_{ij\ell}\, s_i s_j s_\ell + \ldots \quad . \tag{3.3}$$

From the condition

$$\frac{\partial H}{\partial x_{i\alpha}} = 0 \qquad i = 1,N \qquad \ell = x,y,z \quad , \tag{3.4}$$

one obtains in the lowest order in J, keeping nearest neighbor terms only,

$$\vec{u}_i^o = \frac{J'(a)}{\omega_o^2} \sum_j \vec{n}_{ij}\, s_j \quad . \tag{3.5}$$

Introducing (3.5) into (3.1), one obtains

$$J_{ij} = J(a) - \frac{J'(a)}{\omega_o^2} + O(J^3) \quad , \tag{3.6}$$

$$J_{ij\ell} = \frac{1}{2}\,\frac{J'^2(a)}{\omega_o^2} \quad , \tag{3.7}$$

where (i,j) and (i,j,ℓ) refer to nearest neighbors. The expansion procedure breaks down if hard-core effects are important. This is the case if

$$2R > a \quad , \tag{3.8}$$

where R is the hard-core radius. In this case it may be appropriate to look for a representation with a density (ρ) dependent effective pair interaction. If we use a molecular field approximation, we can choose $J_{ij}(\rho)$ in (3.4) such as to give the same average energy for an additional particle as obtained from (3.1). The result is

$$J_{ij}(\rho) = (\frac{1}{3}(2\rho + \rho^2) + 1))\, E_2 \quad , \tag{3.9}$$

where

$$E_2 = \frac{1}{4}\, m\, \omega_o^2\, a^2\, (\frac{2R}{a} - 1)^2 \quad , \tag{3.10}$$

is the energy of two nearest neighbor particles.

4. Discussion and applications

In Refs. [7-10] it has been shown that the model of section 2 is able to explain various properties of hollandite: $K_{2x}Hg_xTi_{8-x}O_{16}$, $0.75 < x < 1$. In particular the x-ray scattering is given by

$$I(s) = \langle \sum_{\ell,\ell'} \exp|-2\pi i s(x_\ell^{(\alpha)} - x_{\ell'}^{(\alpha)})|\rangle \quad . \tag{4.1}$$

Here $x_\ell^{(\alpha)}$ is given by Eq. (2.13) and the average is taken with respect to the configurational Hamiltonian (2.7). The comparison with experimental data showed that the model very well reproduces the rather complex diffuse scattering and permits an unambiguous determination of the parameters ω_o and A.

The excitation spectrum of the configurational Hamiltonian (2.7) is given by [10]

$$\Delta_n = J\gamma^{3n+1}(\gamma + 1/\gamma - 2) \qquad n = 0,1,\ldots \quad , \tag{4.2}$$

where J and γ are given by Eqs. (2.7) and (2.14), respectively. From the x-ray analysis of [8] it has been determined that $\alpha \sim 0.34$, $J \approx 0.7$eV. This results in $\Delta_1 \sim 150^oK$, $\Delta_2 \approx 6^oK$, $\Delta_3 = 0.23K$, etc. The specific heat is particularly sensitive to this excitation spectrum. Approximately the specific heat can be written as [10]

$$C(T) = \sum_{n=1,..} f_n(\rho) \frac{\Delta_n^2}{(k_BT)^2} \frac{e^{\Delta_n/k_BT}}{1 + e^{\Delta_n/k_BT}} \quad . \tag{4.3}$$

The weight functions $f_n(\rho)$ reflects (ρ density of holes) strongly the commensurability of the system [10].

A particularly interesting example for the application of the two dimensional model is the layered compound $AgCrS_2$ [11]. The silver ions occupy half of the tetrahedrally coordinated sites of a puckered honeycomb sublattice. From x-ray data one finds the first peak in the pair correlation function at 2.05Å. This value is larger than the nearest neighbor site distance (2Å). Information about the framework potential has been obtained from the FIR spectrum [11]. A very low frequency optical phonon mode is observed at 30cm^{-1} at room temperature. In this mode the silver ions are more parallel to the layers in a direction opposite to the chromium and sulfur ions.

The frequency of this mode decreases linearly with temperature from 34 cm^{-1} at 138°K till 22 cm^{-1} at 720°K. The silver content of the material can be reduced to a considerable extent by substituting part of the Cr^{3+} ions by ions having about the same size, but a different valency, such as Ti^{4+}, Sn^{4+}. The most prominent effect of such a change in silver concentration (ρ) is on the phase diagram. For $\rho = 1/2$ a second-order phase transition to a two sublattice honeycomb structure occurs at $T_c = 673°K$. The transition temperature shifts according to [11]

$$\frac{1}{T_c}\frac{dT_c}{d\rho} \simeq 3 \quad . \tag{4.8}$$

If we use the theoretical model represented by Eq.(3.10) and (3.11), the mean field transition temperature is given by ($\rho = \frac{1}{2}$)

$$T_c = \frac{1}{k_B}\frac{17}{32} m\, \omega_o^2\, a^2 \left(\frac{2R}{a} - 1\right)^2 \quad , \tag{4.9}$$

where the values of the various parameters are: $m = 1.07M_A$, $\omega_o = 0.41*10^{+13}$ sec^{-1}, $a = 2$Å, $2R = 3$Å.

We then obtain from (4.9)

$$T_c \approx 1200°K \quad . \tag{4.10}$$

For the concentration dependence (3.9) predicts

$$\frac{1}{T_c}\frac{dT_c}{d\rho} = \frac{12}{17} \quad . \tag{4.11}$$

The rather unsatisfactory agreement with respect to the theoretical values are due to (i) molecular field approximations and (ii) the too simplified model. From preliminary calculation it seems that going beyond the molecular field approximation does not significantly improve the results.

The excitation spectrum at low energies is of the same complexity as for the one dimensional system. The longer range forces can always be represented by an expansion as given in (3.4). As was the case for the one dimensional system, we expect no linear specific heat but some specific structure for the ideal system. The linear specific heat is caused by addition inhomogeneities.

References

1. P.Bak, D.Mukamel, J.Villain, and K.Wentowska, Phys.Rev. B19, 1610 (1979).
2. H.Shiba, J.Phys.Soc.Jpn. 46, 1852 (1979), and 48, 211 (1980).
3. J.Villain, in Ordering in Strongly Fluctuating Condensed Matter Systems, edited by T.Riste (Plenum, New York, 1980), p. 221, J. Phys. (Paris), Ltt. 44, L267 (1980), and Surf.Sci 97, 219 (1980).
4. V.L.Pokrovsky and A.L.Talapov, Phys.Rev.Lett. 42, 65 (1979), and Zh.Eksp.Theor.Fiz. 78, 269 (1980) [Sov.Phys.JETP 51, 134 (1980)].
5. H.J.Schulz, Phys.Rev.B22, 5274 (1980).
6. H.J.Schulz, Phys.Rev.Lett. 46, 1685 (1981).
7. L.Pietronero and S.Strässler, Phys.Rev.Lett.42, 188 (1979).
8. H.U.Beyeler, L.Pietronero, and S.Strässler, Phys.Rev.B22, 2988 (1980).
9. H.v.Löhneysen, H.J.Schink, W.Arnold, H.U.Beyeler, L.Pietronero and S.Strässler, Phys.Rev.Lett. 46, 1213 (1981).
10. L.Pietronero, W.R.Schneider and S.Strässler, Phys.Rev. B24, July 15 (1981).
11. P.Brüesch, W.Bührer, T.Hibma and S.Strässler, to be published.

Melting and Nucleation of a Two-Dimensional Electron Solid

R.K. Kalia and P. Vashishta

Argonne National Laboratory
Argonne, IL 60439, USA

Abstract

The method of molecular dynamics has been used to study dynamical and structural properties of a two-dimensional electron system. In the temperature variations of internal energy and self-diffusion constant there are evidences of hysteresis as well as superheated solid and supercooled liquid. These features suggest that the 2-D electron solid melts via first order transition; the melting occurs between Γ = 126 - 132, and the change in entropy on melting is 0.3 k_B per particle. The homogeneous nucleation of a solid from a supercooled liquid has also been studied. Results on the dispersion of longitudinal excitation obtained from dynamic structure factors explain the occurrence of well-defined oscillations in the velocity auto-correlation function in terms of a coupling between single particle motion and longitudinal excitation of the wave vector on the order of inverse interparticle separation.

Introduction

It has been known for nearly fifty years that two dimensional (2-D) solids lack long-range positional order.[1, 2] In 1935 Peierls[1] had shown that the mean-square displacement of atoms in 2-D solids is proportional to the logarithm of the size of the system; the larger the system, the larger is the mean-square displacement. In 1968 Mermin[3] demonstrated that there is another kind of long-range order, namely, the orientational order which does exist in 2-D solids. Kosterlitz and Thouless,[4-6] in 1972, proposed a theory of melting for the 2-D solids. According to this theory, two-dimensional systems have bound pairs of dislocations in the solid phase. With increase in temperature the system stays in the solid phase until it develops free dislocations, whose presence leads to a viscous flow. Around 1972 Feynman[7] had independently arrived at the same conclusion regarding the mechanism of melting for 2-D solids. In 1978, Halperin and Nelson,[8, 9] and Young[10] suggested a very interesting theory according to which the melting of 2-D solids as envisioned by Kosterlitz and Thouless may drive the system into an intermediate liquid crystal-like phase instead of an isotropic liquid. If the solid from which the liquid crystal phase emerges on heating has a hexagonal symmetry, the corresponding liquid crystal phase is called a "hexatic" phase, whereas if the solid has a square symmetry, the liquid crystal phase is called a "tetratic" phase.[9] In the liquid crystal phase, Halperin and Nelson show that there are weakly bound pairs of disclinations which dissociate on heating at a certain temperature T_i. The phase that results from the heating of the liquid crystal phase is an isotropic liquid. Schematically, the transition from a solid to an isotropic liquid is shown in Fig.1; at T_m the solid undergoes a continuous transition into a liquid crystal phase by dissocation of dislocation pairs,

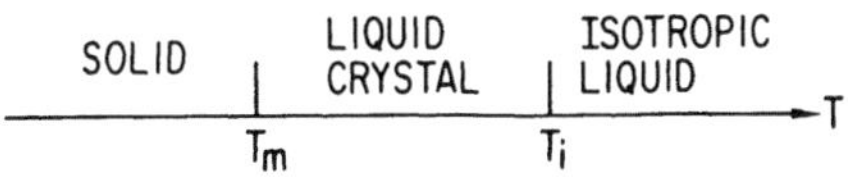

Figure 1: The sequence of transitions in the melting process of two-dimensional solids, according to the theory of Halperin and Nelson (Ref. 8).

and at a higher temperature T_i the latter undergoes another continuous phase transition by dissociation of disclinations leading to an isotropic liquid. According to Halperin and Nelson the two continuous transitions with an intervening liquid crystal phase may be pre-empted by a first-order transition.

One of the simplest two-dimensional systems in nature is a monolayer of electrons on the surface of liquid Helium.(11) The electrons are bound perpendicular to the surface of liquid Helium by their image forces, while along the surface they can move freely. Typically, the surface electron density is between 10^5-10^9 cm^{-2}, which corresponds to a Fermi energy in the range of 10^{-5}-10^{-1} K. This is much less than the liquid He temperature around which experiments are carried out, and the surface electrons can therefore be regarded as a classical system. Because electrons interact via Coulomb interaction $1/\vec{r}$ ($\vec{r}$ being a 2-D vector), the properties of this classical system can be characterized by a single dimensionless variable, Γ, which is a ratio of potential to kinetic energy:

$$\Gamma = \frac{e^2/r_o}{k_B T}$$

where $\pi r_o^2 = \rho^{-1}$, ρ being the number of electrons per unit area.

Grimes and Adams(12) in 1978 showed that the 2-D electrons on the surface of liquid He form a hexagonal solid at sufficiently low temperatures. In terms of the dimensionless variable Γ, they find that the electron lattice melts in the range $\Gamma_m = 131 \pm 7$.

Thouless(13) has estimated Γ_m using the dislocation mediated melting theory of Kosterlitz and Thouless,(4) and Feynman.(7) The energy and configuration entropy of a single isolated dislocation are given by

$$E = \frac{b^2}{4\pi}\,\frac{\mu(\mu+\lambda)}{2\mu+\lambda}\,\ln\left(\frac{A}{A_o}\right) \qquad A_o \gtrsim a^2 , \tag{1}$$

$$S = k_B \ln\left(\frac{A}{a^2}\right), \tag{2}$$

where μ and λ are the Lamé coefficients for a 2-D hexagonal solid, A the area of the system, a the lattice spacing and $\vec{b}$ is the Burger's vector of a free dislocation. At temperatures where E dominates TS the system will not have a free dislocation because its energy increases logarithmically with the area of the system. With increase in temperature the entropy term becomes increasingly more important and when T exceeds the melting temperature T_m, where

$$T_m = \frac{b^2}{4\pi k_B}\,\frac{\mu(\mu+\lambda)}{2\mu+\lambda} , \tag{3}$$

Table 1. A summary of computer calculations and the experimental result for a 2-D electron system.

	Method	Remarks	Γ_m
Hockney and Brown[14]	MD	Melting transition is second order.	95±2
Thouless[13]	K-T Theory	Melting is due to dissociation of dislocations.	79
Totsuji[15]	MC	Investigated static properties in the liquid phase.	-
Grimes and Adams[12]	Expt.	Study of melting transition.	131±7*
Gann et. al.[17]	MC	Study of thermodynamic properties in the solid and liquid phases.	125±15
Morf[18]	MD	Investigated the temperature dependence of shear modulus.	120-140
Hansen, Levesque and Weis[24]	MD	Studied the single particle motion in the liquid phase with and without a magnetic field.	-
Kalia, Vashishta, and deLeeuw[19]	MD	Study of structural and dynamical properties in liquid and solid phases. Melting transition is first order.	126-132
Totsuji and Kakeya[16]	MD	Dynamic structure factor and transverse current correlation function are studied in the liquid phase.	-
Kalia, Vashishta, deLeeuw, and Rahman[26]	MD	Calculation of velocity auto-correlation function and dynamic structure factor in the liquid phase.	-

*In Ref. 12, Grimes and Adams reported that Γ_m = 137 ± 15. But the value reported here is a later estimate; see, e.g., D. S. Fisher, in the proceedings of the conference on "Ordering in Two Dimensions", ed. S. K. Sinha, North Holland, 1981, p. 189.

it is possible to have a free dislocation in the system. Since the Burger's vector connects particles in their equilibrium positions, Thouless[13] has taken $|\vec{b}| = a$. Using the zero temperature values for μ and λ, namely,

$$\lambda = \infty, \; \mu = 0.245065 \; e^2\rho^{3/2}, \text{ and } \rho = \frac{2}{\sqrt{3}a^2} ,$$

Thouless has estimated that

$$\Gamma_m \cong 79. \tag{4}$$

<u>Molecular Dynamics (MD) and Monte Carlo (MC) Calculations</u>

The methods of molecular dynamics and Monte Carlo are very suitable for studying the properties of the classical system of electrons. Using MD method, Hockney and Brown[14] were first to study the melting transition in the electron system. The authors have investigated the melting of a square lattice of 10^4 electrons, and they find that the melting transition is second order and it occurs for $\Gamma_m = 95 \pm 2$. It should be pointed out that a square lattice is dynamically unstable and also it seems that Hockney and Brown did not give sufficient time for the system to acquire equilibrium.

A couple of years after the work of Hockney and Brown, Totsuji[15] used the MC method to study the thermodynamic properties of 2-D electrons in the liquid phase, away from the region of melting. Recently, Totsuji and Kakeya[16] have reported MD calculations for the dynamic structure factor and transverse current correlation functions in the liquid phase.

Gann, Chakravarty, and Chester[17] have studied the melting of the hexagonal electron lattice using the Monte Carlo method. These authors find that the transition is perhaps weakly first order and it occurs in the range of $\Gamma_m \lesssim 125 \pm 15$. Morf[18] has used the MD method to calculate the temperature dependence of the shear modulus μ. From the dependence of μ on T, Morf estimates Γ_m from Eq.(3) to be in the range of 120-140. Table 1 provides a summary of computer calculations and experiment on the 2-D electron system.

Recently we have used the molecular dynamics method to investigate structural and dynamical properties of electrons in the liquid and solid phases.[19, 20] The calculations are performed for N=100, 256, and 576 electrons with a rigid neutralizing, uniform, positive background in a rectangle of area A with periodic boundary conditions. The sides of the rectangle have a ratio of $\sqrt{3}/2$ so that it can accomodate $4M^2$ (M is an integer) electrons on a perfect hexagonal lattice. The calculations are carried out at a fixed density $\rho = 1.477 \times 10^8$ cm^{-2} and various temperatures in the range of $0.2 < T < 1$ K, which corresponds to $36 < \Gamma < 180$. At $\Gamma = 36$ the system is a hot liquid while at $\Gamma = 180$ it is a low-temperature hexagonal solid. The evaluation of forces and potential energy requires special care because of the long-range nature of the $1/r$ interaction. This is carried out by using Ewald's summation for forces and the potential energy. The Newton's equations of motion for electrons are solved using a predictor-corrector algorithm.[21] In the integration of equations of motion, the time step is chosen so that the energy is conserved to at least 1 part in 5×10^4 even after 15,000 time steps. The system is thermalized for several thousand time steps at each temperature and the results reported here are based on long (15,000 time steps) MD calculations after thermalization. When the system is close to the melting region, the MD runs extending up to 40,000 time steps were carried out.

Results

The temperature dependence of internal energy, E, is shown in Fig.2. The system in the solid phase is represented by squares and in the liquid phase by triangles. The points corresponding to the solid phase are obtained by monotonically heating a low-temperature hexagonal lattice, while the liquid states for the system are obtained by monotonically cooling from a hot liquid at $\Gamma = 36$. The distinction between the solid and liquid phases is evident from the behaviour of self-diffusion coefficient D; in the solid phase D is zero, while in the liquid phase it is finite. At $\Gamma = 36$, D is around 5 cm^2/sec and on cooling D decreases continously up to the state marked L_2 where D is around 0.5 cm^2/sec. The state L_2 is the last supercooled liquid and on further cooling the system comes to the state marked S_2 where D = 0. In the solid phase, S_1 is the last superheated state with D = 0. On heating, the state at S_1 melts and goes into the state marked L_1 where D acquires a value of about 0.7 cm^2/sec. The region between the vertical dotted lines is thus the hysteresis region having both supercooled liquid and superheated solid phases. To further illustrate the

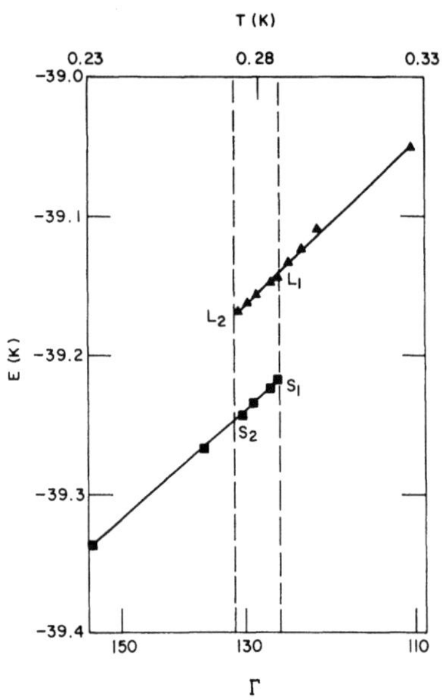

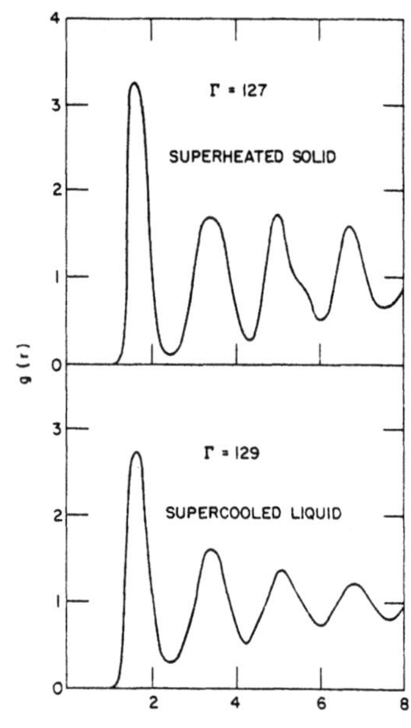

Figure 2: The dependence of internal energy E of the 2-D electron system on the dimensionless variable Γ and temperature T.

Figure 3: Pair-correlation function, g(r), for a superheated electron solid and supercooled liquid. In the liquid, g(r) is smoother but less sharp than in the solid.

difference between a supercooled liquid and a superheated solid, we show in Fig.3 the pair-correlation function, which in the superheated solid is sharper than in the supercooled liquid even though the latter phase is at a lower temperature. In addition, g(r) has a distinct hump between the third and fourth peaks in the superheated solid, whereas in the supercooled liquid g(r) is smooth.

As for the nature of the phase transition, the presence of hysteresis, supercooling, and superheating found in the temperature variation of internal energy are strong indicators of a first-order transition. The latent heat of melting, L, which is the difference in energy of the liquid at L_1 and superheated solid at S_1 is 0.09 K. Taking the transition temperature T_m to be in the middle of the hysteresis region of Fig.2, we find that the transition entropy per particle $\Delta S = L/T_m = 0.3\ k_B$. The results we have presented above are independent of the size of the system; for N = 100, 256, and 576 particle systems, we find supercooling, superheating, release of latent heat on melting and a transition entropy of $0.3\ k_B$. It is therefore clear that for all three sizes of the system the melting transition of a 2-D electron solid is first order. Our calculations for the 576 particle system show that the melting occurs in the range of Γ_m = 126 -132.

The formation of a solid from a supercooled liquid constitutes a very interesting study of the nature of the freezing process. Taking a 576 particle system, we cool it from a supercooled liquid, at time t = 0, by lowering the energy by an amount slightly greater than the latent heat. The time dependence of instantaneous temperature $T(t)$ $\left(= \frac{1}{2Nk_B} \sum_{i=1}^{N} m\left|\vec{V}_i(t)\right|^2\right)$ and mean-square displacement are then monitored over a long period of time. During the first several thousand time steps, T(t) fluctuates around a mean temperature T_1. The temperature T_1 and the corresponding energy, say E, lie above the lower line for the solid phase in Fig.2. Subsequently, in a relatively small interval of time ($\simeq$ 1000 time steps), T(t) rises and the temperature fluctuates around a value of T_2. The system shows zero diffusion in this state. Since our MD calculation is done in a microcanonical ensemble, the energy E stays constant with time. The system with energy E and temperature T_2 lies on the lower line corresponding to the solid phase in Fig.2. We have monitored the system for many thousand time steps after it acquires the mean temperature T_2; the state of the system remains unchanged. Such a behaviour has been seen in other systems which undergo first-order phase transition, for example in the MD work of Hsu and Rahman(22) on liquid rubidium, and in the MC calculation of Creutz, Jacobs, and Rebbi(23) on a gauge invariant Ising system.

In 1979, Hansen, Levesque, and Weis (24) reported an interesting MD calculation for 2-D electrons in the liquid phase at Γ = 36 and 90. Instead of placing electrons on a plane, Hansen et al. performed the calculation on the surface of a sphere of sufficiently large radius, thereby avoiding periodic boundary conditions and Ewald summation. The authors found a series of well-defined oscillations in the velocity auto-correlation function (vaf), defined by

$$Z(t) = \frac{\left\langle \sum_i \vec{V}_i(t)\cdot\vec{V}_i(0)\right\rangle}{\left\langle \sum_i \vec{V}_i(0)\cdot\vec{V}_i(0)\right\rangle} . \qquad (5)$$

The oscillations in $Z(t)$ have a period of 5.2τ where τ (= $(mr_0^3/e^2)^{1/2}$) is a temperature-independent unit of time for the electron system. The corresponding oscillatory frequency is $1.2\ \tau^{-1}$, and it is independent of the value of Γ. Such oscillations in vaf which are independent of Γ were earlier noticed by Hansen et al.(25) in the case of a 3-D system of electrons. In the 3-D system the oscillatory frequency in vaf is very close to ω_p, the long wavelength plasmon frequency, $(4\pi\rho e^2/m)^{1/2}$, and therefore the occurrence of the oscillations in 3-D is attributed to a coupling between the single particle motion and the long wavelength plasmons.

A similar explanation will not hold in the 2-D electron liquid because the dispersion of 2-D plasmons is of the form $\omega_p = (2\pi\rho e^2 k/m)^{1/2}$, and therefore ω_p is zero at $k = 0$. In order to investigate the nature of the coupling between the single particle motion and longitudinal collective excitation, we calculated the vaf and dynamic structure factor, $S(\mathbf{k},\omega)$, at $\Gamma = 36$ and 90 from our MD data.(26) Figs. 4 and 5 show the time dependence of vaf and the dispersion of longitudinal excitations from $S(\mathbf{k},\omega)$. Because the dispersion at $\Gamma = 36$ and 90 are very close to each other, the vertical dotted lines in Fig.5 entail both the dispersions and are indicators of peak positions in $S(\mathbf{k},\omega)$; full widths at half maxima are also shown. The dispersion of plasmons, $\omega_p = (2\pi\rho e^2 k/m)^{1/2}$ is shown by the dashed curve. Note, the period of oscillations in vaf (= $5.2\ \tau$) is the same at $\Gamma = 36$ and 90. From Fig.5 it is also evident that the longitudinal

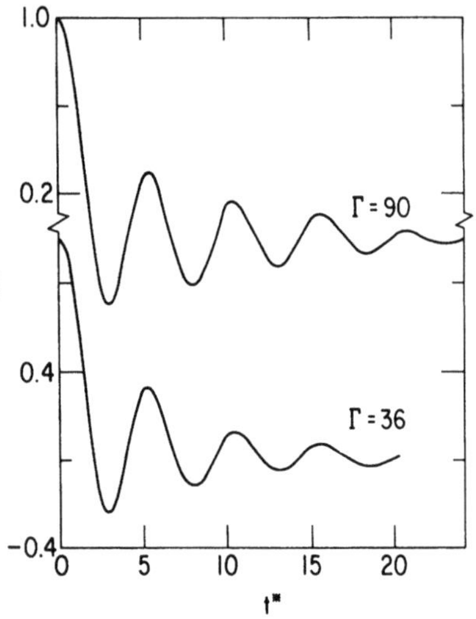

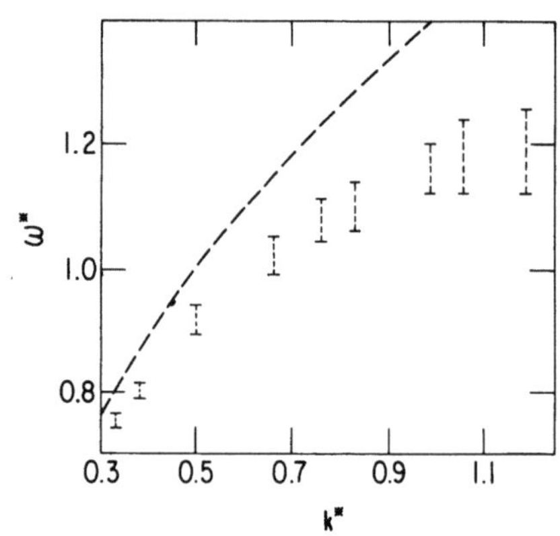

Figure 4: Velocity auto-correlation function (vaf) in liquid phases at $\Gamma = 36$ and 90 as a function of reduced time t^* (= t/τ, where $\tau = (mr_0^3/e^2)^{1/2}$). Note, the period of oscillations in vaf is independent of Γ.

Figure 5: Dispersion of longitudinal excitations (vertical dotted lines) obtained from dynamic structure factor, $S(\mathbf{k},\omega)$. The broken curve denotes the dispersion of long wavelength plasmons, $\omega_p = (2\pi\rho e^2 k/m)^{1/2}$. The frequencies ω^* are measured in unit of time τ and the reduced wave vector $k^* = kr_0$.

excitation tends to become constant when $kr_0 \sim 1$. The frequency of excitation around $kr_0 \sim 1$ is 1.2 τ^{-1}, which is the same as the frequency of oscillations in vaf. Hence, the coupling between the single particle motion and the collective modes is appreciable only for $kr_0 \sim 1$.

References

1. R. E. Peierls, Ann. Inst. Henri Poincare 5, 177 (1935).
2. L. D. Landau, Phys. Z. Sowjetunion II, 26 (1937).
3. N. D. Mermin, Phys. Rev. 176, 250 (1968).
4. J. M. Kosterlitz and D. J. Thouless, J. Phys. C5, L124 (1972).
5. J. M. Kosterlitz and D. J. Thouless, J. Phys. C6, 1181 (1973).
6. J. M. Kosterlitz and D. J. Thouless, Progress in Low Temperature Physics, ed. D. F. Brewer (North-Holland, Amsterdam, 1978), vol. VIII B, Chap. V.
7. R. P. Feynman, quoted by R. L. Elgin and D. L. Goodstein in Monolayer and Submonolayer Helium Films (J. G. Daunt and E. Lerner, eds.), Plenum Press, New York, 1973.
8. B. I. Halperin and D. R. Nelson, Phys. Rev. Lett. 41, 121 (1978).
9. D. R. Nelson and B. I. Halperin, Phys. Rev. B19, 2457 (1979).
10. A. P. Young, Phys. Rev. B19, 1855 (1979).
11. C. C. Grimes, Surf. Sci. 73, 379 (1978).
12. C. C. Grimes and G. Adams, Phys. Rev. Lett. 42, 795 (1979); see also D. S. Fisher, B. I. Halperin, and P. M. Platzman, Phys. Rev. Lett. 42, 798 (1978).
13. D. J. Thouless, J. Phys. C11, L189 (1978).
14. R. W. Hockney and T. R. Brown, J. Phys. C8, 1813 (1975).
15. H. Totsuji, Phys. Rev. A17, 399 (1978).
16. H. Totsuji and H. Kakeya, Phys. Rev. A22, 1220 (1980).
17. R. C. Gann, S. Chakravarty, and G. V. Chester, Phys. Rev. B20, 326 (1979).
18. R. H. Morf, Phys. Rev. Lett. 43, 931 (1979).
19. R. K. Kalia, P. Vashishta, and S. W. deLeeuw, Phys. Rev. B23, 4794 (1981).
20. R. K. Kalia and P. Vashishta, to be published in Nato Advanced Study Institute on Nonlinear Phenomena at Phase Transition and Instabilities, Geilo, Norway, March 29-April 9, 1981.
21. C. W. Gear, ANL Report #7126, Argonne National Laboratory (1966); Numerical Initial Value Problems in Ordinary Differential Equations, Prentice Hall, Englewood Cliffs, N.J. (1971).
22. C. S. Hsu and A. Rahman, J. Chem. Phys. 70, 5234 (1979).
23. M. Creutz, L. Jacobs, and C. Rebbi, Phys. Rev. Lett. 42, 1390 (1979).
24. J. P. Hansen, D. Levesque, and J. J. Weis, Phys. Rev. Lett. 43, 979 (1979).
25. J. P. Hansen, I. R. McDonald, and E. L. Pollock, Phys. Rev. A11, 1026 (1975).
26. R. K. Kalia, P. Vashishta, S. W. deLeeuw, and A. Rahman, to be published.

*Work supported by the U. S. Department of Energy.

Melting of the Two-Dimensional Electron Solid. A Kosterlitz Thouless Phase Transition?

R. Morf

Laboratories RCA Ltd.
CH-8048 Zürich, Switzerland

1. Introduction

Previously it was believed that melting is quite generally a first-order phase transition. Recently, however, HALPERIN and NELSON [1] developed a theory which leads to a higher-order melting transition in two-dimensional (2-d) systems. This theory is based on the idea, proposed by KOSTERLITZ and THOULESS [2], that melting is caused by an instability in the 2-d solid for the creation of free dislocations. It predicts an upper bound for the melting temperature T_m

$$k_B\, T_m \leqslant \frac{1}{4\pi}\, \frac{\mu(\mu+\lambda)}{2\mu+\lambda}\, a_o^2 \,, \tag{1}$$

where μ and λ are the Lamé coefficients measured at T_m and a_o is the lattice constant. If the solid does melt via the dislocation unbinding mechanism,the inequality becomes an equality and, in this case, striking predictions are made: At temperatures $T > T_m$ the system has lost its resistance to shear and translational order. However, orientational order persists up to a temperature $T_i > T_m$, at which a second phase transition associated with the unbinding or another type of defect (disclinations) destroys the orientational order. The intermediate phase is characterized by an orientational order parameter

$$\Psi_p(\vec{r}) = e^{ip\,\Theta(\vec{r})} \,, \tag{2}$$

where $\Theta(\vec{r})$ is the angle relative to an arbitrary reference direction of the bond between two nearest-neighbor particles at position $\vec{r}$. The constant p reflects the type of symmetry: for melting of triangular lattices p = 6 and for square lattices p = 4,corresponding to the persisting hexagonal or cubic symmetry of this phase for which the names "hexatic" and "tetratic" have been coined [1]. Defining the orientational correlation function

$$g_p(\vec{R}) = \langle \Psi_p(\vec{R})\, \Psi_p^*(o) \rangle \tag{3}$$

which, in the solid, for large separations R tends to a positive constant [3], Halperin and Nelson showed that the presence of free dislocations for temperatures $T > T_m$ leads to a power law decay

$$g_p(\vec{R}) \sim R^{-\eta_p} \tag{4}$$

with a temperature-dependent critical exponent η_p which increases monotonically with temperature from zero at T_m up to 1/4 at T_i. This is in marked contrast to the behavior of a conventional fluid for which all correlations decay exponentially at large distances.

The theory developed by HALPERIN and NELSON [1] is based on continuum elasticity theory, in which long-wavelength phonons, dislocations, and disclinations are taken into account. The quantities which determine the physical behavior are the Lamé coefficients μ and λ and the core energy of dislocations E_c and disclinations E_d, which plays the role of a chemical potential for these defects. The theory is based upon an expansion in the fugacity of dislocations $y = \exp - E_c/k_B T$ and similarly for disclinations. This expansion is well behaved if the core energies are large, i.e. for small defect densities. The core energies depend in a very nontrivial way on the interparticle potential [4]. If the fugacity of dislocations is not small at T_m, implying a high density of dislocations, a first-order phase transition from solid to isotropic fluid is likely to occur. Particularly in these cases, the possibility must be considered that special types of arrays of dislocations (e.g. grain boundaries) will drive the (first-order) transition [4, 5].

The striking predictions of the Halperin-Nelson theory - the existence of an orientationally ordered fluid (4) and the universal relation (1) between melting point and elastic constants - lend themselves to an experimental test. To summarize the present situation: No experimental evidence of an hexatic phase has been obtained so far in melting of monolayer systems [6]. Also attempts to test relation (1) in experiment have not been successful so far [7].

Computer simulations of several different systems have recently led to considerable controversy. While there is strong evidence that the hard disk system undergoes a first-order phase transition [8] directly from solid to isotropic fluid, the situation in systems with softer interactions (e.g. Lennard-Jones or inverse power r^{-n}) is not so clear. In 3d systems it has been found that the "first-orderness" [9] (measured e.g. by the change of volume ΔV at melting or by the entropy of melting ΔS) is decreasing as the interaction becomes softer (i.e. for decreasing n) and one would expect a similar trend for 2d systems [10, 11].

Earlier Monte Carlo and molecular dynamics simulations [10] focused mainly on thermodynamic properties (e.g. specific heat, Helmholtz free energy, equation of state) and located the phase transition by means of fits to high and low temperature data, respectively, which leads to a bias towards a first-order interpretation. Recently, a molecular dynamics simulation of the Lennard-Jones system by FRENKEL and McTAGUE [12] was analyzed in the spirit of the Halperin-Nelson theory. From the behaviour of the orientational correlation function $g_6(r)$ they concluded that there exist two distinct phase transitions at T_m and T_i with an hexatic phase in between. Subsequently, this interpretation was challenged by TOXVAERD [13], by ABRAHAM [14], and by VAN SWOL, WOODCOCK and CAPE [10], who suggested that owing to the fact that this simulation was done at constant density, the system actually underwent a phase separation and that the temperatures identified with T_m and T_i rather correspond to limits of stability of liquid and solid, respectively.

Another simulation of this system (constant density Monte Carlo (MC)) was carried out by TOBOCHNIK and CHESTER [15]. They suggest that the system shows a distinctly different behaviour depending on the density. At high density they find a first-order phase transition with elastic constants at the melting point considerably greater than the Kosterlitz-Thouless values (1) and a steep rise in the energy in the vicinity of T_m. At lower densities, however, they find elastic constants consistent with (1) as an equality and a much more gradual increase of energy.

Recently, McTAGUE, FRENKEL and ALLEN [16] studied the soft disk r^{-6}-system in a constant density MC simulation. Using a clever microscopic definition of disclinations and dislocations, they find evidence of a two stage dislocation - disclination unbinding process. However, the order of the transition could not be determined. In a recent paper, ABRAHAM [17] discusses the possibility that quite generally in computer simulations, melting may take place at the Kosterlitz-Thouless temperature as an upper limit for the stability of a metastable solid but that the thermodynamic melting temperature is lower but unobservable in a simulation unless surfaces are present.

In this work we discuss this controversial situation on the example of the 2-d electron system. This system has two major advantages: (i) It is experimentally accessible (electrons trapped at the surface of liquid Helium) and results from simulations can be compared with experiment. (ii) The system does not change area on melting [11] and thus no phase separation can be induced by simulation at constant density. Furthermore, like in all soft disk systems (with r^{-n} interactions), the thermodynamic state depends on one single parameter Γ which is for this system defined by

$$\Gamma = \frac{(\pi n_s)^{1/2} e^2}{k_B T} \qquad (5)$$

and measures the ratio between potential and kinetic energy. The areal electron density is denoted by n_s. We first summarize briefly the results of previous work on this system.

In an early molecular dynamics simulation HOCKNEY and BROWN [18] observed a λ-like anomaly in the specific heat at $\Gamma \approx 95$ and identified this with the melting point. This result was first cast in doubt by a MC simulation by GANN, CHAKRAVARTY and CHESTER [19] who did not find any anomaly at $\Gamma = 95$. However, they observed melting at $\Gamma_m = 125 \pm 15$ evidenced by the change in the behaviour of the positional fluctuation $\langle r^2 \rangle$ which is finite for $\Gamma > \Gamma_m$ and grows indefinitely for $\Gamma \lesssim 120$.

A minor anomaly was also found at Γ_m in the free energy. Using high- and low-temperature fits to the internal energy and calculating the free energy, the high-T and low-T branches intersect at $\Gamma \approx 125$ although with a minute difference in slope of 0.03 percent. Using a double-tangent construction, they also estimated the width $\Delta\Gamma$ of the two-phase region to be $\Delta\Gamma \approx 0.1$.

Since this type of analysis is biased toward a first-order interpretation, such a saml1 discontinuity cannot help to decide upon the order of the transition, although it is pointed out in [19] that in the thermodynamic quantities no evidence for divergences exists. This, however, is consistent with the Halperin-Nelson theory in which the specific heat and free energy have only an unmeasurable essential singularity [1].

An experimental value for the melting point Γ_m was obtained by GRIMES and ADAMS [20] in their experiments with electrons trapped at the surfaces of liquid helium. They found

$$\Gamma_m^{exp} \approx 131 \pm 7 , \tag{6}$$

consistent with [19]. At temperatures below the melting point ($\Gamma > \Gamma_m$) they found evidence that the system forms a triangular lattice [21].

A theoretical estimate for the melting point Γ_m was given by THOULESS [22]. Based upon the Kosterlitz-Thouless criterion (1) and using the exactly known T=0 values $\lambda = \infty$ and $\mu = 0.245065\, e^2 . n_s^{3/2}$ he obtained $\Gamma_m = 78.71$.

The large difference between this value and the experimental result (6) may cast doubt on the applicability of this theory. In fact, the shear modulus u would have to decrease by 40% from its T=0 value if (1) is to hold at the experimental melting point $\Gamma_m^{exp} = 131$.

A first attempt to test the Kosterlitz-Thouless criterion (1) was made by myself [23]. The temperature dependence of the shear modulus μ of the 2d-electron solid was investigated in a MD simulation and found to be consistent with predictions of the Halperin-Nelson theory (see Section 2). More recently, KALIA, VASHISHTA and LEEUW [24] reported results of a MD simulation which they interpret as evidence for a first-order melting transition (cf. the article by KALIA in this volume).

However, numerical simulations of 2d-melting are hampered by serious equilibration problems: While the high-temperature fluid state can be obtained from any initial particle configuration, the low-temperature (defect-free) solid state is not obtained upon cooling the melt, but generally grain boundaries and microcrystalline structures persist down to low temperatures at which they are not expected to exist in thermodynamic equilibrium. Consequently, it is not clear that one is able to simulate the melting transition as it occurs in thermal equilibrium. Moreover, even if the problem of equilibration can be overcome, the determination of the order of the transition remains a difficult task, particularly if this transition is at most weakly first order (e.g. near a tricritical point), as suggested by the results of ref. [19] discussed above. For the case of Ising spin systems this problem has been addressed recently by LANDAU and SWENDSEN [25].

In this work, we present new results of a MD simulation of the 2d-electron system which show the existence of premelting effects in the solid at temperatures significantly lower than those at which the fluid starts to show prefreezing effects. This behaviour differs from the one discussed by ABRAHAM [17] who relates the premelting effects observed in the 2d-Lennard-Jones system to the apparent "criticality" of a superheated solid.

In Section 3 we draw conclusions and discuss open problems.

2. Premelting and Prefreezing Effects

Our MD calculation is done on a system with 780 (= 26 x 30) electrons in a uniform positive background and subject to periodic boundary conditions. The forces and energy are calculated by means of the method described in ref. [4]. The summation over interactions between particles with separation less than a radius R_c is carried out exactly, while the contribution from particles with larger separations is calculated by means of a mesh method using Fast Fourier transforms.

For the present calculation we use $R_c = 1.83\, a_0$ and a mesh with 128 x 128 points which assures that the small errors in the calculated forces are incommensurate (to the extent possible in a finite system) with the lattice containing 26 x 30 particles. All periodic images of the particles are included in the computation by means of Ewald summation.

For the dynamic simulation we use a canonical procedure [26] in which each particle is coupled to a heat bath. The equations of motion for particle i at position $\vec{r}_i$ are

$$m\ddot{\vec{r}}_i = \vec{F}_i + \vec{\phi}_i(t) - m\gamma\dot{\vec{r}}_i, \tag{7}$$

where $\vec{F}_i$ is the Coulomb force and $\vec{\phi}_i = (\phi_i^{(1)}, \phi_i^{(2)})$ is a Gaussian white noise force representing the heat bath at temperature T and obeys the conditions $\langle\vec{\phi}_i\rangle = 0$ and $\langle\phi_i^{(k)}(t)\,\phi_j^{(\ell)}(t')\rangle = 2m\gamma\, k_BT\,\delta(t-t')\delta_{ij}\,\delta_{k\ell}$. The coupling to the heat bath is controlled by the parameter γ. It is used to heat or cool the system. A value of γ was chosen such that the associated relaxation time is about twice the period of the slowest transverse phonon mode in the system.

The numerical integration of (7) is done by means of the finite difference method of ref. [26] with a timestep $\Delta = 2\pi/20\omega_0$, where $\omega_0^2 = 8e^2/ma_0^3$. This time step corresponds to 1/16 of the period of the fastest longitudinal mode at T = 0, which guarantees sufficient numerical accuracy.

We first discuss the results of a simulation of the solid started at T=0 with a perfect triangular lattice. The shear modulus was determined from the frequencies $\omega_t(q)$ of the transverse phonons which for small wavenumbers q are given by

$$\omega_t^2(q) \approx \mu q^2/mn_s\ . \tag{8}$$

Details about the determination of ω_t are given in [23] . In Fig. 1 we show the shear modulus $\mu(\Gamma)/\mu(\Gamma=\infty)$ as a function of temperature (or Γ^{-1}) . We observe a reduction of μ which is linear in T up to a temperature corresponding to $\Gamma \approx 170$. It is the result of phonon-phonon interactions [23, 27] . A stronger decrease for $150 > \Gamma > 130$ is followed by a sharp drop of μ for $130 > \Gamma > 120$ accompanied by a sudden rise of particle diffusion at $\Gamma \approx 130$ (inset in Fig 1). This suggests a melting point $\Gamma_m^{MD} \approx 130$, consistent with the experimental value (6). In [23] the reduction of the shear modulus by the polarizability of dislocation pairs has been calculated by integrating the Halperin-Nelson-Young renormalization group equations starting with the (with respect to dislocations) "bare" values $\lambda = \infty$, $\mu(\Gamma)/\mu(\Gamma=\infty) = 1 - 30.8\,\Gamma^{-1}$ (dashed line in Fig. 1) and a core energy $E_c = 0.46\,\mu(\Gamma)\,a_o^2$ [4, 23] . The result is shown as a solid line in Fig 1. A melting point $\Gamma_m^{th} = 128.2$ is predicted, consistent with experiment. Also, the calculated reduction of μ is consistent with the MD results for $\Gamma > 130$, and we may interpret this as evidence for a premelting effect caused by the presence of dislocation pairs. We note that the starting value of the fugacity [23] $y(o) = \sqrt{\frac{4}{3}} \exp -E_c/k_BT_m$ at $\Gamma = 128$ is y(o) 0.00075 and thus the fugacity expansion should be reliable [1].

We now turn to the behavior of the high-temperature fluid phase. For this purpose a configuration obtained at $\Gamma = 126$ was used as a starting configuration. The system was then coupled to a heat bath at $\Gamma = 60$ ($T \approx 2.2\, T_m$) during 4000 time steps. Kinetic and potential energy showed no drift during the last 2000 time steps. The system is then cooled down again in 7 steps by coupling to a heat bath at Γ = (66, 74, 85, 91, 99, 108, 118). At each temperature the system was monitored for 1500 time steps. In order to look for prefreezing effects in the fluid we determine the susceptibility χ_6 of the orientational order parameter Ψ_6 (2), which according to the fluctuation-dissipation theorem we define by

$$\chi_6 = \frac{2\pi}{k_BT} \int_0^R (g_6(r) - g_6(\infty)) \frac{r dr}{A_c} \quad . \qquad (9).$$

Here, $A_c = n_s^{-1}$ is the area per particle and R is the radius of a circle containing N ($\approx$ 780) particles ($\pi R^2 = N \cdot A_c$). Since in the fluid phase there is no long-range orientational order, $g_6(\infty) = 0$.

If a hexatic phase exists, χ_6 is expected to diverge as the hexatic to isotropic phase transition at T_i is approached from high temperatures

$$\chi_6 \sim \exp a/(T-T_i)^{1/2} \quad . \qquad (10).$$

In Fig. 2 we show the results obtained in this cooling simulation. The points represent averages of χ_6 over 6 configurations at 200 time-step intervals and the bars represent the standard deviation. It can be seen that while χ_6 varies smoothly at high temperatures ($\Gamma < 90$), there is a dramatic increase both in χ_6 and in its standard deviation, which is the result of a strongly varying correlation length ξ_6 from configuration to configuration.

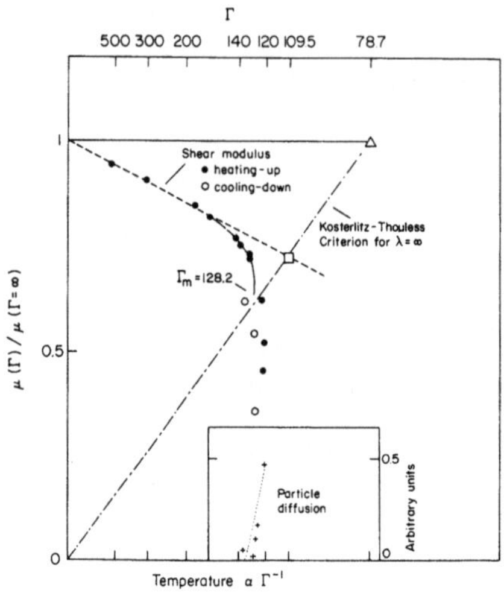

<u>Fig. 1</u>. Temperature dependence of shear modulus and particle diffusion (inset). The circles represent the MD results for $\mu(\Gamma)/\mu(\Gamma=\infty)$ as function of Γ^{-1}. The full line, ending at Γ = 128.2 displays the RG results. The Thouless value Γ = 78.7 (triangle) is obtained with use of T = 0 elastic constants.

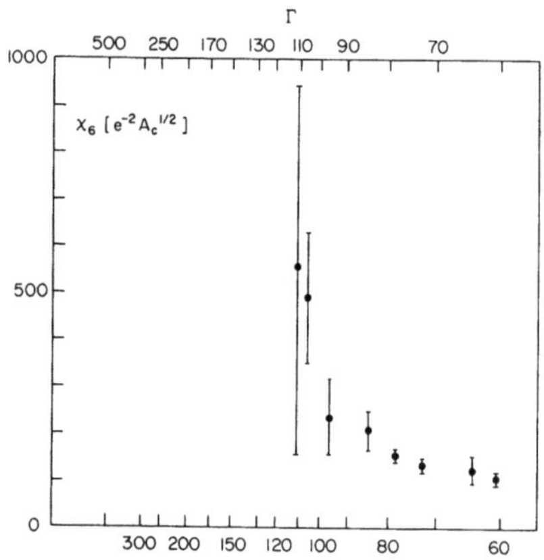

<u>Fig. 2</u>. The susceptibility χ_6 as a function of Γ^{-1}.

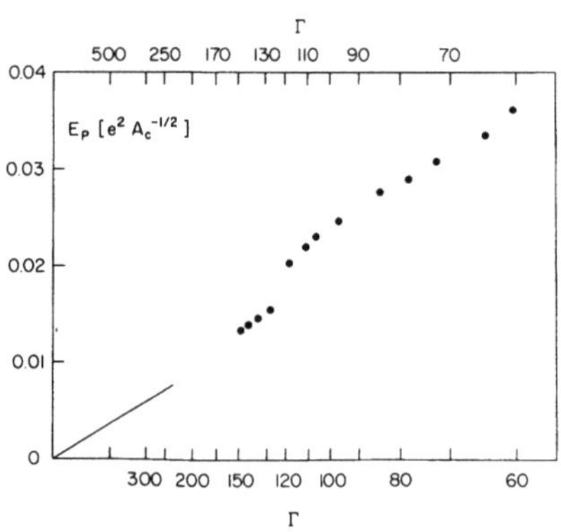

<u>Fig. 3</u>. The potential energy E_p as a function of Γ^{-1}.
The full line starting at $\Gamma = \infty$ represents the harmonic contribution k_BT. Values for $\Gamma < 120$ are obtained from cooling the high-T fluid. Values for $\Gamma > 120$ are obtained from heating the solid.

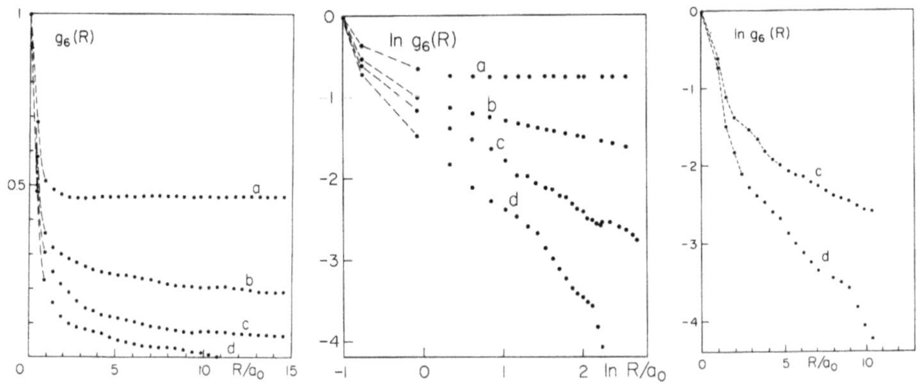

Fig. 4. The orientational correlation function g_6 (R) as function of separation R in linear and logarithmic scales. The letters a, b, c, d refer to the 4 configurations displayed in Fig. 5 below.

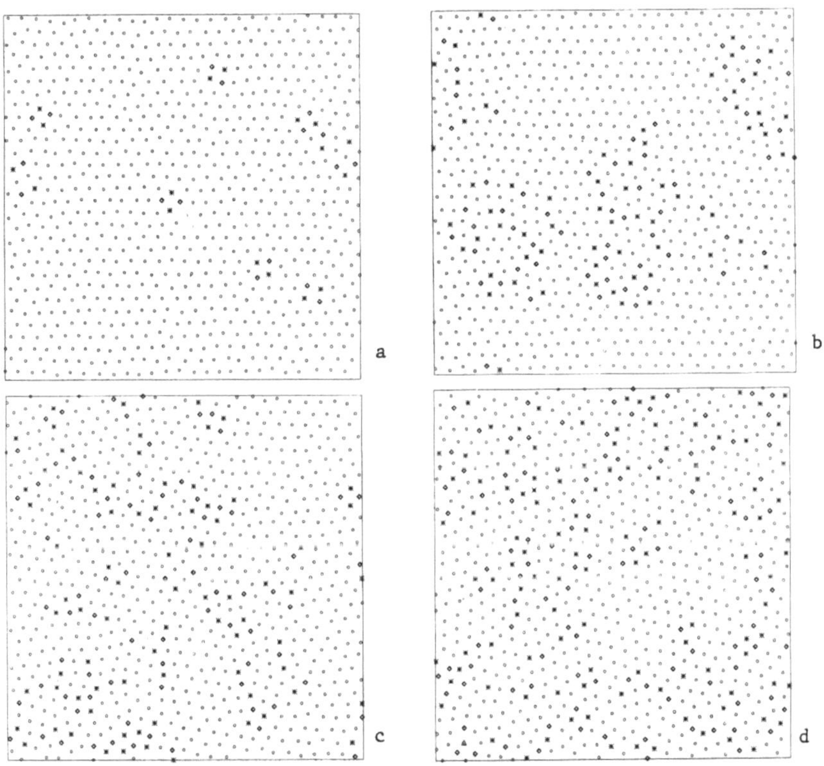

Fig. 5. Configurations from solid to isotropic fluid.
The symbols ×, ◇, o, ✱, Δ designate particles with 4, 5, 6, 7, 8 neighbours. Configurations a, b, c, d are at Γ = 135, 132, 131, 125.

While these results do not allow a determination of T_i and consequently cannot be used to confirm the existence of a hexatic phase, they clearly show the existence of prefreezing effects in the fluid at temperatures 20 to 30 percent above T_m (compare Figs. 1 and 2). A more accurate determination of χ_6 is very desirable but will require simulations during much longer times and the discussion of finite size effects. In Fig. 3 we show the behavior of the potential energy E_p (from which the T=0 energy is subtracted). The harmonic energy k_BT is indicated as a solid line. Owing to equilibration problems we are unable to determine reliable values for E_p in the interesting range $130 > \Gamma > 120$. This difficulty can be illustrated by the fact that upon cooling the fluid below T_m ($\Gamma \approx 135$) and annealing for 6000 time steps only a small fraction (less then 1/4) of the apparent discontinuity ΔE_p of the potential energy could be removed from the system. However, we may use the results of Fig. 3 to derive an upper limit for the melting entropy ΔS

$$\Delta S < 0.3\ k_B \quad . \tag{11}$$

This result is not inconsistent with the value of ΔS quoted by [24]. However, in view of the equilibration problems and the clear indication of premelting and prefreezing effects, the question about the order of the melting transition cannot yet be answered reliably.

To conclude this section we look at the orientational correlation function $g_6(R)$ and at the defects that appear as the solid melts. In Fig. 4 the orientational correlation function $g_6(R)$ is plotted as a function of separation R in linear, log-log, and log-linear scale for the four configurations a, b, c, and d displayed in Fig. 5. They correspond to values of $\Gamma = 135$, 132, 131, and 125 and potential energy E_p/k_BT of 1.15, 1.26, 1.35, and 1.42, respectively. Configuration a clearly shows long-range orientational orders with $g_6(\infty) \approx 0.46$ and an orientational order parameter $\tilde{M}_6 = |\langle\Psi_6\rangle| = \sqrt{g_6(\infty)} \approx 0.68$. In configuration b the long-range orientational order seems to have disappeared and $g_6(R)$ has a well defined power law decay with an exponent $\eta_6 \approx 0.22$. This type of behavior, which is characteristic for the hexatic phase, has been observed during 6000 time steps with η_6 varying between 0.1 and 0.22 and Γ determined from the kinetic energy (averaged over 100 time steps) varying between $134 > \Gamma > 130$. As the system is heated further (by coupling to a heat bath at $\Gamma_B = 126$), at first configurations like c appear which are characterized by a much steeper, but possibly still algebraic, decay of $g_6(R)$. However, this state is not in equilibrium ($\Gamma = 131 > \Gamma_B$) and after about 1000 time steps configurations like d are obtained with exponentially decaying $g_6(R)$. To complete the picture the configurations a, b, c, and d are shown in Fig. 5. To make the defects visible we follow [16] and mark the particles with different symbols according to the number of neighbors as defined by the Wigner-Seitz criterion. The cross, diamond, circle, star, and triangle denote particles with 4, 5, 6, 7, and 8 neighbors, respectively. In this picture, disclinations correspond to particles with a number of neighbors unequal to six surrounded by particles which all have six neighbors. Dislocations, as bound pairs of disclinations, correspond to pairs of particles with 7 and 5 neighbors. In the solid phase, configuration a, consistent with the Kosterlitz-Thouless picture, only bound dislocation pairs ("5-7-5-7-units") are visible and only in one pair a dislocation separation greater than one lattice constant can be found. Configuration b shows a number of free dislocations as is characteristic for the hexatic phase. The system then goes through an unstable phase (configuration c) with a number of isolated disclinations into the isotropic fluid (configuration d) where more than one quarter of the area contains defects. These observations are consistent with the Halperin-Nelson picture. However, owing to equilibration problems, we are unable to determine the width in temperature of the hexatic phase.

The jump $\Delta \tilde{M}_6$ of the orientational order parameter $\tilde{M}_6 = |\langle \psi_6 \rangle|$ at the melting point can be estimated from the behaviour of $g_6(R)$. Since configuration a is still in the solid and is characterized by $\tilde{M}_6 \approx 0.68$, the jump $\Delta \tilde{M}_6$ is at most 0.68. From the results of our simulation we estimate

$$0.5 < \Delta \tilde{M}_6 < .65 \quad (12).$$

3. Conclusions

We have presented results of an MD simulation which show the existence of premelting effects in the solid and prefreezing effects in the fluid. As the melting point is approached from low temperatures we observe a reduction of the shear modulus which is accompanied by the appearance of dislocation pairs and is consistent with predictions based upon the Halperin-Nelson-Young theory. On the other hand, upon cooling the high temperature fluid we find a sudden increase in the orientational correlation length at temperatures about 20 percent above the melting point. While a phase characterized by algebraic decay of the orientational correlation functions has been observed, owing to equilibration problems it is still not clear if the proposed hexatic phase exists in thermal equilibrium.

Since the hexatic phase is characterized by an infinite susceptibility χ_6, it will be interesting to study the effect of putting the system on an incommensurate substrate with hexagonal anisotropy, which corresponds to a field H_6 that couples to the orientational order parameter ψ_6. Preliminary results [28] indicate that the equilibration problems are greatly reduced if such a substrate is present. In particular, it is found that the melting transition in the simulation becomes reversible. Also the observed melting curve $\Gamma_m(H_6)$ is consistent with the Halperin-Nelson theory [28].

Acknowledgements

The author would like to thank B.I. Halperin, D.R. Nelson, and R.G. Petschek for many helpful and stimulating discussions and S.A. Orszag for invaluable help with the computations. He also thanks B.I. Halperin for his hospitality at Harvard University where most of this work was done. It was supported in part by the U.S. National Science Foundation through grants DMR-77-10210 and ATM-78-17092 and through the Harvard Materials Research Laboratory.

References

1. B.I. Halperin and D.R. Nelson, Phys. Rev. Lett. 41, 121 (1978), Phys. Rev. B 19, 2457 (1979). See also A.P. Young, Phys. Rev. B 19, 1855 (1979), who did however not consider an hexatic phase.
2. J.M. Kosterlitz and D.J. Thouless, J. Phys. C 6, 1181 (1973).
3. N.D. Mermin, Phys. Rev. 176, 250 (1968).
4. D.S. Fisher, B.I. Halperin and R. Morf, Phys. Rev. B 20, 4692 (1979).
5. S.T. Chui, Grain boundary theory of 2d melting (to be published).
6. A 3d-stacked hexatic has been observed by R. Pindak, D.E. Moncton, S.C. Davey, and J.W. Goodby, Phys. Rev. Lett. 46, 1135 (1981).
7. R. Pindak, D.J. Bishop, W.O. Sprenger, Phys. Rev. Lett. 44, 1461 (1980).
8. B.J. Alder and T.E. Wainwright, Phys. Rev. 127, 359 (1962), and W.G. Hoover and F.H. Ree, J. Chem. Phys. 49, 3610 (1968).
9. W.G. Hoover, S.G. Gray, and K.W. Johnson, J. Chem. Phys. 55, 1128 (1971).
10. F. van Swol, L.V. Woodcock, and J.N. Cape, J. Chem. Phys. 73, 913 (1980).
11. J.D. Weeks, Volume Change on Melting... (to be published)
12. D. Frenkel and J.P. McTague, Phys. Rev. Lett. 42, 1632 (1979).
13. S. Toxvaerd, Phys. Rev. Lett. 44 1002 (1980).
14. F.F. Abraham, Phys. Rev. Lett. 44, 463 (1980).
15. J. Tobochnik and G.V. Chester in Ordering in Two Dimensions, edited by S.K. Sinha (North Holland, New York, 1980), p. 339.
16. J.P. McTague, D. Frenkel and M.P. Allen, in Ordering in Two Dimensions, (op cit.) p. 147.
17. F.F. Abraham, Phys. Rev. B 23, 6145 (1981).
18. R.W. Hockney and T.R. Brown, J.Phys C 8, 1813 (1975).
19. R.C. Gann, S. Chakravarty, and G.V. Chester, Phys. Rev. B 20, 326 (1979).
20. C.C. Grimes and G. Adams, Phys. Rev. Lett. 42, 795 (1979).
21. D.S. Fisher, B.I. Halperin, and P.M. Platzman, Phys. Rev. Lett. 42, 789 (1979)
22. D.J. Thouless, J. Phys. C 11, L189 (1978).
23. R.H. Morf, Phys. Rev. Lett. 43, 931 (1979).
24. R.K. Kalia, D. Vashishta, S.D. Leeuw, Phys. Rev. B 23, 4794 (1981).
25. D.P. Landau, R.H. Swendsen, Phys. Rev. Lett. 46, 1437 (1981).
26. R.H. Morf and E.P. Stoll, in Numerical Analysis, ed. by J. Descloux and J. Marti (Birkhäuser, Basel, 1977), Vol. 37, p. 139.
27. D.S. Fisher calculated the linear term of $\mu(T)$ analytically. His result agrees with ours (private communication).
28. R.H. Morf, to be published.

Part V

Magnetic Resonance

Magnetic Resonance in Graphite Intercalated Compounds

J. Conard, H. Estrade-Szwarckopf, P. Lauginie

C.R.S.O.C.O. - C.N.R.S.
F-45045 Orléans, France

and

G. Hermann

Bruker Spectrospin, F-67160 Wissembourg, France

Magnetic resonance has been used in various ways to obtain information about G.I.C.

A first series of papers gives us some light about the chemical species present in the inserted layer of binary acceptor compounds by observing the chemical shift of inserted nuclei. As an example KLEINBERG et al. [1], MILLER et al. [2] convinced us of a complicated situation with a chemical equilibrium between various species of graphite AsF_5 compounds. In these experiments the implicit assumption, for the observed shift, is to be essentially chemical in nature,i.e. with no paramagnetic electron upon the observed species.

Actually WEINBERGER et al. [3] estimated the presence of a Curie-like paramagnetic electron to be less than 0.1 $\mu\beta$ in AsF_5-graphite compounds. But the presence, for a limited time, of a Pauli electron remains a question quite open. In a recent paper, F. BEGUIN et al. [4] concluded that the presence of a Pauli electron upon C_6H_6 in the ternary compound $KC_{24}(C_6H_6)_2$ was necessary to explain a ^{13}C shift too high to be chemical only.

By EPR, WEINBERGER et al. [5] and DELHAES et al. [6] developed investigations about the origin and nature of magnetic susceptibility observed in these compounds comparable to results coming from static susceptibility measurements and low temperature specific heat obtained by DI SALVO [7] or ALEXANDER et al. [8] and MIZUTANI et al. [9]. The conclusions of these experiments and theoretical predictions are very important for interpreting in a detailed manner the NMR results discussed later.

Our group devoted its work to the understanding of the metallic character of these compounds. For that purpose we used EPR, NMR of inserted species mainly in donors, and graphitic ^{13}C NMR.

1. Electron Paramagnetic Resonance

EPR gives characteristic line shapes of conductive solids from which we extracted physical information listed below [10,11].

1.1 The mean free path along the c-axis is very large in lithium compounds and exceptionnally small in binary acceptor compounds yet studied. For these compounds the electron wave along the c-axis appears to be captive in a box of length $\sim$ 3.5 Å. Thus, the c-axis residual conductivity could be due to hopping only or to the use of bridges produced by defects.

1.2 The free electron goes, for a fractional time, upon the inserted species in donor compounds. The role of the atomic number Z in the ESR linewidth proves this to be true both in binary compounds and in ternary ones such as graphite) or (K, H, graphite) compounds [12].

1.3 The density of states at the Fermi level N_{EF} given by EPR, provided one uses questionable assumptions, is affected by e-e correlations and electron-phonon coupling relative to the one given by γ, low tpt specific heat coefficient and important differences are noted. Nevertheless, it is clear that this density of states is high in donor compounds, even higher than in connected metal, whereas it is low in acceptor G.I.C.

1.4 We were looking for the reappearance of the graphite g anisotropy when going to higher stages of K G.I.C. and the answer is that, even in the fifth c-plane from the inserted layer, the g anisotropy is very low. It proves an electron doping larger than 10^{-3} e/C, in agreement with ^{13}C NMR results given later and DI SALVO measurements of static susceptiblity [7].

2. Inserted Species NMR: Origin of the Lines, Quadrupolar Coupling, Linewidth.

We have studied Li [13] and Cs [14] NMR, while the first studies of Cs and Rb were made by CARVER [15] and BERTHIER et al. [16], respectively.

2.1 A first surprizing result given by ^7Li NMR is that the IInd stage and perhaps higher ones, prepared by solid state diffusion at 400°C by GUERARD et al. [17], are not actually stable at room temperature: they seem to change slowly towards a mixture of graphite and Ist stage compound.

2.2 We obtain information from the quadrupolar coupling constant as deduced from the structure of ^7Li NMR spectra [13]. We find an axial symmetry for the electric field gradient (E.F.G) in agreement with the Li ne site symmetry. A divergence from the precise stoichiometry LiC_6, as an example $LiC_{6.5}$, gives wings to the line, due to non-axial symmetry [13]. The theoretical E.F.G. computed from six first-neighbour point charges as in LiC_{12} structure gives 43,2 kHz, whereas the experiment yields 20 kHz. The role of π-orbitals can therefore be assimilated in an external screening factor of two in the a,b direction. But going from LiC_{12} to LiC_6, the E.F.G. change is 6 kHz, while the E.F.G. computed from the two new first neighbours Li atoms is 40 kHz. Then the π-screening factor in the c-direction appears to be 6.

Various ways are offered to improve the agreement between experiments and calculation i.e.:

i) a reduction of the Li ionic charge: not evident.
ii) the criticism of the used point charge model, not at all convenient.
iii) the self polarization of the p-surrounding orbitals by the Li ion.
iv) the complete calculation from the best available charge distribution map.

2.3 The linewidth narrows when the temperature is growing, showing a Li diffusion motion. From that change [13], activation energy for the motion can be calculated, not very accurately, and shows an easier motion in LiC_{12} than in LiC_6. We feel the lithium layer to be lacunar in this LiC_{12} sample and may be lacunar in all LiC_{12} samples. At high temperature when the lines are narrow, the spectrum shows a small line of 10% relative intensity. If we decide this line to be due to lithium atoms neighbouring a vacancy, it indicates that the amount of vacancies is about 1.6%.

Now we come to the specific interactions between a nucleus and a conduction electron: we devote the next section to this important analysis.

3. Electron-Nuclei Interactions in NMR of G.I.C.

The relative shift δ_{ppm} is the sum of three terms $\delta = \delta_{PAULI} + \delta_{INTERNAL} + \delta_{ORBITAL}$

δ_{PAULI} $\propto N(E_F)$		$\delta_{INTERNAL}$ $\propto$ internal e susceptibility	$\delta_{ORBITAL}$ $\propto$ all electrons in the band
Same origin as χ_{Pauli} ($\propto$ paramagnetic shift in molecules) TWO CONTRIBUTIONS:			Same origin as $\chi_{orbital}$ (chemical shift in molecules)
CONTACT (KNIGHT)	DIPOLAR	ISOTROPIC, ASSUMED TO BE A CONSTANT.	= 0 for s electrons
$K = \frac{8\pi}{3} \lvert\psi_0\rvert^2 \chi_{Pauli}$	$\propto \langle \frac{1}{r_{ij}^3} \rangle N(E_F)$		
$= \frac{8\pi}{3} P_a \, \xi \, \mu_B^2 \, N(E_F)$	(would be zero)		
P_a = e density upon the nucleus in the isolated atom.	for s electrons or cubic sites		ANISOTROPIC, NON-ZERO TRACE.
	ANISOTROPIC		
ξ specific reduction factor.	TRACELESS		

3.1 Alkaline Knight Shift and Relaxation

If we look at alkaline inserted metal, dipolar and orbital terms are zero; we see only the contact shift:

$$\delta H/H_0 = K = \frac{8\pi}{3} P_a \, \xi \, \chi_{Pauli}$$

3.1.1 For independant electrons:

$$\chi_{Pauli} \propto N(E_F) \propto \gamma \text{ mJ/mole of observed nuclei}$$

where γ is the low temperature specific heat.

So $K = A\xi\gamma$ or $K/\gamma = A\xi$ (A = Cte). K/γ is a measurement of the electron density upon the nucleus compared to the atomic one. We assume now that the transferred charge is related to this electron density: $K/\gamma \,/\, (K/\gamma)_M = \xi/\xi_M$ is the fraction of electron left upon the alkaline atom and $1 - \xi/\xi_M = f$ is the non-s character of the conduction electron wave.

Experimentally, $\gamma_{MC_x} > \gamma_{Metal}$ probably because the alkaline-like part of the band is narrower than the metallic one (cf. PFLUGER [18] VOLPILHAC [19]) $K_{MC_x} < K_{Metal}$ so K/γ is greatly reduced. In the next table we summarize information obtained from alkaline inserted atoms.

Table 1 Values of K/γ and f as deduced from alkali studies.

	LiC_6 [6] [25]	LiC_{12} [24]	RbC_8 [8]	CsC_8 [8] [9]	CsC_{24} [9]
$(\frac{K}{\gamma})$ G.I.C.	$\frac{38}{3.00} = 12{,}7$	$\frac{45}{5.27} = 8{,}52$	$\frac{1450}{10.4} = 139$	$\frac{2579}{5.68} = 454$	$\frac{200 \pm 50}{6.24} = 32 \pm 8$
$(\frac{K}{\gamma})$Metal	$\frac{260}{1.67} = 159$		$\frac{6520}{2.41} = 2705$	$\frac{14900}{3.20} = 4656$	
$\xi/\xi_M = 1-f$	0,078	0,054	0,051	0,097	0.008 ± 0.02
f	0.92	0,95	0,95	0,90	0,99 - 1
$(x+1)\xi/\xi_M$	0,56	0,70	0,46	0,88	0,15 - 0.22

Note: $\xi_M = 0.44$ in Li metal. Assuming the electron to be univormly shared among all metal and C nuclei, $(x + 1)\xi/\xi_M$ would be 1.

Two anomalies appear in this table :

i) It is surprizing to find $\gamma LiC_{12} > \gamma LiC_6$: as we noticed in 3.1.1, the bandwidth was decreasing from metal to LiC_6. So, the decrease of Li-Li coupling from LiC_6 to LiC_{12} might be able to narrow the bandwidth, giving rise to a higher density of states at the top. ESCA experiment would be able to check this property.

ii) The γ value for RbC_8 is very high without any appearant reason. New measurements of Rb compounds γ values would be very useful.

3.1.2 For Non-Independant Electrons

Now we are able to give a more detailed description with the introduction of the two main interactions which break the independant electrons model: the electron-phonon interaction and the e-e correlation. Table 2 gives the formalism which describes these interactions.

Table 2 List of quantities relevant to the study of metallic character.

	N_{E_F}	χ_p	γ	K	K/γ	$s = \|I_1\|^T K^2$
independant electrons	N_o	$\chi_o = \mu_B^2 N_o$	$\gamma_o = \pi^2 k^2 N_o/3$	k_o $\pi P_a \xi\, \mu_B^2 N_o/3$	$k_o/\gamma_o = A\xi = \frac{8}{\pi} \frac{P_a \mu_B^2 \xi}{k^2}$	$s_o = \frac{\hbar}{4\pi k} (\frac{\gamma e}{\gamma n})^2$
+ e-phonons	$N_o(1+\lambda)$	$\sim \chi_o$	$\gamma_o(1+\lambda)$	K_o	$\frac{K_o}{\gamma_o} \frac{1}{1+\lambda}$	s_o
+ e-e correlations	$N_o \frac{1+\lambda}{1+x}$	$\chi_o \frac{1+\mu}{1+x}$	$\gamma_o \frac{1+\lambda}{1+x}$	$K_o \frac{1+\mu}{1+x}$	$\frac{K_o}{\gamma_o} \frac{1+\mu}{1+\lambda}$	$s_o(1+\mu)^2$

where $0 < \lambda < 1$ characterizes the e-phonon interaction; $x > 0$ and $\mu > 0$ are relative to e-e correlation: the first one describes the changes in N_{EF} and the second one, the change in the ratio χ_p/N_{EF}.

This formalism perfectly agrees with ^{7}Li NMR in the pure metal where $\mu = 0.31$; $(1+\mu)/(1+\lambda) = 1.21$; $\xi(1+\mu/1+\lambda) = 0.53$ (coming from K/γ experimental value) and $\xi = 0.44$ in agreement with direct determination by OVERHAUSER effect.

In LiC_6 it gives $\mu \sim 0$ (poor correlation), whereas CsC_8 shows a high degree of correlation with a value: $\mu = 1.32$. These results agree with the idea of an expansion of the lithium lattice in LiC_6, relative to metallic lithium, while in CsC_8 the cesium lattice is compressed.

3.2 ^{13}C NMR: Lines Assignment to C Sites and Planes

The ^{13}C lines appear with a shift and an axial anisotropy, typical of the π orbitals symmetry [20],[26],[27].

3.2.1. In first stage donor compounds, only one kind of sites is seen, as expected in LiC_6. In the MC_8 compounds [23] , it confirms the existence of the $\alpha\ \beta\ \gamma$ rotation rules of vicinity for the alkaline atoms in adjacent layers.

3.2.2 The II^d stage donor compounds studied so far show a very similar structure of two lines, with the anisotropy not reversed relative to graphite as it was for the Ist stage compounds. We assign these two lines of same intensity to two kinds of sites in the C plane: those neighbouring an alkaline atom to which the C atom is covalently bounded and those in second or third rank position.

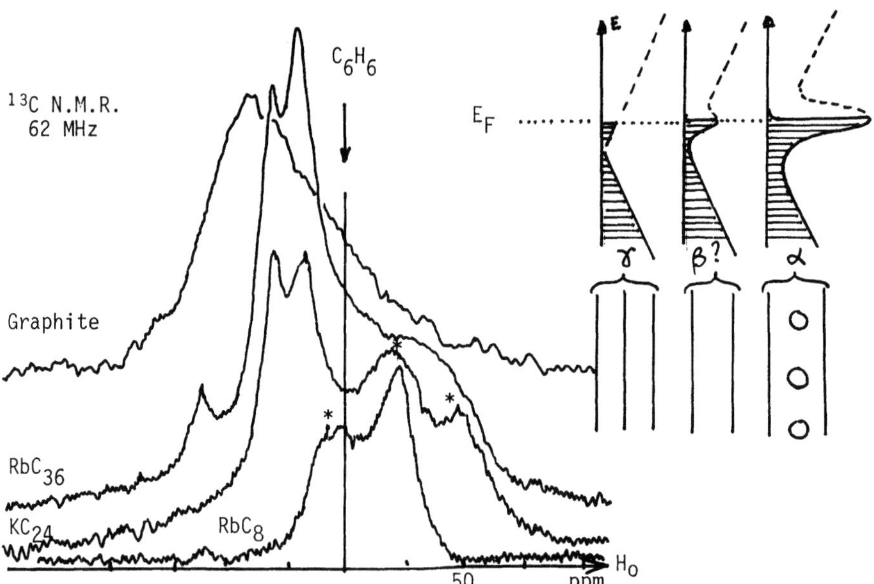

Fig.1 RbC_8: one site, axial; KC_{24}: two sites, axial (α band shape) RbC_{36}: KC_{24} line structure more a narrow line at the H_0 // a,b graphite position (i.e. $\alpha + \gamma$ band shape). The β band shape may be exists in K series. N.B. Starred peaks result from orientation [21].

3.2.3 The IIInd, IVth and higher stage lines can be resolved at high field very clearly into

- the group of two lines assigned to the sandwich MC_{24} described in 3.2.2 and
- a line similar to graphite, with except two features:

i) it has a T_1 shorter than that of graphite

ii) this line is at the position of the a,b // H_0 graphite's component but it seems to be symmetric (like boronated coke) and narrower than the graphite line. We assign this line to a remote e-doped carbon planes.

3.3 ^{13}C NMR: Charge Sharing

At the Nijmegen conference, we proposed a model of bandshape with a shape change one C layer to the other [20].

The sandwich (donor layer plus the two adjacent carbon layers) has a peak of density of states at E_F issued from the alkaline level, whereas the remote carbon layers are more accurately described by the rigid band model. Perhaps the samll π-π coupling between C layers is able to produce a small peak in the density of states of the second neighbour plane.

So, the high stage G.I.C. is a solid with a density of states at E_F varying from one C plane to the other. That peculiar situation explains how a conductive material may have an extended inhomogeneous distribution of charges.

A first independant check of this model was obtained from the γ values of II^{d}, III^{d} and IVth stages in the K and Cs series. It gives a value of density of states divided by a factor of 2 to 3 when going from the first C plane to the second one.

Table 3 N_{E_F} [states/eV C] vs rank of C plane/intercalated layer, from γ [8,9]

	1st plane	2nd plane	self test [21]
KC_{12n}	0.106	0.034	($C_{48}K$, γ value missing)
CsC_{12n}	0.11	0.069	$\sim$ 0.0695

In order to check in another way this charge sharing, we obtained at very high field ($\sim$ 100 kG) a series of experiments at variable repetition time. If the trigger time is shorter than the relaxation time T_1, the line is rubbed out by saturation effects. Figure 2 relates these experiments with the K-graphite ninth stage compound.

i) Due to the high resolution obtained at that field, the graphite-like line at low field is clearly kept apart from the KC_{24} line. Moreover, it appears to be quite symmetric. The specific orbital contribution to graphite lineshape c-component has disappeared. That means the e-doping of the C-plane is larger than 10^{-3} e/c [22]. This idea has been checked by observing the ^{13}C NMR of a 0.5% boronated graphitized coke shown in the same picture. It gives a narrow symmetric line at the position of the a,b peak of graphite.

ii) The relative intensity of the graphite-like line is clearly growing when the repetition rate goes from 2 to 10 s, and keeps constant from 10 to 100 s. So, the lower T_1 value for remote planes seems to be of the order of 10 seconds, i.e. 9 times shorter than graphite T_1. It proves that the 5th

plane has enough e-doping. The known Fermi level shift (0.17 eV) in a rigid band model for CsC_{36} would give about $2.4\cdot10^{-3}$ e/C, in good agreement.

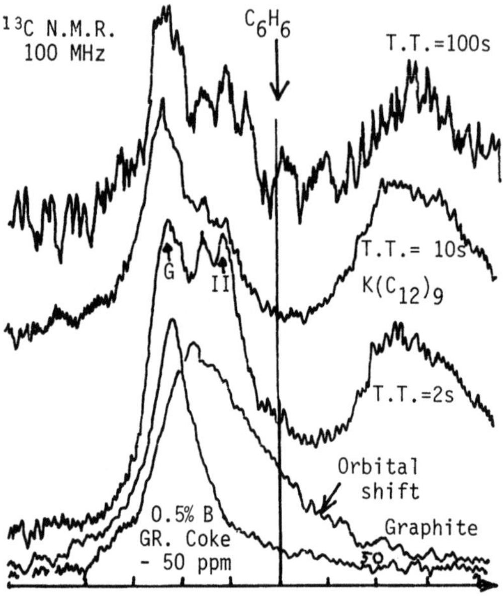

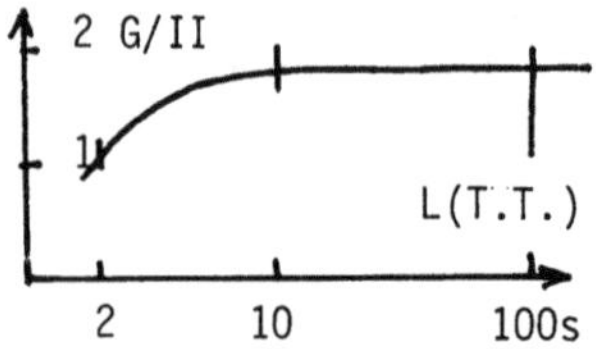

Fig.2 Very high field ^{13}C NMR of $K(C_{12})_9$ at variable repetition rate (T.T.). We estimate the T_1 of remote planes to be $\sim$ 10 s.
Their line is quite well resolved and comparable to pure and 0.5 boronated graphite.

If we add to these experiments the χ_p vs stage measurement by DI SALVO et al. [7] and the EPR results by LAUGINIE et al. [10], we have a set of five independant measurements giving information about the charge sharing along the c direction. They agree showing an e-doped C-plane at long distance.

A consequence of this model is that for a high stage G.I.C. the charge along the remote planes would be approximately constant, if we except the first two or three planes, which may show a residual peak in the density of states.

3.4 ^{13}C NMR: Interpretation of the Shifts

All the effects predicted in §3 are now acting simultaneously. The indirect contact and dipolar terms can only be separated easily, using the traceless property of the later one. For each component (// a,b) or (// c) of the axial spectrum, we can write

$$K_{a,b} = K_{ind} + K_{dipolar} = (A\,\xi_0 + B)(1 + \frac{f}{x})\,\gamma$$

$$K_c = K_{ind} + K_{dipolar} = (A\,\xi_0 - 2B)(1 + \frac{f}{x})\,\gamma\ .$$

But the orbital contribution δ_{ORB} is not so easy to analyze. It has the same origin as the orbital susceptibility χ_{ORB} but with a different average. DI SALVO et al [7] found a positive $\chi_{c,ORB}$, while on the contrary ^{13}C shifts are all upfield. An indication of the χ_{ORB} value is the EPR $g-g_{free}$ value.

As an example, graphite with $N(E_F)$ = 0,014 mJ/mole $\sim$ 0 is a purely orbital case for the H_0 // c component. The a,b // H_0 component is assumed to be very near the absolute reference. N_{EF} = 0, $\delta_{orbital}$ = 0 , as confirmed by the presence at this position of the peak of boronated graphite and the a,b peak of acceptors and frozen benzene.

In G.I.C. two orbital components can occur, introducing too many variables in the problem to be solved. So, we decided to postulate $\delta_{a,b\,ORB} \sim 0$ as a general hypothesis, although it would not be true in the first stage donor compounds. Then the shifts are :

$$\delta_{a,b} = (A\ \xi_0 + B)(1 + \frac{f}{x})\ \gamma$$

$$\delta_c = (A\ \xi_0 - 2B)(1 + \frac{f}{x})\ \gamma + \delta_{(ORB)c}$$

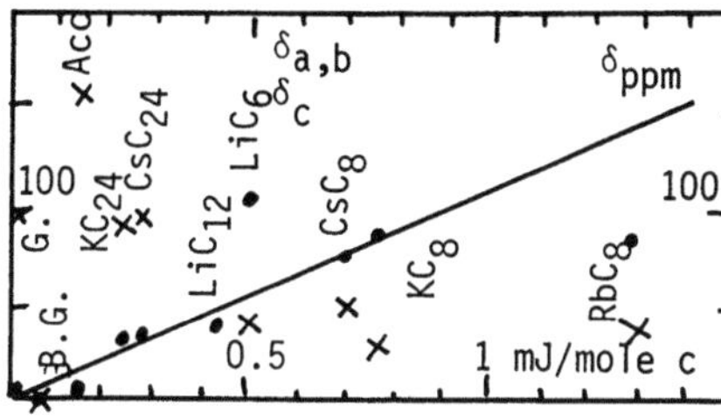

Fig.3 Plot of $\delta_{a,b}$ and δ_c vs γ. We can check the proportionality between $\delta_{a,b}$ and γ. We see a quite good alignment except for LiC_6 and RbC_8.

LiC_6 might have a non-zero orbital contribution to the $\delta_{a,b}$ component. EPR gives a diffusion constant D = 18 $cm^2\ s^{-1}$ (Li metal D = 20; LiC_{12} D = 0,95; KC_8 D < 0,1 $cm^2\ sec^{-1}$). So, LiC_6 is too conductive along the c-axis to support the hypothesis: $\delta_{a,b\,ORB} \sim 0$. If that is true, the orbital contribution can be estimated to be $\sim$ 50 ppm by difference between the $\delta_{a,b}$ experimental shift and the expected one on the right line $\delta_{a,b}$ vs γ.

The δ_c positions vs γ show a quite large misalignment, in agreement with an orbital species-dependant contribution. The "anomaly" of RbC_8 could be easily "solved" by a change in the γ value by a factor of about two.

At this point of our analysis, what is clear is that we cannot obtain much more information without detailed knowledge of the orbital contribution. This is probably available only from theoretical calculation. With the drastic hypothesis $\delta_{c\,ORB}$ = 0, we can obtain A, ξ and f. But the proportionality between $\delta_{a,b}$ and γ shows that f/x and B could be approximately constant: at least the sum (A ξ_0 (1+f/x) + B) is a constant.

Actually, (1 + f/x) is known to be between 1 and 1.1, and we would need a very high precision in the γ and δ measurements to obtain more accurate f determinations. In donors, the use of the a,b shift to deduce values of f [20] is not a bad approximation. In acceptors compounds, where γ and K are small, there is no hope in obtaining very precise f measurements.

Nevertheless, this discussion proves the boronated graphite position to be an adequate reference for the shift. Its distance to the a,b component of graphite is no more than 1 ppm (i.e. inside the measurement precision).

Of great interest would be a comparison with the K/γ values for the IInd rank planes, but as for graphite and acceptors, we have not enough precision to give significant values now. From Table 4 we infer the charge transfer at the carbon level to be small and related to the 1 + f/x value.

Table 4 ^{13}C K/γ values in I^{st} and IInd stages compounds: It evidences also anomalies for LiC_6 and RbC_8 (shift reference was $\delta_{a,b}$ in boronated graphite)

^{13}C	Graphite	LiC_6	LiC_{12}	KC_8	KC_{24}	RbC_8	CsC_8	CsC_{24}	Acceptors
K/γ	not precise	216	98	112	123	58	106	104	ε/ε

Conclusion

At the beginning of this redaction, we had the idea to conclude with the successes of magnetic resonance:

- identification of chemical species and sample composition
- localization map of free electron waves
- description of charge sharing along the c-axis
- proposal of a now well-tested band-model
- bondings clearly different in donor and acceptor compounds
- rough charge-transfer estimation.

Though illuminating answers to these problems were obtained, we summarize areas in which work is yet to be done :

- Obtain more information on acceptors, best described at present as decoupled graphitic planes.
- Obtain precise estimation or calculation of orbital ε-component of the NMR shift.
- Obtain a more accurate calculation of $(\psi_0)^2$ upon the alkaline atom as did colleagues in this conference.
- Obtain clear understanding of the origin of the NMR linewidth in relation with the electronic correlation time.
- Understand the thermal variation of T_1 and the experimental values of the Korringa constant.
- Achieve precise measurements of γ values in a complete table of I^{st} to VIIIth stage compounds if possible.
- Calculate core polarization hyperfine field with carbon π_z orbitals.

We apologize for having centered our paper on the use of magnetic resonance as a tool for the analysis of the conduction electron system and we thank many colleagues for past and future cooperation and discussions and more specifically F.J. DI SALVO, for χ_p values and the Nancy group who prepared the great majority of our samples.

References

1 R.L.Kleinberg, L.B. Ebert, Synthetic Metals, 2 (1980) 245.
2 G.R. Miller, H.A. Resing, P. Brant, M.H. Moran, F.L. Vogel, T.C. Wu, D. Billaud and A. Pron, Synthetic Metals, 2 (1980) 237
3 B.R. Weinberger, J. Kaufer, A.J. Heeger, E.R. Falardeau and J.E. Fischer Sol. St. Com., 27 (1978) 163.
4 F. Beguin, L. Facchini, G. Hermann, M.F. Quinton, A.P. Legrand and R. Setton, 15th Biennal Conf. on Carbon, Philadelphia (1981).
5 B.R. Weinberger, J. Kaufer, A.J. Heeger, J.E. Fischer, M. Moran, N.A.W. Holzwarth, Phys. Rev. Let., (1978) 1417.
6 P. Delhaes, Mat. Science Eng., 31 (1977) 225.

7 F.J. Di Salvo, S.A. Safran, R.C. Haddon, J.V. Waszczak and J.E. Fischer, Phys. Rev. B, 20, (1979), 4883.
S.A. Safran and F.J. Di Salvo, Phys. Rev. B, 20, (1979), 4889.
8 M.G. Alexander, D.P. Goshorn and D.G. Onn, Phys. Rev. B, 22, (Nov. 1980) or Synthetic Metals, 2, (1980), 203.
9 U. Mizutani, T. Kondow and T.B. Massalski, Phys. Rev. B, 17, (1978), 3165.
10 P. Lauginie, M. Letellier, H. Estrade, J. Conard and D. Guerard, Vth Int. Carbon Graphite Conf. London (1978), 645. Soc. Chem. Ind. Ed.
11 P. Lauginie, H. Estrade, J. Conard, D. Guerard, P. Lagrange and M. El Makrini, Physica 99B, (1980), 514.
12 J. Conard, H. Estrade-Szwarckopf, P. Lauginie, M. El Makrini, P. Lagrange and D. Guerard, to be published in Physica B (1981).
13 J. Conard and H. Estrade, Mat. Science and Eng., 31, (1977), 173.
14 H. Estrade-Szwarckopf, P. Lauginie and J. Conard. (This conference).
15 G.P. Carver, Phys. Rev., B2, (1970), 2284.
16 C. Berthier, Y. Chabre and P. Segransan, Physica, 99B, (1980) 107.
17 D. Guerard, A. Herold, Carbon, 13, (1975), 337.
18 P. Pflüger, P. Oelhafen, H.U. Künzi, R. Jeker, E. Hauser, K.P. Ackermann, M. Müller and H.J. Güntherodt, Physica, 99B, (1980), 395.
19 G. Volpilhac and J. Hoarau, J. Phys. C, 13, (1980), 2281.
20 J. Conard, H. Estrade, P. Lauginie, H. Fuzellier, G. Furdin and R. Vasse, Physica, 99B, (1980), 521.
21 J. Conard, M. Gutierrez-Le Brun, P. Lauginie, H. Estrade-Szwarckopf and G. Hermann, Synthetic Metals, 2, (1980), 227.
22 M.P. Sharma, L.G. Johnson and J.W. Mc Clure, Phys. Rev. B, 9, (1974), 2467.
23 D.E. Nixon and G.S. Parry, Brit. J. Appl. Phys., 1, (1968), 291.
24 P. Delhaes, J.P. Manceau and D. Guerard, Synthetic Metals, 2, (1980), 277.
25 C. Ayache, E. Bonjour, R. Lagnier and J.E. Fischer, Physica, 99B, (1980), 547.
26 D.T. Haworth and C.A. Wilkie, J. Sol. St. Chem., 31, (1980) 343.
27 M. Suganuma, U. Mizutani and T. Kondow, to be published.

^{133}Cs NMR Studies of Cesium G.I.C.

H. Estrade-Szwarckopf, J. Conard, P. Lauginie, C.R.S.O.C.I. - C.N.R.S.
Orléans, France and

J. Van der Klink, Laboratoire de Physique Expérimentale
Ecole Polytechnique, Lausanne, CH and

D. Guérard, P. Lagrange, L.C.M.A., Université de Nancy, France

NMR seems to be an adequate tool for investigating local electronic properties, by means of nuclear spin-lattice relaxation time, line-shape, line position and their temperature dependence.

In graphite lamellar compounds, much information could be obtained from ^{13}C NMR because of its quite general applicability in all kinds of compounds (acceptor, donor and ternary ones): the model of local electronic band structures could be built by this means [1].

Moreover, especially in alkaline compounds, some inserted nuclei can also be observed by NMR and this situation is even more interesting when

a) the mobility of this inserted species and its interactions with the lattice, neighbouring nuclei, electrons, electric field gradient,... can be measured (ex. Li [2] and Rb [3]);

b) the comparison between ^{13}C NMR shift and that of the inserted species allows a better understanding of the electronic space distribution as seen by both species [4]. Results from ^{133}Cs NMR of CS graphitides have been published as early as 1967 [5] and, some later on; in 1970, CARVER [6] performed in his PHD thesis quite a complete study at both very low and room temperatures.

In this paper, we will give new results about this ^{133}Cs NMR in stages I to IV cesium-graphitides, obtained with accuracy higher than previous studies.

1. Experimental Conditions

The powdered samples have been prepared from ground natural graphite in Nancy using the classical preparation mode. Those air-sensitive samples are kept in NMR usual sealed tubes (pyrex glass, Ø 10mm, 40 mm height). The free space above the sample has been limited as much as possible when sealing in order to avoid the destruction of the sample when heat treated at 500°C: in that configuration, nearly the whole tube is at the same temperature and the thermal stability can be insured.

NMR pulsed Bruker Spectrometer SXP working at 11.8 MHz (21 kG) has been used. The spectra are obtained by Fourier transform of several thousands of accumulated free induction decays of the magnetization following a 90° pulse. The highest explored frequency sweep width is 100 kHz: with this frequency scale, both the reference CsCl solution (Cs^+) and the CsC_8 signals could be observed, but the whole quadrupolar line structure of the 7/2 spin could not.

The variable temperature device is a nitrogen circulation system with electronic regulation. The stabilization is quite good (± 1°K) but the accuracy for its determination is not (± 10°K) because of a temperature gradient near the measuring thermocouple. This device allows the temperature to be varied between 100 to 500 K. Some measurements were performed at lower temperatures at the L.P.E. in Lausanne from 15 to 100 K. However, the electromagnet was there not very homogeneous and the lines were broadened. In this spectrometer the whole quadrupolar structure has been detected.

2. Experimental results.

2.1 CsC_8

The ^{133}Cs line is 2-3 kHz broad at room temperature and its observation quite difficult: about 10000 accumulations are necessary for the signal-to-noise ratio to be usable.

Its position is about 30 kHz low-field shifted with regards to the reference (Cs^+ in CsCl solution). When temperature is increased, this shift increases regularly (Fig. 1a) as does the linewidth, varying from 2.75 kHz at 300 K to 3.2 kHz at 500 K.

2.2 CsC_{24}

Two CsC_{24} samples have been studied: they gave both the same results. At 100 K the line looks like a dissymetric doublet whose shape is characteristic of a powder pattern of axial second-order quadrupolar splitting [7]. The distance between the doublet peaks is 1.6 kHz and the structure very well resolved (Fig.2). At low temperature, the line is similar down till 5 K, although broadened by the magnet inhomogeneity of the then used spectrometer. The doublet splitting is temperature independant from 15 K up till 290 K.

In Lausanne, the whole quadrupolar structure of the 7/2 spin has been observed. The distance between the first satellites is 73 ± 4 kHz at 40 K and 76 ± 2 kHz at 115 K.

The line position has been estimated through the theoretical formula [8] at $9/25\ \Delta\nu$ from the highest peak, that is to say from the up-field one. It is about 4 kHz high-field shifted with respect to the Cs^+ solution. When temperature is increased, the whole doublet shifts towards low fields, slowly till 290 K and very abruptly above, crossing the reference position near room temperature (Fig.1a).

The linewidth has been estimated through the up-field peak's halfwidth at half maximum. We observe a line broadening when temperature is increased (Fig.1b) but this broadening does not really follow the shift variation: although it also presents an abrupt change, this one occurs at temperature 50° higher than that of the shift. This broadening is so great that at 305 K, the line structure is no longer observable and that at $T > 350$ K, the whole line disappears almost completely.

The relaxation time T_1 decreases very abruptly from 27 minutes observed by CARVER at 1.3 to 0.1s at 300 K (Fig 1b) and no simple law can be inferred.

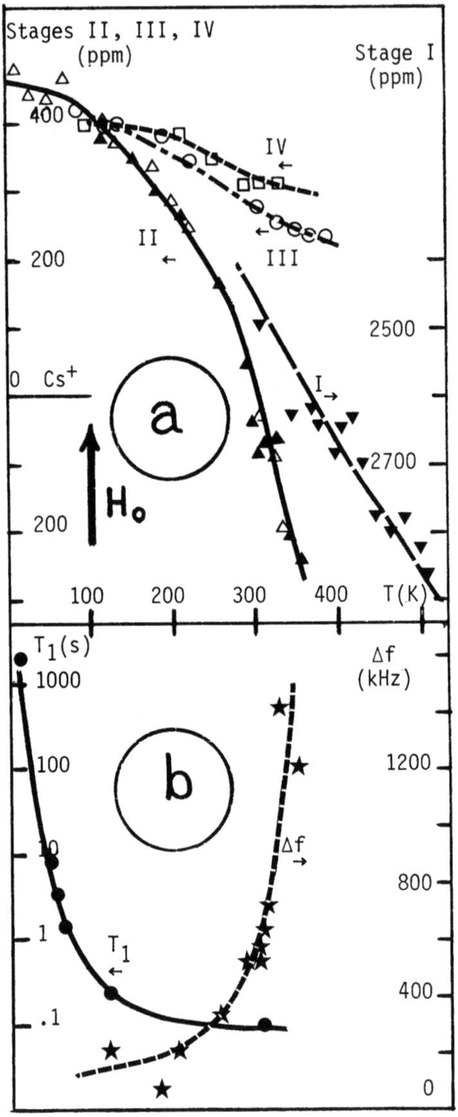

Fig.1

a) Line positions vs T
▼ CsC_8 (right scale)
▲△ CsC_{24}, ○ CsC_{36}, □ CsC_{48} (left scale)

b) Relaxation times T_1 (left scale) and half linewidth Δf (right scale) vs T in CsC_{24}.

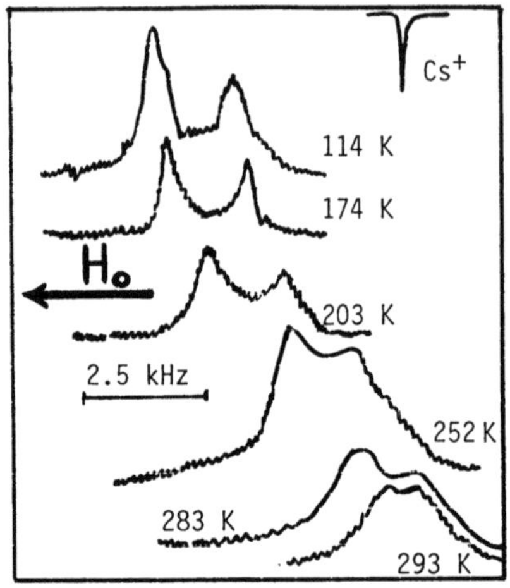

Fig.2

Central-component shape and positions in CsC_{24}.

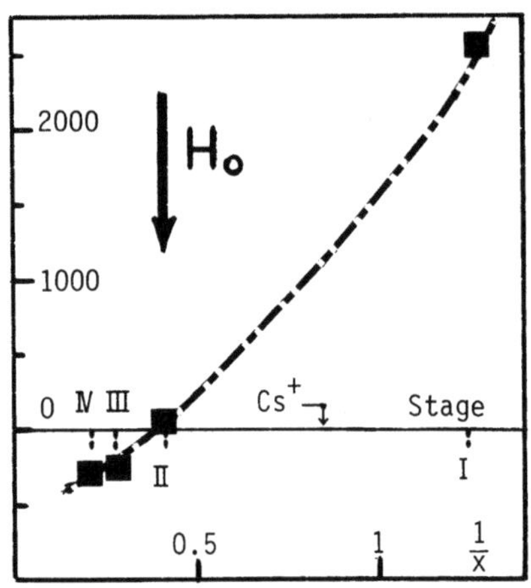

Fig.3

Line positions vs metal Cs/C = 1/x at room temperature.

2.3 CsC_{36} and CsC_{48}

At low temperatures, these lines are very similar in shape and position to that of the second stage compound, but when temperature is raised, they slowly shift toward low fields, though without showing any transition (Fig.1a).

The relaxation times are short (< 100 ms) for both compounds.

3. Discussion

3.1 Relaxation Times.

CARVER did observe at 1.3 K very long relaxation times T_1 in the three studied stages I, II, III: respectively, 7.5 s, 27 min and 48 min [6]. At this temperature, the relaxation is probably essentially due to the interaction with the conduction electrons. This interaction is important for the first stage and very weak for the other ones, as will be confirmed further on when discussing the Knight shift.

At higher temperature, and even at 40 K for the CsC_{24}, the relaxation is much more effective, probably because of quadrupolar interaction of the 7/2 spin with the electric field gradient.

3.2 Knight Shifts

The room temperature results are treated in terms of charge distributions in [4]. In the following, we will only discuss the main aspects of the line positions (Fig.3).

In the first stage CsC_8, the measured Knight shift (K) is about the same as that proposed earlier [5 and 6], confirming the metallic state of the alkaline atoms in graphitides. When compared to that of metallic cesium, the K value is reduced by a factor of 6, the same order of magnitude as observed in other alkaline graphitides: 7 and 4.5 in Li and Rb, respectively [8,3].

For the higher stages, II, III, IV, at low temperature, the shift is much smaller and opposite in sign to that usually found with s-electrons. CARVER, who measured 0-shifts, because of lower accuracy of his spectrometer, observed at 1.3 K unusually long relaxation times (27 and 48 minutes, respectively, in CsC_{24} and CsC_{36}). This is in qualitative agreement with the Korringa relation T_1TK^2 = cte, indicating a very weak interaction between the nuclei and the conduction electrons. We suggest that the cesium atoms are not very far from complete "ionization" [1].

In this case, the Knight shift being close to zero, we might observe another contribution to the shift, usually neglected, for example a chemical shift which may be opposite in sign to the s-electron Knight shift (For Cs ∿ 200 ppm [9]).

With increasing temperature, all the four stages lines are low-field shifted, i.e. in opposite direction to the metallic Cs whose Knight shift decreases, as can be expected from regular thermal expansion [10]. The knowledge of the thermal expansion coefficients (known to be "normal" in graphitides contrary to those in graphite) is necessary to relate the observed line shift with structural behaviour. However, because of its sign, we would infer that the thermal evolution of the line position is due to an until now not determined contribution which would be temperature sensitive.

Lastly, around room temperature, the second stage Cs graphitide undergoes an abrupt shift towards low fields. Here too, this shift is opposite in sign to that observed upon melting in metallic cesium [10]. We must, however, remind that this melting occurs precisely at 300 K.

3.3 Quadrupolar Constant

In II, III and IV stages at 110 K, the line structure is due to a second-order quadrupolar interaction which splits the central component into such a doublet. Using the $\Delta\nu = 1.6$ kHz distance between both peaks, we can calculate the quadrupolar interaction constant ν_q through the theoretical formula given by COHEN and REIF [8].

$$\Delta\nu = \left[I\,(I+1) - \frac{3}{4} \right] \frac{\nu_q^2}{16\nu_0} \cdot \left(1 + \frac{16}{9}\right)$$

with ν_0 = Larmor frequency, here: 11.8 MHz
$I = 7/2$

We obtain $\nu_q = 85.21$ kHz. This quadrupolar interaction constant, ν_q, has also been measured from the first-order quadrupolar structure in CsC_{24}: 73 ± 4 kHz at 40 K and 76 ± 2 kHz at 115 K. The agreement with the first value is fairly good. Moreover, those values are very close to that measured from the same spectra in CsC_8 at 1.3 K: 73.9 kHz (in agreement with CARVER, this value has been recalculated by us from the published spectrum because of a mistake in its printed value). So, the electric field gradient (E.F.G.) is of the same order of magnitude in CsC_8 and in higher stages.

If the E.F.G. measured by ν_q is due to neighbouring atomic charges, we may expect two modifications when going from CsC_{24} to CsC_8 structure:

1. each Cs atom has two metallic neighbours in the successive planes, following the $\alpha\beta\gamma$ stacking rule and whose presence must distort the axially symmetry of the E.F.G.: indeed the doublet structure disappears in CsC_8 where the line is quite broader than in stages II, III and IV.
2. each Cs atom has twice as many neighbours in its plane in CsC_8 as in CsC_{24}: we expect an E.F.G. two times greater. Our opposite result could perhaps be explained by a decrease of the Cs "ionization" by a factor of 2 (assuming the π-screening effect to be constant). In fact, the point charge model has to be used cautiously in these iono-covalent compounds.

The comparison with the E.F.G. values in Li compounds would also be interesting. Unfortunately, Li and Cs quadrupolar constants and antishielding factors found in the literature appear to be too erratic.

4. Conclusion

In higher stages, the line structures are very well resolved at least up till 400 K in CsC_{36} and CsC_{48} and up till the "transition" at 300 K in CsC_{24}. That means that the Cs atoms undergo an axially symmetric electric field gradient. This electric field gradient is well defined, as are the Cs sites.

Moreover, the unique value found for E.F.G. is about the same in all the stages from I to IV. The only explanation is that this axial electric field gradient is imposed by the neighbouring graphite lattice, the Cs atom positions being above the centers of the C hexagons. That means that if the Cs lattice is not well ordered as shown by the R.X. diffraction, this disorder is due to a random occupation of the sites but not to random sites. This is also the case in LiC_6 and LiC_{12} where we observed by NMR the mobility of Li atoms hopping from one site to the other [9], the disorder being in that case dynamical.

Moreover, in CsC_{24}, a transition-like behaviour is observed around room temperature: apart from the sudden line shift, the linewidth increases until the structure and finally the whole line vanishes. This new situation may be that of randomly distributed Cs atoms that no longer occupy hexagonal sites, undergoing a wide distribution of E.F.G. If it was a dynamical disorder with great atomic mobility, this E.F.G distribution would be time averaged and the line would be narrow and observable. So, if the Cs planes are "liquid", the correlation times of the atomic mobility are relatively long ($\tau_c > 1.5 \cdot 10^{-8}$ s).

The observed transition is perhaps to be related to the "melting" described from X-ray diffraction results by CLARKE et al. [11], unless it occurs at higher temperature : 350 K instead of 150 K where we don't observe any significant variation of the NMR characteristics.

In conclusion, we can say that NMR of inserted species in graphitides provides a lot of information about both structural and electronic aspects. However, their total interpretation implies results from other techniques. In Cs compounds, for instance, the Knight shift thermal evolution cannot be interpreted without measurements of expansion coefficients, especially in the (a,b) plane. As well, the real nature of the observed "transition" in CsC_{24} would be confirmed when observing the eventual modification of the X-rays diffraction pattern on the same sample. We hope those measurements are soon performed.

References

1 J. Conard, H. Estrade, P. Lauginie, H. Fuzellier, G. Furdin and R. Vasse, Physica, 99B, (1980), 521.
2 H. Estrade, J. Conard, P. Lauginie, P. Heitjans, F. Fujara, W. Buttler, G. Kiese, H. Ackermann and D. Guerard, Physica, 99B, (1980), 531.
3 C. Berthier, J. Chabre and P. Segransan, Physica, 99B, (1980), 107.
4 J. Conard, H. Estrade-Szwarckopf, P. Lauginie and G. Hermann (This issue).
5 V. Jensen, D.E. O'Reilly and Tung Tsang, J. Chem. Phys., 47, (1967), 1195.
6 G.P. Carver, Phys. Rev. B, 2, (1970), 2284.
7 M.H. Cohen and F. Reif, Solid State Physics, 5, (1957).
8 P. Lauginie, M. Letellier, H. Estrade, J. Conard and D. Guerard, Proc. Vth London Intern. Carbon and Graphite Conf., (1978), 645.
9 D.E. O'Reilly, Prog. NMR Spectroscopy, 2, 7
10 B.R. Mc Garvey and H.S. Gutowsky, J. Chem. Phys., 5, (1958) 241.
11 R. Clarke, N. Caswell, S.A. Solin and P.M. Horn, Phys. Rev. Lett., 43, (1979), 2018.

The Electronic Structure of Donor- and Acceptor-Type Intercalation Compounds by Means of Low Temperature Specific Heats and ^{13}C Nuclear Magnetic Resonance

U. Mizutani and M. Suganuma, Department of Crystalline Materials Science
Faculty of Engineering Nagoya University, Nagoya, Japan and

T. Kondow, Department of Chemistry, Faculty of Science
The University of Tokyo, Tokyo, Japan

1. Introduction

The characteristic properties of graphite intercalation compounds arise from their layered structure, in which a spacial inhomogeneity in the charge distribution could be present in the direction along the C-axis as the result of charge transfer between the carbon and intercalant layers. A variety of studies deal with this charge distribution and different methods of measurements sometimes give contradictory conclusions [1-5].

In order to gain insight into this problem, low-temperature specific heats and ^{13}C nuclear magnetic resonance (n.m.r.) were measured concurrently as a function of intercalant concentration. The samples studied were the intercalation compounds of potassium, rubidium and cesium and of bromine. Since the specific heats not only give the density of states at the Fermi level but also phonon behaviors and ^{13}C n.m.r. reveals local electronic structure in the vicinity of the carbon atom, information on the charge distribution along the C-axis and the degree of ionization of the intercalants can be determined more reliably by combining these two techniques.

2. Experimental

First and second stage compounds of alkali metals were synthesized by the dual furnace method [6] from spectroscopic grade graphite powder (Union Carbide SP-1). Higher-stage compounds were prepared by diluting a corresponding stage-1 compound with the necessary amount of the pure graphite. Every sample was annealed for 3 days at 280°C. The bromine compound, $C_{16}Br_2$ was formed when the degassed graphite was exposed to bromine vapor. Compounds with smaller concentrations of bromine were produced by leaving the sample in an argon atmosphere for several hours to 100 days. The sample, sealed in a Pyrex tube, was transfered under pure argon atmosphere (15 ppm of H_2O and $\sim$ 1 ppm of O_2) to a copper container and a glass tube for the specific heats and the ^{13}C n.m.r. measurements, respectively.

The heat capacity measurements were made in a conventional calorimeter in the temperature range between 15 and 6K. The sample weighed approximately 5 g. ^{13}C n.m.r. spectra were taken by use of Varian Wideline CW spectrometer WL-112. A tuned cross coil of WL-230 and a Blume-type head were employed in room temperature and 77K measurements, respectively. The rf magnetic field was calibrated by the rotary saturation method. A Varian C-1024 time averaging computor was used to improve the signal-to-noise ratio. The signal averager was triggered by the proton n.m.r. signal so as to minimize the effect of any magnetic field drift. Typical accumulation time required was 10 - 20 hrs. The n.m.r. signals were observed in magnetic fields of 8.4, 5.6 and 2.8KG so as to examine the field dependence of the signals.

3. Specific Heats

As demonstrated in Fig.1, the temperature dependence of the specific heat, C,was expressed as

$$C = \gamma T + \alpha T^2 + \Delta C \tag{1}$$

$$\Delta C = AN_A \cdot k_B (\frac{\Delta}{T}) \frac{\exp(\Delta/T)}{[\exp(\Delta/T) - 1]^2} , \tag{2}$$

where the terms γT and αT^3 represent the electronic and lattice specific heats and the additional term ΔC arises from a dispersionless optical mode of lattice vibration. In Eq.(2), Δ is the characteristic temperature, N_A is the Avogadro number and k_B is the Boltzmann constant. The coefficient A is given as [7]

$$A = (N/N_A) f , \tag{3}$$

where N is the total number of intercalants and f represents the degree of freedom of each intercalant contributing to ΔC. The parameters thus obtained are summarized in Table 1.

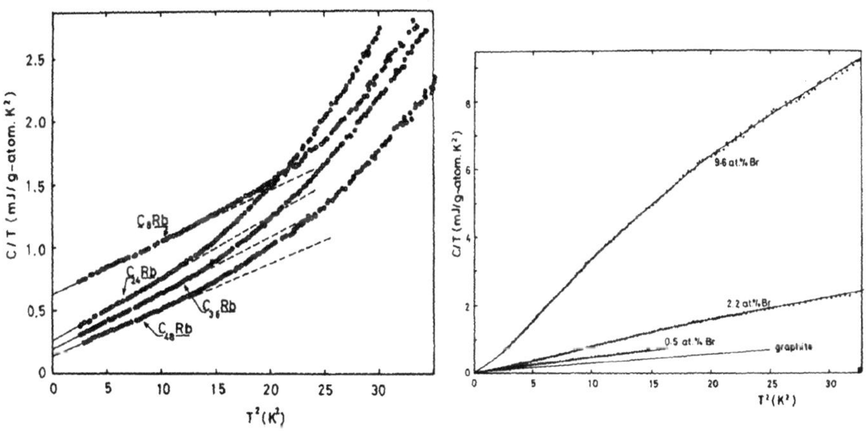

Fig.1 C/T is plotted against T^2 for the rubidium (left panel) and the bromine (right panel) graphite intercalation compounds in the temperature range of 1.5 to 6K

Figure 2 shows the coefficient γ of the electronic specific heats as a function of the intercalant concentration. The γ values of the first stage compounds of alkali metal were approximately 0.5 mJ/g-atom$\cdot K^2$ after the correction due to the electron-phonon interaction [7]. These experimental values agrees fairly well with the theoretical prediction of Inoshita et al. [8]. The γ values of the higher stage compounds of the alkali metals depended linearly on the alkali metal concentration. This fact is indicative of non-uniform charge distribution along the C-axis. In addition, the γ values were independent of the intercalant species, irrespective of their

Table 1 Specific heat parameters of the alkali metal and bromine graphite intercalation compounds

Sample	γ [mJ/g-atom·K^2]	α [mJ/g-atom·K^4]	Θ_D [K]	Δ [K]	A
C_8K	0.697 ±0.006	0.1502 ±0.0013	234.8 ±0.7	——	——
C_8Rb	0.646 ±0.002	0.0410 ±0.0002	361.9 ±0.6	71 ±1	0.8 ±0.1
C_8Cs	0.63	0.049	341	47.6 ±1.0	0.126
$C_{24}K$	0.241 ±0.003	0.0375 ±0.0007	372.9 ±2.1	——	——
$C_{24}Rb$	0.255 ±0.001	0.0510 ±0.0001	336.5 ±0.2	51 ±1	0.070 ±0.008
$C_{24}Cs$	0.25	0.085	284	35.1 ±1.0	0.0647
$C_{36}K$	0.189 ±0.003	0.0339 ±0.0007	385.2 ±2.5		
$C_{36}Rb$	0.192 ±0.001	0.0440 ±0.0002	353.5 ±0.5	51 ±1	0.057 ±0.006
$C_{36}Cs$	0.19	0.072	300	33.4 ±1.0	0.0444
$C_{48}Rb$	0.143 ±0.001	0.0376 ±0.0001	372.5 ±0.3	51 ±1	0.042 ±0.004
$C_{48}Cs$	0.16	0.060	319	33.9 ±1.0	0.0320
C – Br (at%Br)					
9.6	0.05 ±0.04	0.21	210	16.7	3.3×10^{-3}
2.2	0.04 ±0.04	0.056	326	19.1	0.9×10^{-3}
0.5	0.06 ±0.03	0.040	365	——	——
Graphite	0.03 ±0.01	0.026 ±0.001	421 ±4	——	——

different ionization potentials. Accordingly, the intercalant layers do not play an important role for the electronic structure in the vicinity of the Fermi level, and besides, the alkali metals should be fully ionized.

Suppose in an extreme case that the valence electron transferred from the alkali metal atoms is localized in the carbon layers (C_b layer) located at the nearest neighbors of the alkali metal layer. In other words, the valence electron does not reside on a carbon layer (C_i layer) bounded only by other carbon layers. Since the C_b and C_i layers can be regarded as the carbon layers in $C_{24}M$ and in pure graphite, respectively, the γ coefficient for C_xM can be expressed as [7]

$$\gamma_{C_xM} = 25M(\gamma_{C_{24}M} - \gamma_G) + \gamma_G \tag{4}$$

Eq.(4), shown in Fig.2 as the dotted line, reproduces the observed data points. This finding leads to the conclusion that the electronic structure in the vicinity of the Fermi level is determined by the C_b layers, while the C_i layers remains graphitic. The values of $\gamma_{C_{24}M}$ can be evaluated on the assumption that the valence electrons on the C_b layers behave as a two-dimensional free electron gas. The density of state per one alkali atom is estimated to be 2.16/eV·M atom since the average area occupied by one electron on the carbon layer is approximately 62 Å^2. The density of state of 2.61 corresponds to the γ value of 0.246 mJ/g-atom·K^2, which agrees well with the observed values (0.24 ∿ 0.25 mJ/g-atom).

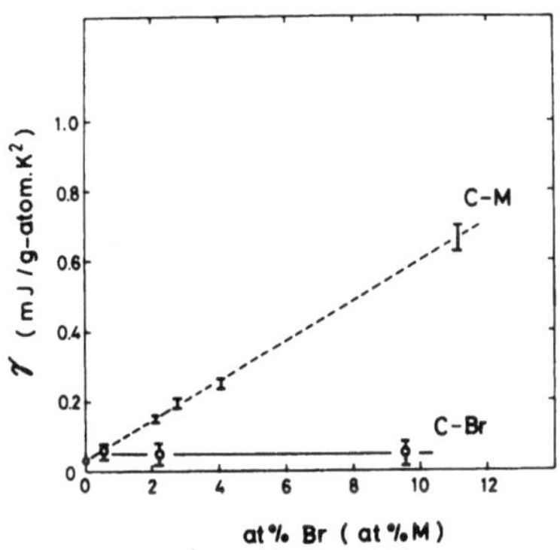

Fig.2 The coefficient (γ) of the electronic specific heats as a function of the intercalant concentration

In the case of the bromine compounds which are typical acceptor-type compounds, the γ values were not enhanced greatly upon intercalation and almost independent of the intercalant concentration. Hence, the density of state at the Fermi level does not change appreciably upon intercalation.

On the basis of this localized picture, the stage dependence of the coefficient α for the compound C_xM ($x \geq 24$) is shown as [7]

$$\alpha_{C_xM} = \left(\frac{1}{1+x}\right)[25\alpha_{C_b} + 12\alpha_{C_i}(n-2)] , \tag{5}$$

where α_{C_b} and α_{C_i} are the α values for the C_b and C_i layers, respectively, and n is the stage number of C_xM. Figure 3 shows the plot of α' (= $(1+x)\alpha_{C_xM}$) against n. α_{C_i}(mJ/mol·K^4) was estimated from the increment ($12\alpha_{C_i}$) to be in the range from 0.25 to 0.033. These values agree fairly well with the α value of pure graphite (0.026 ± 0.001 mJ/mol·K^4).

The localized nature also reflects on the specific heats arising from the dispersionless optical phonon. As shown in Fig.4, the parameter A was proportional to the intercalant concentration. It should be pointed out that the energy gap Δ was independent of the intercalant concentration. In the case of the C_xM compounds ($x \geq 24$), this optical mode should be associated

Fig. 3

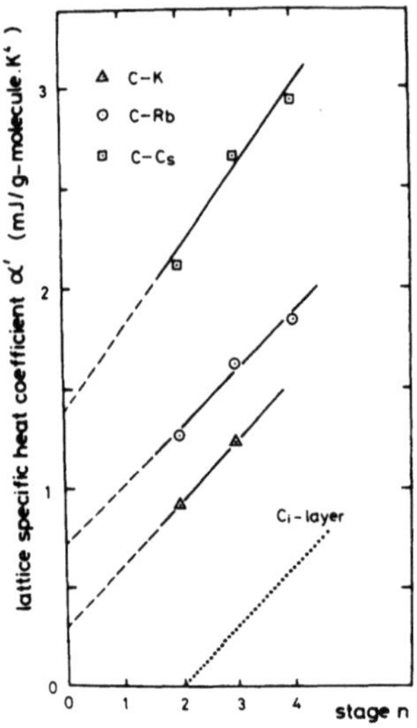

Fig.4

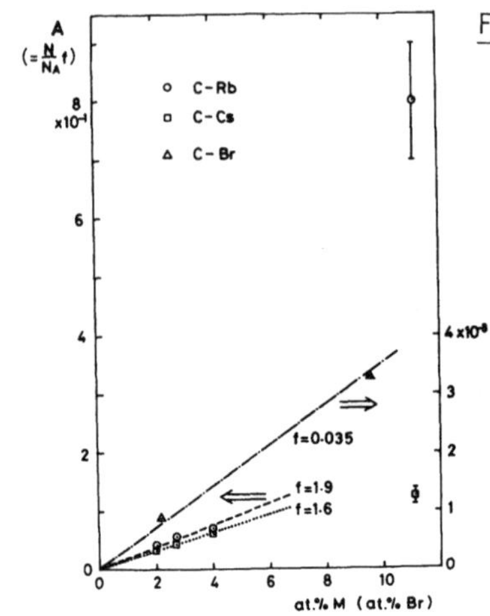

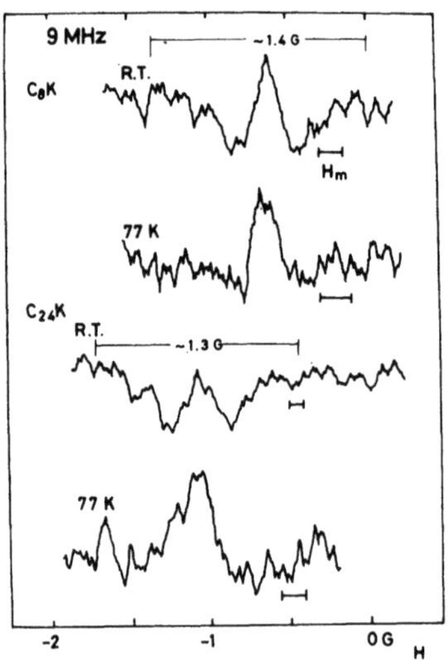

Fig.3 The plots of α' (=(1+x)α) vs. the stage number of C_xM

Fig.4 The parameter A is shown as a function of the intercalant concetration.

Fig.5 Typical ^{13}C n.m.r. spectra of C_8K and $C_{24}K$ at room temperature and 77K

with the two-dimensional lattice vibration of the alkali metal layers because both Δ and f are independent of the stage number and the values of f are about 2. The first stage compounds are the exception in the case of bromine compounds, the f value was 0.035. If only an ionized portion of bromine contribute to the specific heats, then the degree of ionization becomes 1.7%, which is in agreement with the value reported [3].

4. ^{13}C n.m.r.

Typical dispersion derivative curves of ^{13}C n.m.r. signals for C_8K and $C_{24}K$ are shown in Fig.5, and the ^{13}C Knight shift and linewidth of all the compounds studied are summarized in Table 2. The dependence of the linewidth on the intercalant concentration and static magnetic field can be explained on the basis of the non-uniform picture. Suppose that the ^{13}C signal observed is the simple superposition of the signal of the ^{13}C nuclei in the C_b layer on the C_i layer. The signal for the C_i layer could be distinguished from that associated with the C_b layer by taking advantage of the different behaviour of these signals to the static magnetic field, H_0, applied. The linewidth of the signal for the C_i layer may increase as increasing H_0 because of the anisotropic magnetic susceptibilities of the C_i layer or pure graphite [9], while the linewidth of the signal associated with the C_b layer may not be dependent on H_0 as reported [10]. In fact, in the alkali metal compounds, line broadening of the signal was observed with the decrease of the intercalant concentration. This phenomenon can be accounted for if the contribution to the central resonance peak is attributed to the C_b layer and the line broadening results from increasing fraction of the C_i layer which provides broader signals [9]. The lack of the line broadening due to H_0 may indicate spillover of the valence electron to the C_i layer to some extent. The bromine compounds which are typical acceptor-type compounds showed a sharp contrast to the alkali metal compounds. The linewidth remained almost the same with changing the bromine concentration, while the width increased of H_0. As discussed above, the degree of ionization of bromine is several percents in the concentration-range up to 6 mol % of bromine so that only a small portion of the compounds can be regarded as the C_b layers substantially and the rest remains to be the C_i layers even at great concentration of bromine. Therefore, the large portion of the graphitic C_i layers should show H_0-dependence as the pure graphite does.

Table 2 The linewidth and Knight shift of the alkali metal and bromine intercalation compounds

Sample		^{13}C shift* Present results	Linewidth (G) 9 MHz	6 MHz (3 MHz)
Graphite (R.T.)		176 ± 10	5	4 (2)
	(77K)	180 ± 15	6.5	——
Donor-typc	C_8Rb	76 ± 5	1.4	——
	C_8K	81 ± 5	1.4	1.3
	$C_{24}Rb$	131 ± 4	1.4	1.4
	$C_{24}K$	128 ± 4	1.3	1.4
	$C_{36}Rb$	122 ± 5	1.6	——
	$C_{48}Rb$	123 ± 5	2.0	(2.1)
Acceptor-type	C-Br (at.%)			
	11.1	154 ± 10	3.2	2.8
	7.9	155 ± 10	3.6	2.8
	3.1	155 ± 10	3.6	——
	0.5	155 ± 10	5	3.7

*The values of ^{13}C shift are shown with respect to that of TMS

The ^{13}C signal did not show noticeable Knight shift on the intercalation, in spite of the definite change in the charge concentration in the C_b layers. This phenomenon indicates that the majority of electrons or holes reside in the p-character band.

References

1. S. Tanuma, Y. Iye, O. Takahashi, Y. Koike: Synthetic Metals 2, 341 (1980)
2. S.A. Solin: Physica (Utrecht) 99B, 443 (1980)
3. M.S. Dresselhaus, G. Dresselhaus, P.C. Eklund, D.D.L. Chung: Mater. Sci. Eng. 31, 141 (1977)
4. F. Batallan, I. Rosenman, C. Simon, G. Furdin, H. Fuzellier: Physica 99B, 411 (1980)
5. J. Conard, M. Gutierrez-le Brun, P. Lauginie, H. Estradeszwarckopf, G. Hermann: Synthetic Metals 2, 227 (1980)
6. D. Nixon, G.S. Parry: J. Phys. D 1, 291(1968)
7. M. Suganuma, T. Kondow, U. Mizutani: Phys. Rev. B23, 706 (1981)
8. T. Inoshita, K. Nako, H. Kamimura: J. Phys. Soc. Jpn. 43, 1237 (1977)
9. M. Suganuma, U. Mizutani, T. Kondow: Phys. Rev. B22, 5079 (1980)
10. F.J. Disalvo, S.A. Safran, R.C. Haddon, J.V. Waszczak, J.E. Fischer: Phys. Rev. B20, 4883 (1979)

Part VI

Transport, Conductivity, Superconductivity

Electrical Conductivity and Superconductivity in Graphite Intercalation Compounds

F. Lincoln Vogel, Richard Wachnik and Louis A. Pendrys

Laboratory for Research on the Structure of Matter
University of Pennsylvania, Philadelphia, PA 19104, USA

ABSTRACT

This paper will provide a current review of the electrical conductivity of the donor and acceptor intercalation compounds of graphite based on the research of the group at Penn and their collaborators.

In the lithium graphitides, which have a different stoichiometry than the graphitides with larger alkali atoms, many interesting characteristics appear: the conductivity is higher for stage one than stage two, the conductivities for both stages is rather high, $\sim 2 \times 10^5 \Omega^{-1}$, and the anisotropy of conductivity is only ten. For the potassium intercalated graphite at higher pressures, startling discontinuities in the c axis resistivity variation with pressure indicate transformations of the interplane ordering and concomitantly the intraplane packing. In other experiments, it was found that alteration of the planar density by alloying permits the synthesis of the unusual ternary series $K_xNa_{1-x}C_g$ where x is between 0.45 and 1. These ternaries were found to have lower in plane resistivities than the corresponding KC_g compounds.

It has been known for some time that the acid acceptor compounds that form liquid-like intercalated layers (HNO_3, UBBELOHDE) will, on cooling below 240 K, exhibit anomalies in the ρ (T) curve. This has been confirmed for compounds synthesized with SbF_5, H_2SO_4 (electrolytically) and NO_2SbF_6. X-ray structural examination of the SbF_5 compound reveals that ordering changes in the intercalated layer coincide with the abrupt changes of resistivity with temperature.

Low-temperature superconductivity measurements have been made on stage 1 and 2 K/Rb mercurographitides and thallographitides. Transitions have been observed in $KHgC_8$ (II) at 1.90 K and $RbHgC_8$ (II) at 1.44 K, but no transitions were observed in $KHgC_4$ (I) or $RbHgC_4$ (I) down to 1.3 K. On the other hand, transitions were observed in $KTl_{1.5}C_8$ (II) at 1.3 K. Suppression of the transition in a magnetic field was observed and anisotropy of the critical field as well.

1. Lithium Graphitides

Stages 1 and 2 lithium intercalated graphite were synthesized by BASU et al. [1] by immersion of highly oriented pyrolytic graphite (HOPG) samples in molten metal at ~350°C. Stage 1 was obtained from pure liquid Li while stage 2 resulted from a mixture of 3.8% Li in Na. Stage 1 has the stoichiometry of C_6Li and a $\vec{c}$ axis identity distance d_i = 3.706 $\pm$ 0.08 Å.

Room temperature electrical conductivity, σ_a, measurements were made parallel to the layer planes using a contactless eddy current technique operating at 1 kHz [2]. The results are shown in Fig. 1 and compared with those of potassium. Whereas potassium-graphite shows a maximum conductivity in the vicinity of stage 3, 4, while the stage 1 stoichiometry was found to be close to the ideal C_6Li, the stage 2 was closer to $C_{18}Li$ than C_{12} and so contains a large fraction of intercalant vacancies which probably explains the anomolous maximum.

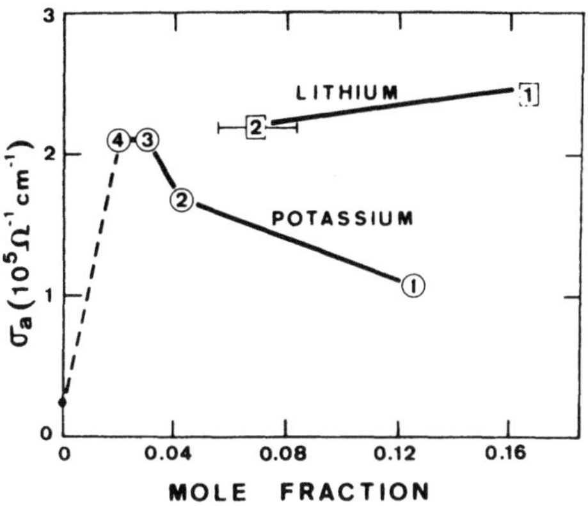

Fig. 1 Conductivity, σ_a vs. mole fraction for potassium and lithium graphitides.

Preliminary values have been obtained for σ_c, the conductivity perpendicular to the layer planes. These were done using a 1 kHz lock-in amplifier technique with gold foil pressure contacts on the 1 cm^2 c faces. The value of $\sigma_c = (1.5 - 2.0) \times 10^{-4} \Omega^{-1} cm^{-1}$ implies a σ_a/σ_c=10-15 for C_6Li, much less anisotropic than C_8K, for instance, where the ratio $\sigma_a/\sigma_c = 30$.

2. Resistivity vs Pressure for C_xK

Interesting effects, of both a continuous and discontinuous nature, were observed in experiments measuring the change of $\vec{c}$ axis resistivity vs. pressure.

The measurements were made in a hydrostatic cell by FUERST, MOSES & FISCHER [3] using pentane as the pressure medium. Resistivity was obtained from a dc four probe bridge with spring contacts. Samples, 5mm x 5mm x 0.2 mm in size were characterized as to stage by their 00ℓ spacings before but not during or after the experiment.

The relative resistivity of HOPG falls in a nearly straight line fashion from 1.00 to 0.63 as the pressure rises from zero to 22k bar. The intercalation compounds show much more detail, however. The odd stages C_8K, $C_{36}K$ and $C_{60}K$ show qualitative differences among themselves as seen in Fig. 2. For C_8K, R/R_0 increases very slowly at first while for $C_{36}K$ and $C_{60}K$ there is an initial small decrease, small effects compared to HOPG. All three compounds show anomalous sharp jumps in R/R_0 at higher pressures: C_8K at 15 k bar, $C_{36}K$ and $C_{60}K$ at 10 k bar. On the other hand, the even stages behave quite differently, $C_{24}K$ showing a marked initial decrease in R/R_0 and $C_{48}K$ showing a marked initial increase in R/R_0. There is an abrupt drop in R/R_0 for $C_{24}K$ at 3.3 k bar and the increase in R/R_0 for $C_{48}K$ reaches a peak at 9 k bar and then declines.

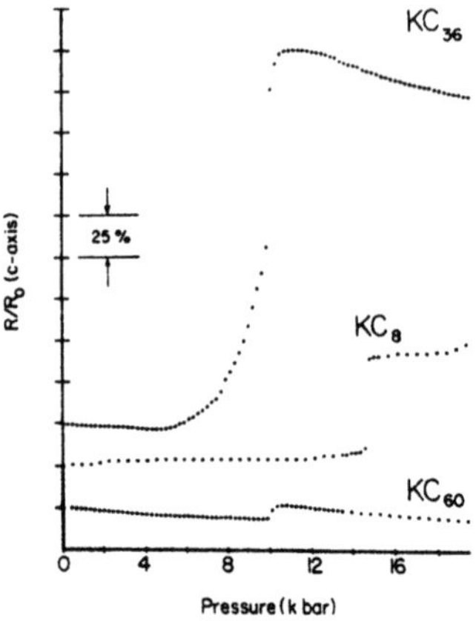

Fig. 2 Relative resistivity change R/R_0 as a function of pressure for stages 1, 3 and 5 C_kK

These results can be interpreted in relation to previously known effects on resistivity, i.e. the stage increase increases R_c by reducing the spatial overlap between screened metal layers, and densification reduces R_c because more charge per unit volume is available for conduction. The relative contribution at a particular transition determines whether R_c increases or decreases. The abrupt drop in R at 3.3 k bar of $C_{24}K$ was observed before by CLARKE, WADA and SOLIN [4] and is attributable to a phase transformation: $C_{12x2}K \rightarrow C_{8x3}K$, a stage 2 compound of metal planar density of one potassium per 12 carbon to a stage 3 compounds with metal planar density of one potassium per 8 carbons. In this case the densification dominates the stage change and we observe a net decrease in R. A small abrupt increase of R/R_0 of $C_{48}K$ at 7.9 k bar is attributable to a $C_{12x4}K \rightarrow C_{8x6}K$ transformation where the stage increase dominates giving a net increase in R.

The following table gives the resistivities of some of the pressure-induced structures.

Table 1

$C_{sxt}K$	C Axis Resistivity ohm-cm x 10^{-3}
$C_{8x1}K$	0.3
$C_{12x2}K$	7.5
$C_{8x3}K$	2.7
$C_{12x3}K$	16.0
$C_{12x4}K$	40
$C_{8x6}K$	88
$C_{12x5}K$	107
HOPG	120

Obviously, further detailed structural work is needed to explain these results.

3. Resistivity of Ternary NaK Compounds

While sodium does not react readily with graphite, it can be made to intercalate when combined with another alkali metal which forms a stable binary compound with graphite [5]. However, in the case where the other metal is potassium, the compounds form in a limited range of composition and are metastable, i.e. sodium is completely eliminated on prolonged contact with the alloy liquid or vapor.

BILLAUD et al. [6] have shown that the resistivities of the NaK ternary alloys, plotted in Fig. 3, are always slightly lower than the potassium compounds of the same stage. This is probably explained by the fact that the Na atom, being smaller than the K atom increases the charge per unit volume of the ternary.

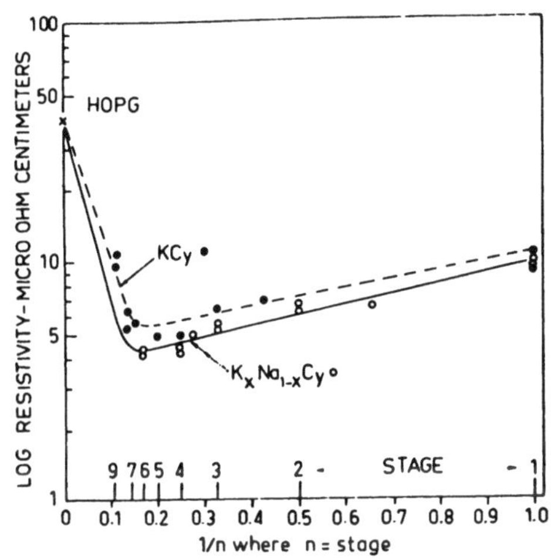

Fig. 3 Resistivity as a function of stage for potassium and sodium potassium graphitides

4. Transport of Acceptor Compounds with SbF_5 and $FeC\ell_3$

Resistivity and magnetoresistance measurements are made using an ac susceptibility technique developed at the University of Pennsylvania [7]. The essential part of the system consists of two transformers in parallel with the secondaries wound in opposition. An alternating current with 1 kHz frequency is passed through the primary coils and the output of the secondary is adjusted to be exactly zero with the help of some extra circuitry. When a sample is inserted between 1 pair of primary and secondaries, the change in mutual inductance causes an output voltage to be generated across the secondaries. The component of the voltage in phase with the exciting voltage is related to magnetic susceptibility, the magnitude of the voltage out of phase with the exciting signal is proportional to the conductivity. Using a phase sensitive detector, the in-phase signal, sensitive to the magnetic properties of a superconductor, and the out-of-phase signal, sensitive to the conductivity of normal state metals, can be observed. By applying a magnetic field one can measure either the critical field of a superconductor or magnetoresistance. Because of the orientation of the sample with respect to the coils the apparatus probes basal-plane conductivity and magnetic moments along the $\vec{c}$ axis. Although resistivity and magnetoresistance are not the most direct probes of the electronic properties of GIC's, they are sensitive to such properties and can provide information on charge transfer defect concentration, structure, etc. We are interested in relating these measure-

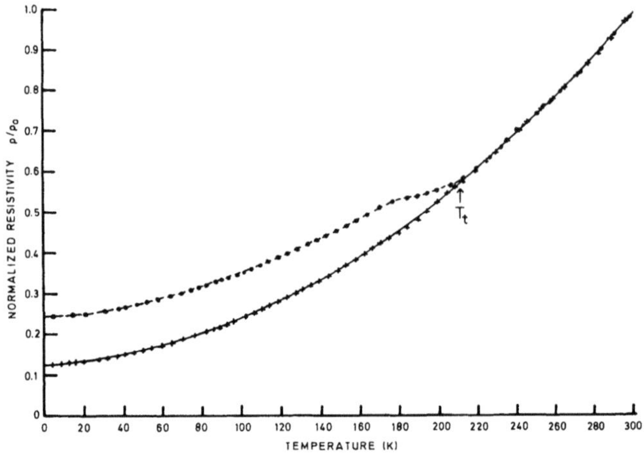

Fig. 4 Normalized resistivity vs. temperature for stage 1 graphite - SbF_5 where ρ_0 is ρ (T) at 300K (ρ_0 = 3.28μ Ωcm) slowly cooled + quenched •

ments to models accounting for low–field galvanomagnetic properties and using these models to interpret the variation of these properties with synthesis.

The resistivity as a function of temperature of SbF_5 intercalation compounds has been measured [8]. The ratio between the room temperature and residual resistivities is much smaller than that of most pure metals; it is approximately 5 to 10. Another interesting feature is the upward curvature in ρ vs. T at temperatures above about 200K where linear behavior is to be expected. The upward curvature has been explained as a result of scattering between various pockets of the Fermi surface.

Most striking, however, is the difference in cooling rapidly and slowly from room temperature to low temperature [9]. A rapidly cooled sample shows a much higher resistivity at low temperatures indicating that a disordered phase has been quenched in. The point at which the two curves meet is approximately the transition temperature and this temperature decreases with increasing stage. Furthermore, the difference between the resistivity of the rapidly and slowly cooled samples is smaller in higher stage samples. Since a quenched–in disordered phase is expected to interfere more with the conductivity contribution of the bounding layers than the interior layers, the smaller difference in resistivity in higher stages is expected. The existence of a transition has been confirmed by differential scanning calorimetry [9] and by x-ray diffraction.

DSC data show a small heat of transformation between 190K and 215K. X-ray diffraction shows the disappearance of several small peaks above the transition temperature. The diffraction data can be indexed using a unit cell in which one dimension is approximately 23 Å above T_t and 5.7 Å below T_t. The x-ray diffraction data did not yield any further information. A tempting idea at first was to assign the structural change to the ordering of a short polymer chains. The molecule SbF_5 is known to polymerize readily. Subsequent measurements by FLANDERS et al. [10] have shown that similar behavior occurs in many acceptor compounds - HNO_3, HSO_3F, AsF_5, and SbF_6^- CH_3NO_2 based compounds, which are not known to polymerize easily.

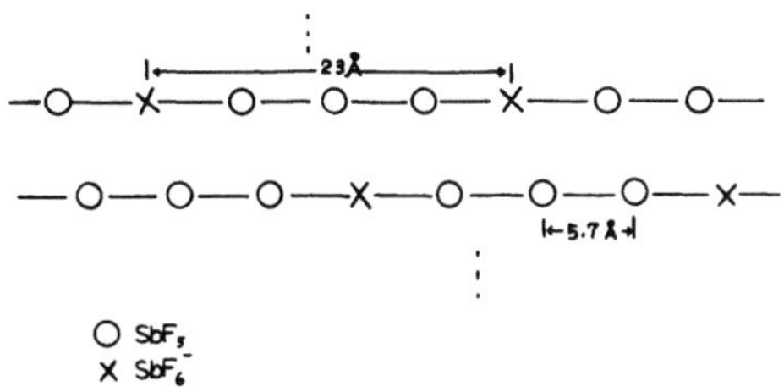

Fig. 5 Proposed ordered structure for SbF_5 graphite

The magnetoresistance of AsF_5 [7], SbF_5 [8], and $FeCl_3$ [11] graphite compounds has been measured; for SbF_5 and $FeCl_3$ graphite compounds several stages have been examined. For $FeCl_3$ graphites, the magnetoresistance decreases with decreasing stage. For SbF_5 graphites, the magnetoresistance decreases from Stage IV to Stage III and increases as one goes to lower stages. These results are not in good agreement with each other. Since the charge transfer is higher in the SbF_5 graphites, one may be measuring the magnetoresistance in a different regime of behavior. Comparison of high stage samples containing different intercalants shows another trend. The magnetoresistance coefficient, μ_m, is smaller in high stage SbF_5 compounds than in high stage $FeCl_3$ compounds. There seems to be a general trend in high stage compounds; higher charge transfer correlates with lower μ_m.

5. Superconductivity of Ternary Graphitides

Superconductivity has been studied in several ternary graphite intercalation compounds using an ac induction technique. The mercurographitides, $KHgC_8$ and $RbHgC_8$, were observed to have superconducting transition temperatures,

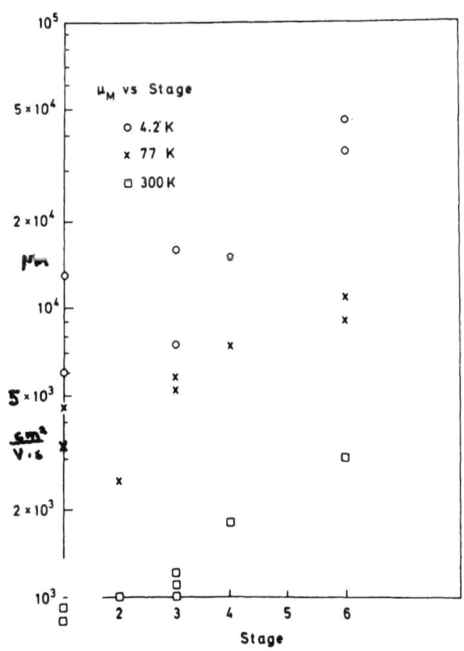

Fig. 6 Magnetoresistance coefficient as a function of stage at three temperatures for C_nFeCl_3

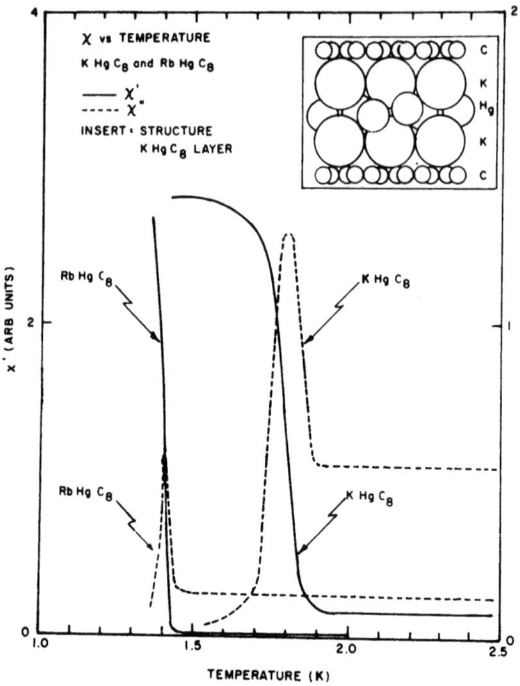

Fig. 7 Magnetic susceptibility vs. temperature for C_8KHg and C_8RbHg. The ac exciting fields were 0.2 Oe and 2.0 Oe for C_8RbHg and C_8KHg, respectively. Insert: proposed structure

Tc's, at 1.90K and 1.44K, respectively. These values are in agreement with specific heat results reported by ALEXANDER et al. [12]. TANUMA [13] has also observed a Tc at 1.9K in $KHgC_8$ in addition to a Tc of 0.723 for $KHgC_4$ [14] using an ac susceptibility technique. The thallographitides, $KT\ell_{1.5}C_8$ and $KT\ell_{1.5}C_4$, were observed by us to have TC's of 1.31K and 2.70K, respectively. Superconductivity in graphite intercalation compounds was first reported by HANNAY et al. [15] in KC_8, RbC_8 and CsC_8. The maximum Tc observed was 550 mK. The recently studied mercurographitides and thallographitides exhibit a considerable degree of anisotropy in their magnetic properties.

The thallographitides are similar to the mercurographitides in that each intercalated layer is composed of two hexagonal layers of alkali metal atoms sandwiching a layer of Hg of Tℓ atoms. This novel structure is shown in the insert of Fig. 7 for a typical mercurographitide.

Mercurographitides are prepared by intercalating HOPG (highly oriented pyrolytic graphite) with K or Rb to form the first stage compounds, KC_8 or RbC_8 [16]. Each sample is transferred to a new tube containing the stoichiometrically required quantity of Hg, evacuated, sealed, and then allowed to react at 100°C. The thallographitides were prepared by heating graphite in the presence of the stoichiometrically required amounts of KTℓ (or RbTℓ) with Tℓ. Samples are air sensitive and therefore were placed in sealed ampoules for susceptibility measurements.

Typical data for both phases of the susceptibility, x' and x'', are shown as a function of temperature in Fig. 7 for $KHgC_8$ and $RbHgC_8$. The signal shown for the $RbHgC_8$ compound was obtained using an ac exciting field of

$\approx$ 0.2 Oe. The signal for the $KHgC_8$ was obtained using an ac field of $\approx$ 2 Oe. The values of the Tc's obtained from susceptibility curves are 1.90K for $KHgC_8$, 1.44K for $RbHgC_8$, 2.70K for $KT\ell_{1.5}C_4$ and 1.31K for $KT\ell_{1.5}C_8$. Specific heat results by ALEXANDER et al. at the University of Delaware leave no doubt that superconductivity in the mercurographitides is a bulk phenomenon.

Typical curves of differential susceptibility versus applied external field are shown for the $KHgC_8$ system at T = 1.41K in Fig. 2 where the magnetic field, $\vec{H}_{app}$, is applied parallel and perpendicular to the $\vec{c}$-axis of the sample.

The upper critical field, H_{c2}, was defined as the intercept of a straight line drawn tangent to the linear region of the diamagnetic curve with the normal state value. Upper critical fields were determined for $KHgC_8$ and $KT\ell_{1.5}C_4$ at several temperatures. Values of critical fields as a function of temperature for $\vec{H}_{app} \parallel \vec{c}$ and $\vec{H}_{app} \perp \vec{c}$ obtained for $KHgC_8$ are shown in the insert of Fig. 8. For the stage II compound $KHgC_8$, H_{c2} versus T exhibits a linear dependence for $\vec{H}_{app} \parallel \vec{c}$ and a positive upward curvature for $\vec{H}_{app} \perp \vec{c}$. For the stage I compound $KT\ell_{1.5}C_4$, the H_{c2} versus T curves exhibit positive upward curvature for both $\vec{H}_{app} \parallel \vec{c}$ and $\vec{H}_{app} \perp \vec{c}$. Although the nature of the H_{c2} versus T curves make it difficult to calculate exactly the values of $H_c(0)$, it is possible to make rough estimates. For $KHgC_8$, the values are $\approx$ 2k Oe and $\approx$ 80 Oe for $\vec{H}_{c2} \perp \vec{c}$ and $\vec{H}_{c2} \parallel \vec{c}$. For $KT\ell_{1.5}C_4$, the values are $\approx$ 20k Oe and $\approx$ 4k Oe for $\vec{H}_{c2} \perp \vec{c}$ and $\vec{H}_{c2} \parallel \vec{c}$.

Specific heat results by ALEXANDER et al. [12] yield a value of the thermodynamical critical field of 78 Oe for $KHgC_8$. Since the experimental values of the critical field are higher than this, the compound is assumed to be a type II superconductor.

In all cases, the Ginzburg Landau parameter, K, appears to be temperature dependant. The ratio of the effective masses, given by $\varepsilon^2 = \frac{H_{c2\perp}}{H_{c2\parallel}} = \frac{m_c^*}{m_a^*}$, of $KHgC_8$ increases with decreasing temperature, equalling 25 at Tc and

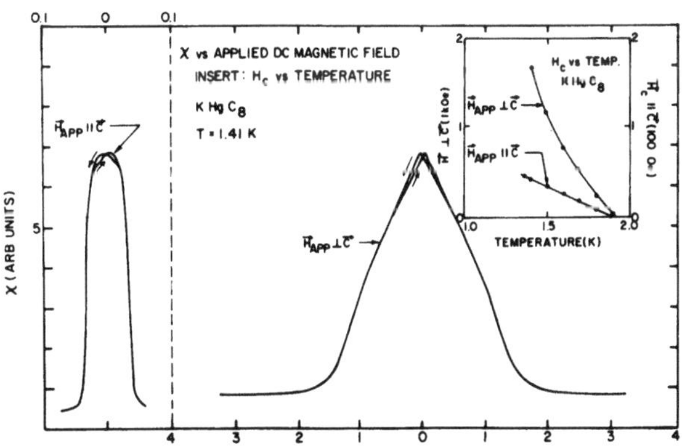

Fig. 8 Differential susceptibility as function of dc field parallel and perpendicular to $\vec{c}$ axis. Insert: critical field vs temperature

equalling 40 at a reduced temperature of 0.74. The effective mass ratio of $KTl_{1.5}C_4$ is temperature independent and is equal to ≈5.

In conclusion, basal-plane superconductivity has been studied by an ac susceptibility technique. The stage I compound $KTl_{1.5}C_4$ was observed to have a Tc of 2.7K, the highest value of a graphite intercalation compound to date. Critical field behavior has been found to be anisotropic in $KHgC_8$ and $KTl_{1.5}C_4$. A temperature dependant anisotropy in critical field was observed for $KHgC_8$ with a value of 25 at Tc which increases with decreasing temperature. A temperature independant anisotropy of critical field was observed for $KTl_{1.5}C_4$ with a value of ≈5.

REFERENCES

1. S. Basu, C. Zeller, P.J. Flanders, C.D. Fuerst, W.D. Johnson and J.E. Fischer, Mat. Sci. Eng. 38 (1979) 275.

2. C. Zeller, A. Denenstein, and G.M.T. Foley, Rev. Sci. Inst. 50 (1979) 602.

3. C.E. Fuerst, D. Moses and J. Fischer, p. 94, Extended Abstracts 15th Biennial Conference on Carbon, June 22-26, 1981, Philadelphia, Pa., American Carbon Society.

4. R. Clarke, N. Wada and S.A. Solin, Phys. Rev. Lett. 44 (1980) 1616.

5. D. Billaud and A. Herold, Bull Soc. Chim 11 (1974) 2407.

6. D. Billaud, A. Herold and F.L. Vogel, Mat. Sci. Eng. 45 (1980) 55.

7. C. Zeller, L.A. Pendrys and F.L. Vogel, J. Mat. Sci. 14 (1979) 2241.

8. T.C. Wu, F.L. Vogel, L.A. Pendrys and C. Zeller, Mat. Sci. Eng. 47 (1981) 161.

9. T.C. Wu and F.L. Vogel, submitted to Carbon.

10. P.J. Flanders, M.J. Moran, J.E. Fischer, D. Billaud, G. Furdin, H. Fuzellier and A. Metrot, p. 98, Extended Abstracts 15th Biennial Conference on Carbon, June 1980, University of Pennsylvania, American Carbon Society.

11. R.A. Wachnik, L.A. Pendrys and F.L. Vogel, Extended Abstracts, 15th Biennial Conf. on Carbon, June 1980, University of Pennsylvania and J.B. Perrachon, Masters Thesis, University of Pennsylvania, December, 1978.

12. M.G. Alexander, D.P. Goshorn, D. Guerard, P. Lagrange, M. El Makrini and D.G. Onn, Synthetic Metals, 2, No. 344 (1980) 203; M.G. Alexander, D.P. Goshorn, D. Guerard, P. Lagrange, M. El Makrini and D.G. Onn, to be published in Solid State Commun.

13. S. Tanuma, to be published in Proceedings of the YAMADA Conference IV on Physics and Chemistry of Layered Materials, September 8-10, 1980.

14. Y. Iye, O. Takahashi, S. Tanuma and Y. Koike, Fifteenth Biennial Conference on Carbon, Philadelphia, (June 22-26, 1981), 54.

15. N.B. Hannay, T.H. Geballe, B.T. Matthias, K. Andies, P. Schmidt and D. MacNair, Phys. Rev. Letters 14, (1965) 225.

16. M. El Makrini, P. Lagrange, D. Guerard and A. Herold, Carbon 18, (1980), 211; M. El Makrini, P. Lagrange, D. Guerard and A. Herold, C.R. Acad. Sci. Paris Serie C 288, (1979) 303.

Superconductivity of Graphite Intercalation Compounds with Alkali and Alkali-Amalgam

Sei-ichi Tanuma, Yasuhiro Iye and Yōji Koike

The Institute for Solid State Physics
The University of Tokyo
Roppongi, Minato-ku, Tokyo 106, Japan

1. Introduction

In the last several years, the research area of "synthetic metals" has enjoyed a rapid growth. The growth is not only of the number of new materials but also of new physical phenomena or concepts. Of the family of "synthetic metals", a good number of members are superconductors, and superconductivity is one of the most intensively studied subjects on these materials. In fact, an expectation for high T_c superconductors has been one of the main forces which have driven physicists into this field of research.

A synthetic superconductor, in its narrowest sense, may be defined as a superconductor synthesized solely from non-superconductive elements. $(SN)_x$, $(SNBr_y)_x$, $(TMTSF)_2PF_6$, $(TMTSF)_2ClO_4$, C_8K, etc. are in this category. But in general, the name may be used in a wider sense, simply meaning a superconductive synthetic metal. Defined in this way, it can include $Hg_{3-\delta}AsF_6$, some of the intercalated transition metal dichalcogenides, and graphite intercalation compounds with alkali-amalgams.

Among them the group of intercalated layered materials is greatest in number. A main point of interest in these materials is the mechanism of T_c-enhancement (or even a transition from a non-superconductor to a superconductor) upon intercalation. This point is important because a solution of this problem will lead to a guiding principle for a designed synthesis of superconductors. Another point of interest is upon the large anisotropy of superconductivity observed in these materials. This point attracts physicists' interest in connection with the problem of dimensionality and superconductivity.

This paper deals with the superconductivity of graphite intercalation compounds (GIC's). One point of great interest about the superconductivity of GIC's is the relation between charge transfer and superconductivity. It is well known that the electronic band structure of a GIC can be regarded in a first approximation as consisting of the graphite rigid band and the intercalant band. In the case of partial charge transfer, carriers exist both in the graphite band and intercalant band. A naive, but fundamental, question may then be presented; which is superconducting, graphite carriers or intercalant carriers?

Before describing the present work, we would like to give a historical review in order to see the present status of the research on superconductive GIC's. Among GIC's, superconductivity was first found by Hannay et al.[1], as early as 1965, in the first stage alkali metal GIC's (C_8K, C_8Rb and C_8Cs).

The transition temperatures they reported were 0.39∿0.55 K for C_8K, 0.023∿ 0.150 K for C_8Rb, and 0.030∿0.135 K for C_8Cs. These superconductivities, however, were not confirmed by other researchers; for instance, Poitrenaud [2] obtained a negative result for superconductivity of C_8K down to 0.30 K. Recently, the superconductivity of C_8K was re-discovered with lower T_c, by Kobayashi et al.[3] and by our group [4,5]. Kobayashi et al. found T_c of 0.080 K for a sample made from graphite powder, and 0.125 K for a sample based on grafoil. We have investigated 13 samples of C_8K synthesized from highly oriented pyrolitic graphite (HOPG) to find transition temperatures ranging from 0.128 to 0.198 K. As for C_8Rb and C_8Cs, recent experiments gave negative results down to 0.09 K for C_8Rb [6] and to 0.06 K for C_8Cs [7].

More recently, a new family of superconductive GIC's were found. Alexander et al.[8] observed an anomaly in the low temperature specific heat of the second stage potassium-amalgam GIC, C_8KHg, at 1.93 K, which was indicative of the onset of superconductivity. The superconductivity of C_8KHg was subsequently confirmed by Pendrys et al.[9] and by our group [10] through observation of the Meissner effect. Pendrys et al.[9] also found a superconducting transition of the second stage rubidium-amalgam GIC, C_8RbHg, at 1.44 K.

In contrast to the above-mentioned second stage alkali-amalgam GIC's, superconductivity has not so far been found in the corresponding first stage compounds, C_4KHg and C_4RbHg. This was very puzzling, because the first stage compounds have higher density of states at Fermi level and hence are expected to be more favorable for superconductivity. We report here, the first observation of superconducting transition of C_4KHg and C_4RbHg. The transition temperatures are, however, lower than the corresponding second stage compounds, so that the above question still remains; why the second stage compounds with lower carrier concentration have higher T_c's than the first stage compounds. We have also studied C_8KHg and C_8RbHg to reconfirm the reported T_c's. We carried out a detailed study on the super/normal transitions under magnetic fields of the four alkali-amalgam GIC's. The anisotropy of superconductivity of these materials will be discussed in comparison with each other and with C_8K.

In order to get a clue to solve the question — which is superconducting, graphite layer or intercalant layer — we investigated the superconductivity of the mother materials of these compounds, KHg and RbHg alloys. Comparison of the results with those on alkali-amalgam GIC's offers a new viewpoint in discussing the superconductivity of GIC's.

The next section is devoted to a description of experimental procedures. Experimental results will be presented in §3, and discussion will be given in §4. As for C_8K, experimental details are fully described in [5], so that only the relevant results will be quoted.

2. Experimental

Highly oriented pyrolytic graphite (HOPG) was used as host material. Synthesis of the first stage potassium-amalgam GIC, C_4KHg, was done by the method reported by Lagrange et al.[11,12]. First, potassium-amalgam KHg was prepared by reacting one-to-one molar ratio of distillation purified potassium and mercury in a sealed pyrex tube at 260°C for a day. The amalgam was sealed in a new tube with one or more HOPG chips, and the tube was kept at 200°C for a week. Formation of the first stage compound C_4KHg was readily judged by its copper-pink luster.

As for the second stage potassium-amalgam GIC, C_8KHg, two different methods of synthesis are reported. One is to react graphite with one-to-two molar ratio amalgam KHg_2 [11]. Samples used in [10] were synthesized by this method. The other method is to react pre-synthesized C_8K with mercury vapor. In the present work, we employed this method of synthesis. The sample of Pendrys et al.'s work [9] was also synthesized by this method. Transformation from C_8K to C_8KHg was witnessed by change in the luster from brass-yellow of C_8K to steel-blue of C_8KHg.

Syntheses of C_4RbHg and C_8RbHg were essentially same as above except that slightly lower reaction temperatures were employed.

Since all the intermediate products as well as the final products were air-sensitive, operations were all done in a glove box which was first evacuated by an oil diffusion pump and then filled with helium gas passed through a cold trap at 77 K. Air-sensitiveness of the alkali-amalgam GIC's, however, is less severe than alkali metal GIC's, because even when directly exposed to air, their surface luster lasts fairly long compared to alkali metal GIC's.

Samples were characterized by (00l) diffraction patterns of Cu-Kα X-ray, as shown in Fig.1. As seen in the figure, the first stage samples show sharp diffraction peaks indicating uniform structure, except that the sample #RF-1 shows small C_8Rb peaks. Compared with these, the X-ray patterns of the second stage samples are less sharp, possibly due to inhomogeneously distributed remnants of C_8K and C_8Rb. Influence of the remnants on the superconductivity is not clear, but since C_8K and C_8Rb are non-superconductive in the temperature range of present work, it can be neglected in a first approximation.

A sample used in the superconductivity measurement was taken out of the reaction tube in the glove box, wrapped with Parafilm (American Can Co.), mounted onto the cryostat and then cooled quickly. A helium-3 evaporation cryostat was used to attain temperatures down to 0.40 K. Temperature was measured by a carbon resister thermometer calibrated against ^3He saturated vapor

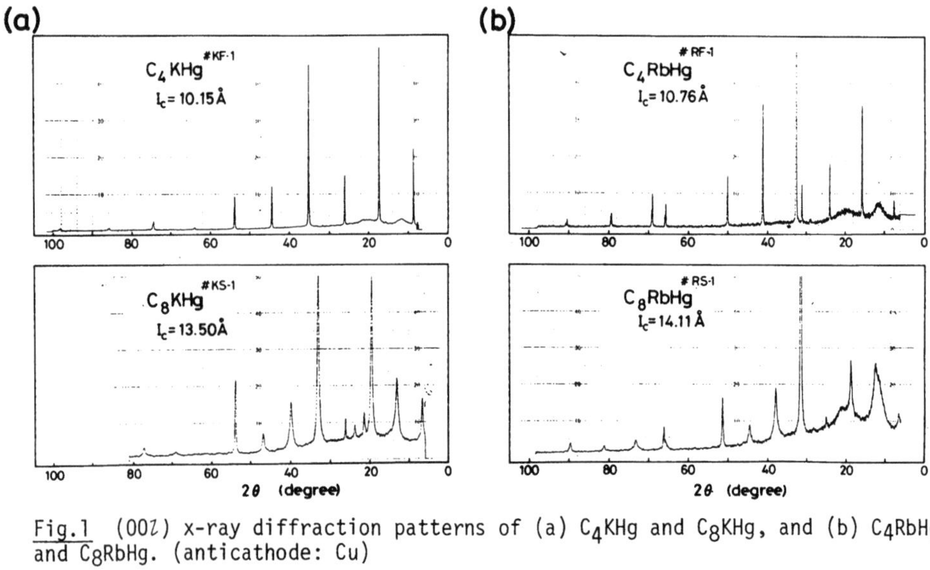

Fig.1 (00l) x-ray diffraction patterns of (a) C_4KHg and C_8KHg, and (b) C_4RbHg and C_8RbHg. (anticathode: Cu)

pressure. The superconducting transition was detected by measurement of a.c. susceptibility by using a simplified Hartshorn bridge. The frequency and amplitude of the modulation field were 27 Hz and 0.2 Oe, respectively.

3. Results

3.1 A.C. Susceptibility

Figure 2-a shows typical recorder traces exhibitting super/normal transition of C_8K under a magnetic field applied parallel or perpendicular to the c-axis. A remarkable feature is that under the parallel field condition, the susceptibility curve shows a clear supercooling effect, while in the perpendicular field geometry no hysterisis is observed. Since the supercooling effect is characteristic of a type I superconductor, C_8K in the parallel field geometry is regarded as a type I superconductor. Hence, the observed critical field

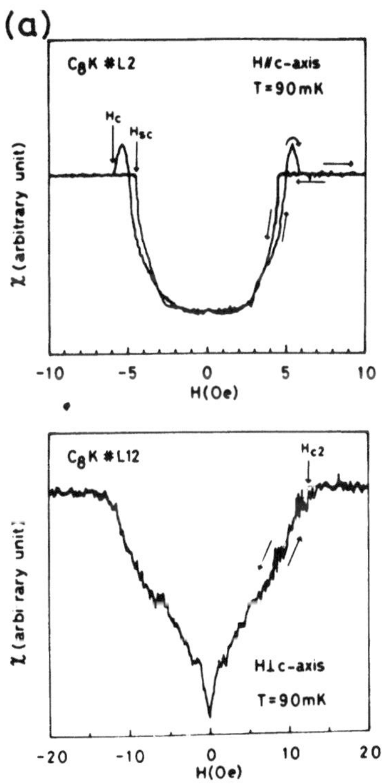

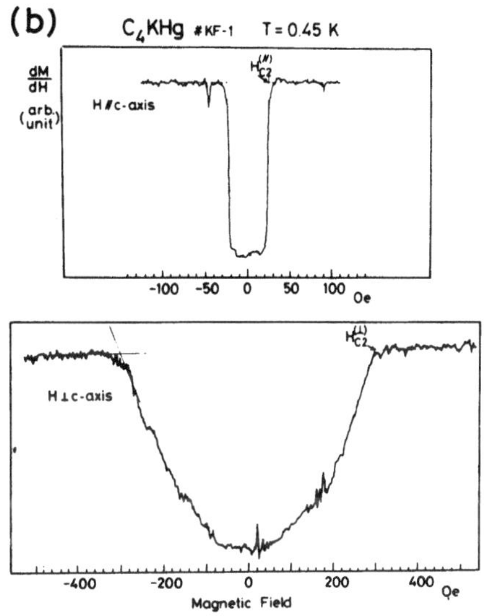

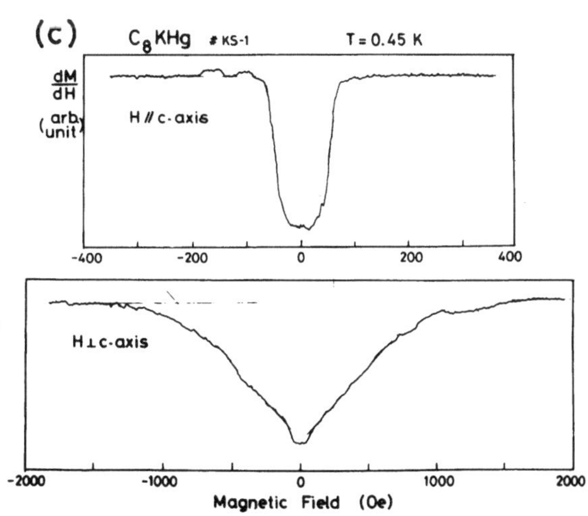

Fig.2 The a.c. susceptibility of (a) C_8K, (b) C_4KHg and (c) C_8KHg, as functions of the static magnetic field applied parallel (the upper figures) or perpendicular (the lower figures) to the c-axis.

in the up-sweep of the magnetic field corresponds to the thermodynamic critical field H_c, and the critical field in the down-sweep is the supercooling field H_{sc}. On the other hand, the behavior under the perpendicular field condition is of type II, with a value of the upper critical field H_{c2} more than twice as large as H_c.

Figure 2-b and c are examples of recorder traces showing super/normal transition of C_4KHg and C_8KHg under magnetic field. The susceptibility curves for C_4RbHg and C_8RbHg are quite similar to those of C_4KHg and C_8KHg, respectively. That the observed change in susceptibility corresponds to approximately full Meissner value is assured by a sensitivity calibration of the detection system by using known superconductors of the same size as GIC samples. In contrast to the case of C_8K, the supercooling effect is absent. Thus, the superconductivity of alkali-amalgam GIC's is considered to be of type II, as are the superconductivities of alloys in general.

We assign the observed critical field to the upper critical field H_{c2}. The way we determined H_{c2}-values from experimental curves is illustrated in the figures. For the first stage compounds, this procedure was obvious as seen in Fig.2-b. In the case of the second stage compounds under the perpendicular field, the susceptibility curve showed a long tail extending towards higher fields, as has been reported by Pendrys et al.[9] on the same compound. Such is the case, the experimental definition of $H_{c2}^{\perp}$ of the second stage compounds has some ambiguities. We adopted here a definition of $H_{c2}^{\perp}$ by the intercept of a straight line drawn tangent to the linear portion of the susceptibility curve, with the normal state level.

3.2 Angular Dependence of Critical Field

Figure 3-a shows the dependence of the critical field of C_8K on the field direction with respect to the c-axis. Figures 3-b$\sim$e are those for C_4KHg, C_8KHg, C_4RbHg and C_8RbHg, respectively.

In the case of a layered superconductor, one of the simplest models which can be used to explain the anisotropy of H_{c2} is to assume an elliptical fluxoid [13]. Ginzburg-Landau (GL) theory [14] for type II superconductors gives the following expression relating H_{c2} to the fluxoid.

$$H_{c2} = \frac{\phi_0}{2\pi\xi_\alpha\xi_\beta}\,, \tag{1}$$

where ϕ_0 is the flux quantum and ξ_α and ξ_β are two principal coherence lengths in the plane perpendicular to H. For a layered superconductor, the coherence length is assumed to be isotropic within the layer ($\xi_a = \xi$) and for the c-direction, it takes a different value ($\xi_c = \varepsilon\xi$). For H parallel to the c-axis, the fluxoid is circular in cross-section, while for other directions it is elliptical. The elliptical fluxoid model with H_{c2} given by (1) results in the following expressions for the angular dependence of H_{c2}.

$$H_{c2}(\theta) = \frac{\phi_0}{2\pi\xi^2}\,\frac{1}{(\cos^2\theta + \varepsilon^2\sin^2\theta)^{1/2}} = H_{c2}^{\parallel}\,\frac{1}{(\cos^2\theta + \varepsilon^2\sin^2\theta)^{1/2}}, \tag{2}$$

where θ is the angle between H and the c-axis. In this model, the coherence lengths are assumed to be large compared with the layer spacing. When the coherence length in the c-direction is on the order of layer spacing, the validity of the model may be questionable.

The critical quantity determining the H_{c2}-anisotropy is the ratio of coherence lengths in the layer plane and in the c-direction. This ratio is often

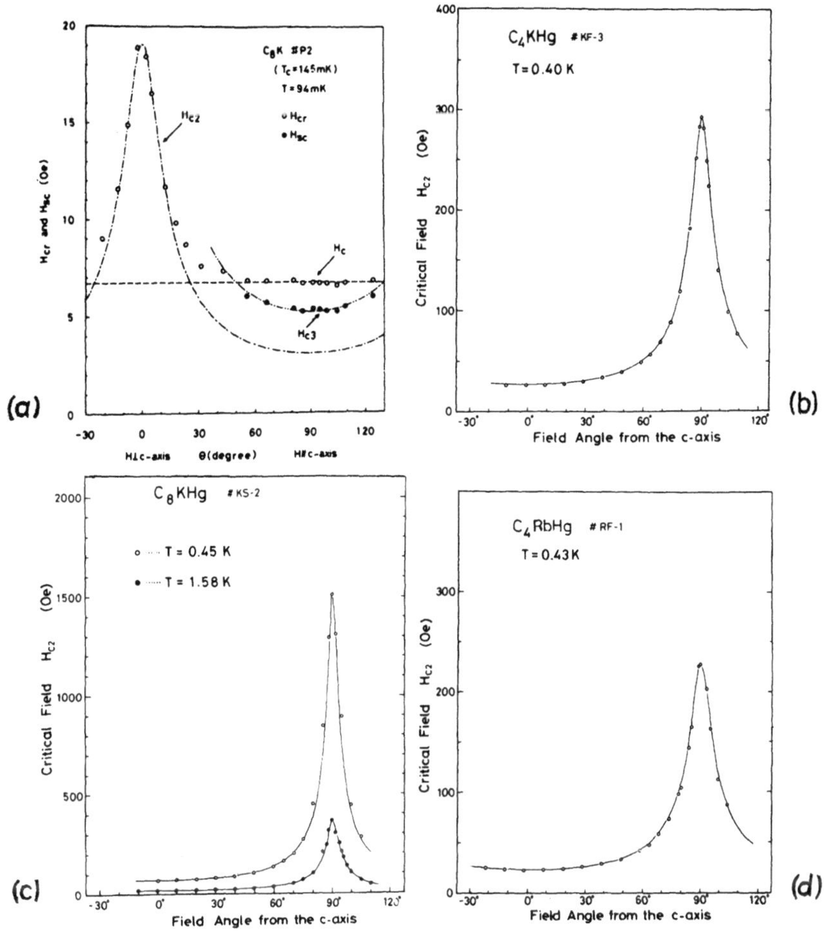

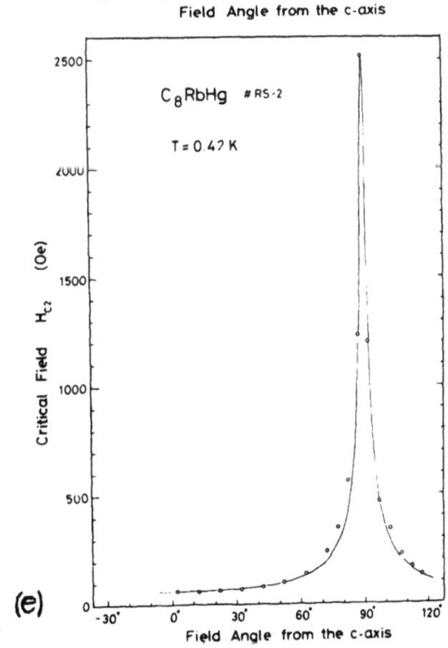

Fig.3 The angular dependence of the critical field of (a) C_8K, (b) C_4KHg, (c) C_8KHg, (d) C_4RbHg and (e) C_8RbHg. The curves in the figures show the fitting of the angular dependence predicted from the effective mass model.

expressed by the ratio of the effective masses which describes the electron conduction parallel and perpendicular to the layers.

$$\varepsilon = \frac{\xi_c}{\xi_a} = \left(\frac{m_a}{m_c}\right)^{1/2} \quad (3)$$

In this sense, the above model for anisotropic H_{c2} is often referred to as the effective mass model. The anisotropic effective mass introduced here is a generalized one and can include contributions from the band effective mass anisotropy as well as anisotropy in the electron-phonon interaction.

In Fig.3, fittings of the angular dependence given by (2) are shown. It is seen that the angular dependences of H_{c2} are well explained for alkali-amalgam GIC's. The model is not directly applicable to C_8K because of the type I behavior near H // c-axis. H_{c2} in this case was evaluated by using a relation $H_{c2} = H_{c3}/1.695$ after assuming the surface nucleation field H_{c3} to be equal to the observed supercooling field H_{sc}. The reasonable fit of the model seen in Fig.3 indicates that these materials are three-dimensional superconductors with varying degree of anisotropy, rather than two-dimensional ones.

3.3 Temperature Dependence of Critical Field

The temperature dependences of the parallel and perpendicular critical fields are shown in Fig.4. From (2) and the temperature dependence of the coherence length near T_c, $\xi(T) = \xi(0)(1-T/T_c)^{-1/2}$, the temperature dependence of the parallel and perpendicular coherence lengths are given by

$$H_{c2}^{/\!/}(T) = \frac{\phi_0}{2\pi\xi^2(0)}\left(1 - \frac{T}{T_c}\right), \quad H_{c2}^{\perp}(T) = \frac{\phi_0}{2\pi\,\varepsilon\,\xi^2(0)}\left(1 - \frac{T}{T_c}\right) \quad (4)$$

Linear temperature dependences are seen in Fig.4, near T_c. For the perpendicular critical field, so-called positive curvatures are observed at sufficiently low temperatures in a few cases. Such positive curvatures have been also reported in $H_{c2}^{\perp}$ of other layered superconductors such as 2H-$NbSe_2$[15].

4. Discussion

Fundamental quantities of the superconductivity of the GIC samples studied in this work are summarized in Table 1. For C_8KHg, Koike et al.'s data [10] and Pendrys et al.'s data [9] are also shown for comparison.

First, the values of T_c are in a reasonable agreement with each other. The problem of T_c will be discussed afterwards. Secondly, as to the H_{c2}-anisotropy ratio, the agreement is fairly good for C_8K, C_4KHg but is poor for second stage GIC's. For example, the anisotropy ratio of C_8KHg ranges from 6.7 to 31 depending on the experimental group. Such a large discrepancy is chiefly due to the scatter of $H_{c2}^{\perp}$ values. One origin of such a scatter in the second stage samples is the ambiguity in the experimental definition of H_{c2} as mentioned in the previous section. This is however not the whole reason for the discrepancy, since Pendrys et al.[9] and we have adopted essentially the same definition of $H_{c2}^{\perp}$.

This problem can be put in a clearer form if we talk in terms of the coherence length calculated by using (4) and listed in Table 1. It is noted that the coherence length in the layer plane is in good agreement with each other, while that in the c-direction shows large sample dependence especially for C_8KHg and C_8RbHg. Since the c-axis coherence length is basically related to the c-axis electron conduction, it is understandable that it is defect-sensitive and hence shows a large sample dependence. It is worth commenting here that #A1 and #A2 (Koike et al.'s sample)[10] were synthesized by reaction of graphite with KHg_2, while other samples were made via C_8K. The difference

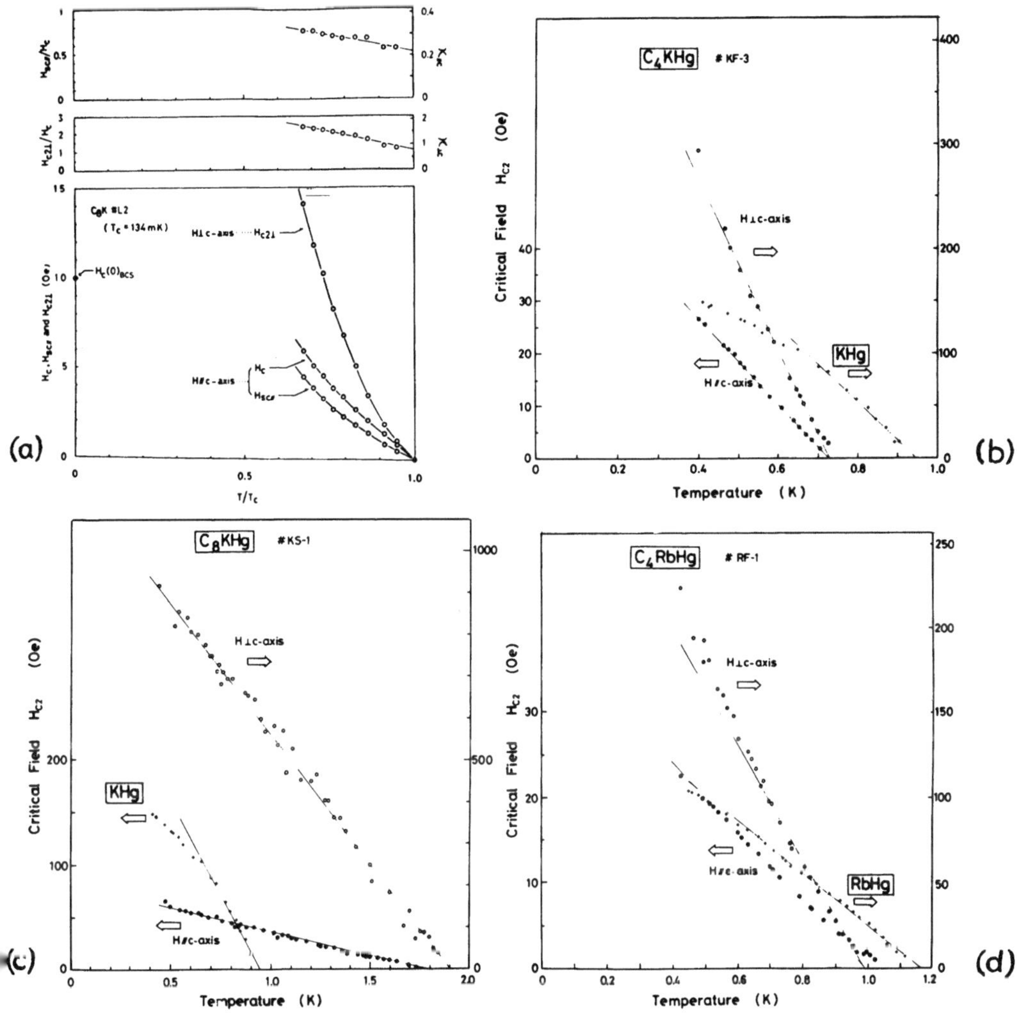

Fig.4 The temperature dependence of the critical field of (a) C_8K, (b) C_4KHg, (c) C_8KHg, (d) C_4RbHg and (e) C_8RbHg. In (b) ∿ (e, see next page), the data on KHg and RbHg are also shown.

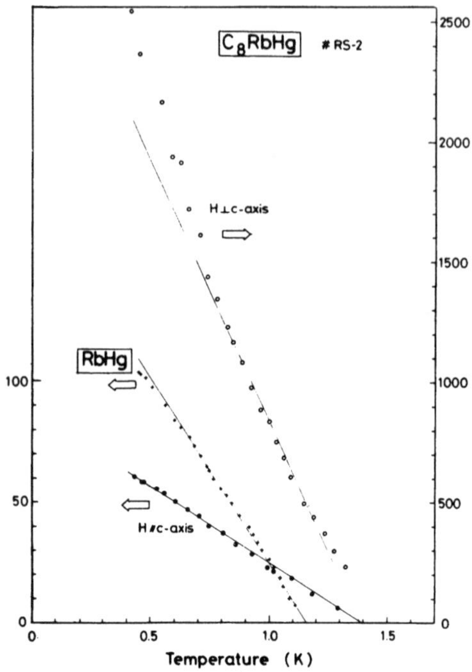

(e)

Fig. 4e.

in the sample preparation method might be related with the observed systematic difference in the anisotropy ratio.

Table 1 Summary of experimental data on the superconductivity of GIC's

Sample	T_c	$H_{c2}^{\perp}/H_{c2}^{\parallel}$	$T_c(\frac{dH_{c2}^{\parallel}}{dT})$	$T_c(\frac{dH_{c2}^{\perp}}{dT})$	$\xi_a(0)$	$\xi_c(0)$
#L12 [5] C_8K	0.134 K	4.7	5.4 Oe	21.4 Oe	7800 Å	2000 Å
#P2 [5]	0.145	6.2	3.7	23.2	9400	1500
#KF1 C_4KHg	0.72	11.5	62.4	715	2300	200
#KF2	0.86	10.4	49.2	654	2600	200
#KF3	0.73	11.0	58.8	575	2400	240
#KS1 C_8KHg	1.90	15.5	81.2	1180	2000	140
#KS2	1.94	21.7	92.2	2050	1900	85
#A1 [10]	1.90	31	131	4370	1600	48
#A2 [10]	1.70	30	126	3910	1600	52
— [9]	1.90	6.7	95	630	1900	280
#RF1 C_4RbHg	0.99	10.2	40.3	359	2900	320
#RH1* $C_4RbHg+C_8RbHg$	1.28	9.7	66.0	564	2200	261
#RS1 C_8RbHg	1.40	20.3	154	2700	1500	83
#RS2	1.40	41.3	87.9	2940	1900	58
— [9]	1.44	—	—	—	—	—

* It is interesting that the sample #RH1, which was proved by the X-ray study to be a mixed crystal of C_4RbHg and C_8RbHg, showed a single-step transition at $T_c = 1.28$ K, instead of a two-step transition corresponding to each stage.

Table 2 summarizes the comparison of superconductivities of C_8K, C_4KHg and C_8KHg. The data of Debye temperature are those reported by Alexander et al. [8]. The values of N(0)V listed in the third column, where N(0) is the density of states at the Fermi level and V is the pairing potential, are calculated by substituting the values of T_c and the Debye temperature Θ_D into the BCS formula [16],

$$T_c = 1.13\,\Theta_D \exp\left(-\frac{1}{N(0)\,V}\right). \tag{5}$$

If we use the McMillan's formula [17]

$$T_c = \frac{\Theta_D}{1.45}\exp\left(-\frac{1.04(1+\lambda)}{\lambda-\mu^*(1+0.62\,\lambda)}\right) \tag{6}$$

with an often adopted assumption, $\mu^* = 0.1$, we obtain the values of the electron-phonon coupling constant λ shown in the fourth column.

Table 2 Comparison of superconductivities of C_8K, C_4KHg and C_8KHg.

	T_c (K)	Θ_D (K)	N(0)V	λ
C_8K	0.14	374	0.13	0.30
C_4KHg	0.73	269	0.17	0.38
C_8KHg	1.90	301	0.19	0.45

The result shown in Table 2 indicates that these GIC's are all in the category of weakly-coupled superconductors, and that the large difference in the transition temperatures should be attributed to the elctron-phonon coupling. On the latter point, it must be added that the simple-minded assumption $\mu^* = 0.1$ may be questionable for these materials. An appropriate treatment of the Coulomb interaction is indispensable for a quantitative estimation of T_c of GIC's.

Let us now return to the question addressed in the first section, i.e. whichever is superconducting, graphite layer or intercalant layer? Although a definite answer to this question cannot be given at present, the following can be pointed out. Table 3 shows the existing data of the superconducting transition temperatures of KHg and RbHg alloys. From the fact that both alkali-amalgam GIC's and their mother alloys have T_c's of the same order of magnitude, it is guessed that superconductivities of these are essentially of same nature. Hence it is believed that electrons in the intercalant layer rather than the graphite layer play a dominant role for the superconductivity of alkali-amalgam GIC's.

Table 3 The transition temperatures of alkali-amalgams.

	Hg†	KHg_8†	KHg_4†	KHg_3†	KHg_2†	KHg*	RbHg*
T_c (K)	4.15	3.42	3.27	3.18	1.20	0.94	1.17

† B.W.Roberts: J. Physical and Chemical Reference Data, 5 (1976)
* present work

In the case of C_8K, such an analogy is not applicable because its mother materials are both non-superconductive. Inoshita and Kamimura [18] calculated the value of λ, on the basis of their theory on electron-phonon interaction in

GIC's, and suggested that the inter-pocket phonon scattering can give a value of λ large enough to make graphite electrons superconductive. Very recently, Takada [19] pointed out that when the inter-pocket Coulomb interaction is properly taken into account, the contribution of the inter-pocket phonon interaction is almost cancelled out, so that the possibility of graphite electrons becoming superconductive is scarce. He made an estimation of T_c by considering a strong polar coupling between the s-like electrons of alkali-metals and the longitudinal optic and acoustic phonons, and concluded that three dimensional electrons of the alkali metal band principally contribute to the superconductivity of the first stage alkali metal GIC's. If this is the case, the partial (i.e. not full) charge transfer is essential for superconductivity in GIC's.

Although the so-far unobserved superconductivity of the first stage alkali-amalgam GIC's has been found, there still remains a question why the second stage compounds have higher T_c's than the first stage. One speculation is as follows. If we regard an alkali-amalgam GIC as a layered alkali-amalgam metal intercalated with carbon layers, the observed T_c vs. stage relation may be viewed as a T_c-enhancement by insertion of another carbon layer, analogous to the T_c-enhancement of transition metal dichalcogenides by intercalation. From this viewpoint, the T_c-enhancement mechanism upon intercalation might be associated with the anisotropy increase by increased separation of superconducting layers. This problem is open to the future reseach.

Acknowledgements

The authors are grateful to D. A.W.Moore of Union Carbide Co., U.S.A., for kindly donating high quality HOPG samples. They are also indebted to Prof. H.Kamimura and Dr. Y.Takada for stimulating discussions. This work was financially supported by the Mitsubishi Foundation.

References

[1] N.B.Hannay, T.H.Geballe, B.T.Matthias, K.Andres, P.Schmidt and D.MacNair: Phys. Rev. Lett. 14 (1965) 225.
[2] J.Poitrenaud: Rev. Phys. Appl., 5 (1970) 275.
[3] M.Kobayashi and I.Tsujikawa: J. Phys. Soc. Jpn., 46 (1979) 1945.
[4] Y.Koike, H.Suematsu, K.Higuchi and S.Tanuma: Solid State Commun. 27 (1978) 623.
[5] Y,Koike, H.Suematsu, K.Higuchi and S.Tanuma: J. Phys. Chem. Solids, 41 (1980) 1111.
[6] Y.Koike and S.Tanuma: unpublished data.
[7] M.Kobayashi and I.Tsujikawa: private communication.
[8] M.G.Alexander, D.P.Goshorn, D.Guerard, P.Lagrange, M.El Makrini and D.G.Onn: Synthetic Metals, 2 (1980) 203.
[9] L.A.Pendrys, R.Wachnik, F.L.Vogel, P.Lagrange, G.Firdin, M.El Makrini and A.Herold: Solid State Commun., to be published.
[10] Y.Koike and S.Tanuma: J. Phys. Soc. Jpn., to be published.
[11] P.Lagrange, M.El Makrini, D.Guerard and A.Herold: Physica 99B (1980) 473.
[12] P.Lagrange, M.El Makrini, D.Guerard and A.Herold: Synthetic Metals, 2 (1980) 191.
[13] R.C.Morris, R.V.Coleman and R.Bhandar: Phys. Rev., B5 (1972) 895.
[14] A.L.Fetter and P.C.Hohenberg: in "Superconductivity" ed. R.D.Parks, Marcel Dekker, Newyork, 1969.

[15] N.Toyota, H.Nakatsuji, K.Noto, A.Hoshi, N.Kobayashi. Y.Muto and Y.Onodera: J. Low Temp. Phys. 25 (1976) 485.
[16] R.Meservey and B.B.Schwartz: in "Superconductivity" ed. R.D.Parks, Marcel Dekker, Newyork, 1969.
[17] W.L.McMillan: Phys Rev. 167 (1968) 331.
[18] T.Inoshita and H.Kamimura: Synthetic Metals, 3 (1981) 223.
[19] Y.Takada: preprint,(submitted to J. Phys. Soc. Jpn.).

Thermal Conductivity: An Old Method and New Tool to Investigate Intercalation Compounds

J.P. Issi, J. Heremans*

Laboratoire de Physico-Chimie et de Physique de l'Etat Solide
Université Catholique de Louvain, B-1348 Louvain-la-Neuve and

M.S. Dresselhaus

Center for Materials Science and Engineering, M.I.T., Cambridge, MA 02139, USA

1. Introduction

Recently measurements on the thermal conductivity of graphite intercalation compounds have been reported [1][2][3] and discussed [4]. We wish to emphasize here the new information that these in-plane and c-axis thermal conductivity measurements provide on the structure and properties of intercalation compounds as well as on the thermal conductivity of quasi-two dimensional systems in general.

In general, the low temperature thermal conductivity κ of solids is due to contributions from the charge carriers κ_e and the phonons κ_g [5][6], so that $\kappa = \kappa_e + \kappa_g$. For both processes the expression of the thermal conductivity is given by the kinetic formula :

$$\kappa = \frac{1}{3} c_V \cdot v \cdot \ell \tag{1}$$

where c_V is the specific heat at constant volume, v the velocity of the particle and ℓ its mean free path. Since these parameters are quite different for the two mechanisms, the temperature variation of κ_e and κ_g will also be quite different. In Fig. 1 we present a summary of typical behaviors for the thermal conductivity of metals (dominated by κ_e) and insulators (dominated by κ_g), including typical low-temperature dependences at low and high T. The in-plane dielectric maximum for pristine graphite is also shown. We have also represented in Fig. 1 the typical behavior of a non-crystalline solid : vitreous silica. In the lowest temperature range the thermal conductivity of the latter is much lower than that of its allotropic crystalline form (quartz). The introduction of defects in a dielectric solid like quartz decreases significantly the low temperature thermal conductivity because of a decrease in ℓ due to phonon-defect scattering [5][6]. In the same way, introducing defects in a pure metal will tend to decrease its electronic thermal conductivity. In both cases for heavily damaged or very impure solids the thermal conductivities around the maximum may be decreased by orders of magnitudes. Thus thermal conductivity measurements should be useful to study lattice imperfections in crystalline solids.

In the case of graphite intercalation compounds there are some additional interesting features. It was shown that in the liquid helium range the electrons or holes dominate in-plane, while phonons carry all the heat in the c-axis direction [4]. Thus it is imperative to treat separately in-plane and c-axis behaviors. As a corollary, it is not surprising to find that the anisotropy of the thermal conductivity and its temperature

---

* Aspirant FNRS

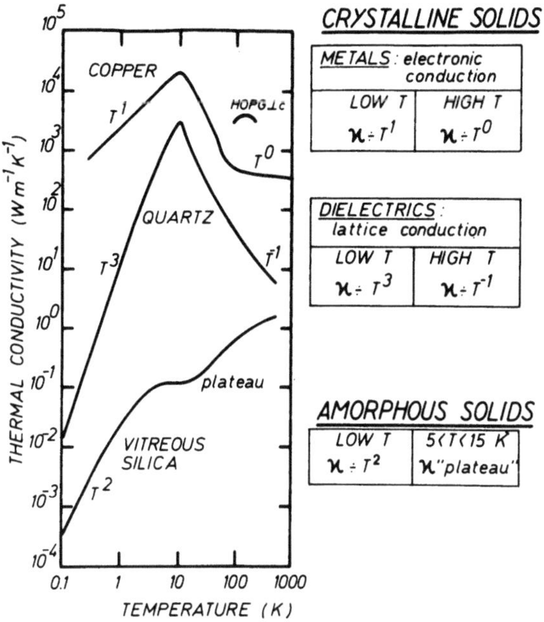

Fig.1. Typical behavior of the temperature dependence of the thermal conductivity of three kinds of solids. In the case of pure *copper* the thermal conductivity is entirely electronic. It starts with a linear temperature behavior at low temperatures in the residual range, reaches a maximum and then decreases to level off at higher temperatures. The constant value of the thermal conductivity corresponds to the range where the electrical conductivity σ is inversely proportional to the absolute temperature T. The electronic thermal conductivity κ_e is directly related to the electrical conductivity via the Wiedemann-Franz law $\kappa_e = L\,T\,\sigma$, where L is the Lorenz number. *Quartz* exhibits a typical dielectric behavior. The low temperature thermal conductivity varies first as T^3, reaches a pronounced maximum and then decreases with temperature. *Vitreous silica*, which is a non-crystalline material, has a much lower thermal conductivity than quartz and its temperature variation is also totally different. The T^2 behavior at low temperatures and the plateau are both universal features of non-crystalline solids. The location of the in-plane dielectric maximum for pristine *HOPG* is also shown.

dependence are much different from that of the electrical conductivity, as is explained below.

2. The Effect of Intercalation on the Host Material

The fact that graphite is the host material distinguishes its intercalation compounds from all the other layered materials as regards the in-plane thermal conductivity for the following reasons. The in-plane Debye temperature ($\theta_D \sim 2500$ K) of this material is, with diamond, the highest one observed in any solid. Usually the dielectric maximum of the lattice thermal conductivity of solids occurs between $\theta_D/20$ and $\theta_D/10$, which for pristine graphite is in the range 120-250 K, in agreement with observations (Fig.1). Defects introduced by intercalation are expected to increase the temperature at which the maximum of the thermal conductivity occurs, which is also verified experimentally [3]. Any other known element or compound will have its dielectric maximum at much lower temperature, say below 30 K. Because the electronic component of the intercalant is expected to be negligible at all temperatures, the contribution of the graphite layers will dominate the higher temperature behavior (Fig. 2). Since for in-plane graphite around room temperature we are still in the low temperature range with respect to the Debye temperature, we have in hand an ideal tool to study the effects of intercalation on the defect structure of pristine graphite. For this purpose we have to compare the behavior of the starting material, i.e. pristine graphite to that of the intercalated material.

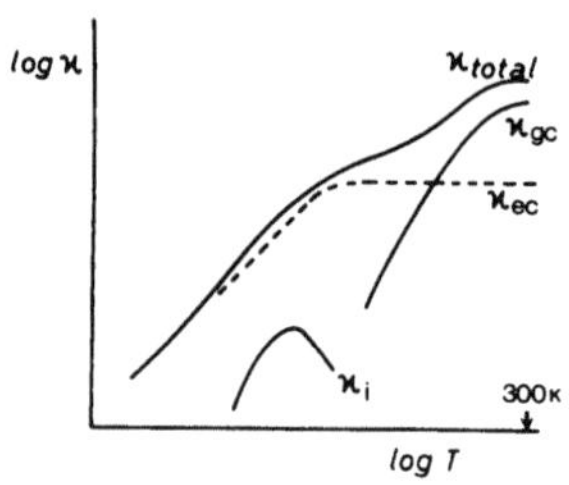

Fig.2. Schematic representation of the temperature variation of the thermal conductivity of a graphite intercalation compound below 300 K plotted on a log-log scale. The maximum thermal conductivity is of the order of a few hundreds $W \cdot K^{-1} \cdot m^{-1}$. At low temperatures the thermal conductivity of the intercalated graphite increases linearily with temperature, then an additional contribution (κ_i), which may be due to the presence of the intercalant, shows up. Around room temperature the thermal conductivity of the intercalated graphite is due to lattice (κ_{gc}) and electronic (κ_{ec}) contributions.

If we make the very crude assumption that the total in-plane lattice thermal conductivity κ_g is the sum of the thermal conductivities of the different types of layers acting in parallel, when the stage $n > 2$, then [4][7]

$$\kappa_g = \frac{d_i \kappa_i + n c_o \kappa_{gc}}{I_c} \qquad (2)$$

where d_i and c_o are the thickness of the intercalate and carbon layers, respectively, I_c the c-axis repeat distance, κ_{gc} the contribution to the lattice thermal conductivity of the carbon layers and κ_i a contribution from the intercalant.

Consider the idealized case where the thermal conductivity of the graphite layers κ_{gc} remains that of pristine graphite after intercalation. Knowing κ_{gc}, c_o and d_i, we may estimate the stage dependence of the lattice thermal conductivity around room temperature from (2) assuming that $d_i \kappa_i$ is negligible. So comparing the experimentally observed values for the lattice thermal conductivity with those computed for the idealized case enables us to estimate the effects due to intercalation. The results of a rough analysis [4] performed on potassium and $FeCl_3$ compounds indicated that the scattering length in these materials is of the order of a few 10^{-7}m, which might be associated with scattering by Daumas-Hérold domain boundaries. Also, point-defect scattering in the host graphite material was found to be more pronounced after intercalation. In some cases, in-plane measurements may reveal a small contribution κ_i resulting from the intercalant (Fig. 2) in a restricted temperature range. A relatively small κ_i contribution was indeed apparent in the case of stage-2 $FeCl_3$ compound. This contribution was even smaller for a stage-3 and too small to be detected with certainty for higher stages [3]. In the stage-5 potassium compound [2] κ_i was not detectable since the electronic thermal conductivity is too high and the ratio d_i/c_0 too small ($\sim$ 0.61 for potassium compared to $\sim$ 1.84 for $FeCl_3$).

In-plane, around room temperature, phonons provide useful tools to probe defects. In the liquid helium range we have another independent tool for this purpose;the electrons. In this temperature range the electronic contribution dominates the scene and may thus be studied directly [3][4]. Using the Wiedemann-Franz law the electrical resistivity may be calculated. At these low temperatures the value of the residual or extrinsic resistivity depends on lattice imperfections. Also, if we substract from the total resistivity the constant residual contribution, the ideal resistivity is deduced, and its temperature variation is an indication of intrinsic electron scattering processes. In that context we believe that residual resistivity determination via four probe thermal conductivity measurements are more reliable than direct measurements at low temperatures [8].

3. Study of the Intercalant

Parallel to the c-axis direction, the situation is entirely different. Here it is the intercalant which will mainly determine the resistance to thermal flow and will thus govern the thermal conductivity. Thermal conductivity studies in that direction, when compared to the behavior of pristine graphite, might then provide information on the intercalate layers. In Fig. 3 the c-axis thermal conductivity of a stage-2 $FeCl_3$ graphite intercalation

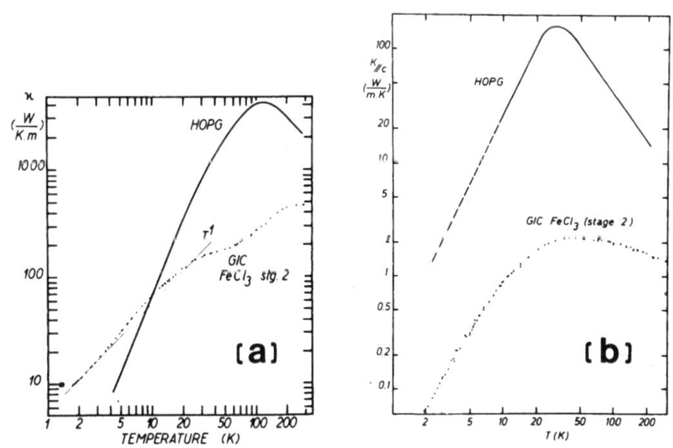

Fig. 3 Typical temperature dependence of the thermal conductivity of a graphite intercalation compound compared to that of pristine HOPG. The data are relative to a stage-2 graphite-$FeCl_3$ intercalation compound in-plane (a) and in the c-axis direction (b). Note that while for pristine graphite the qualitative behavior is not essentially different for both directions, it is quite different for the intercalation compound.

compound is compared to that of pristine graphite in the same direction. It may be seen that the presence of the intercalant decreases by at least an order of magnitude the lattice thermal conductivity. The effect is even more pronounced in the case of the potassium compound [4]. Studies are now in progress to investigate the stage-dependence of the c-axis thermal conductivity of graphite intercalation compounds.

4. Concluding remarks

The first studies of the temperature variation of the in-plane conductivity of graphite intercalation compounds revealed that the main effect of intercalation is to increase the electronic thermal conductivity and decrease the lattice thermal conductivity relative to pristine graphite. c-axis measurements revealed that contrary to the case of the electrical conductivity the anisotropy of the thermal conductivity is comparable to that of pristine graphite in acceptor compounds.

At the lowest temperatures heat transport is entirely due to the electrons or holes in-plane and to the phonons in the c-axis direction. Around room temperature both electrons and phonons carry the heat in-plane. At intermediate temperatures there is eventually a small additional contribution which is tentatively ascribed to the presence of the intercalant.

We have shown that, because of the unique features of graphite intercalation compounds, in-plane thermal conductivity studies provide two independent tools to study defects introduced by intercalation. These probes are the electrons at low temperatures and the phonons at higher temperatures. Also, c-axis thermal conductivity measurements are expected to yield information about the intercalant layers.

Acknowledgements

The authors are thankful to Dr. G. DRESSELHAUS for useful discussions. The help of H. MAZUREK, M. SHAYEGAN, I. ZABALA-MARTINEZ and B. POULAERT concerning the experimental work is also greatly appreciated. This work was partially supported by an AFOSR grant number F 49620-81-C-0006.

References

1 J. Boxus, B. Poulaert, J-P. Issi, H. Mazurek and M.S. Dresselhaus, Solid State Communications, 38, 1117 (1981)

2 J. Heremans, J-P. Issi, I. Zabala-Martinez, M. Shayegan and M.S. Dresselhaus, Physics Letters (in press)

3 B. Poulaert, J. Heremans, J-P. Issi, I. Zabala-Martinez, H. Mazurek and M.S. Dresselhaus, Extended Abstracts 15th Biennal Carbon Conf. (1981),p.92

4 J. Heremans, J-P. Issi and M.S. Dresselhaus, to be published.

5 R. Berman, Thermal Conduction of Solids, Clarendon Press, Oxford (1976)

6 P.G. Klemens,"Theory of the Thermal Conductivity of Solids" in Thermal Conductivity , vol. 1, R.P. Tye Editor, Academic Press, New York (1969)

7 We will not speculate here about the case where n = 1, since there are no measurements available yet for this stage.

8 J-P. Issi, J. Boxus, B. Poulaert and H. Heremans, Proceedings of the 17th International Thermal Conductivity Conference, Gaithersburg (1981)

The Effect of Hydrogen Sorption on the Electronic Properties of $C_{24}K$

T. Kondow

Department of Chemistry, Faculty of Science
The University of Tokyo, Tokyo, Japan and

K. Ando and Y. Tomono

Department of Physics, Sophia University, Tokyo, Japan

1. Introduction

Graphite intercalation compounds possess a peculiar layered structure and the carbon layers intercalated are spaced by the intercalant layers in parallel. Thus the slit-shaped vacancy between the intercalated carbon layers enables a variety of molecules to condense into it at low temperatures [1]. Accordingly the compounds exhibit the molecular sieve-type sorption which differs by an intercalant material inserted because each intercalant gives different slit sizes of molecular dimension. However, this sieving effect is markedly contrasted to sorption into a rigid cage (zeolite, for example) with respect to the extensive swelling ability of the intercalation compounds. It should be emphasized that hydrogen molecules are absorbed at low temperatures by the intercalation compounds of alkali metals except for the first stage compounds, and in addition, deuterium molecules are more readily absorbed (isotope effect) [1]. In the system C_{24}-D_2, heat of the deuterium sorption exceeds that of hydrogen sorption by 10% and a slight increase in the C-axis lattice constant is reported resulting from sorption [2].

In the present study, the electrical conductivity and Hall effect of $C_{24}K$, $C_{24}K(H_2)_2$ and $C_{24}K(D_2)_2$ were measured in the temperatures between 4 and 77K so as to clarify the mechanism of the hydrogen sorption and its isotope effect.

2. Experimental

Fig.1 shows the schematic diagram of the sample cell made of pyrex glass. The sample in the cell was prepared in situ from HOPG (Union Carbide ZYA) by employing the dual furnace method [3]. After the sample was prepared, the glass cell having a breakable seal in which the sample was placed was connected to a hydrogen gas reservoir by glass blowing. The pressure of hydrogen was measured by a Pirani guage which was calibrated against a MKS Baratron manometer. The pressures of hydrogen or deuterium gases (99.99% supplied by Takachiho Co.) were kept sufficiently high (more than 200 Torr) to form at 77K the saturated compounds, $C_{24}K(H_2)_2$ or $C_{24}K(D_2)_2$. As shown in the inset of Fig.1, the 6 platinum wires were adhered to the sample by utilizing a Hanovia platinum paste after the contact areas were platinum-plated. The sample was mounted on a ceramics plate which was held inside the glass tube by applying a stainless steel spring. The electrical resistivity measurements were preformed by a conventional 4-terminal method in the temperature range from 4.2 to 300K. Static magnetic field of 16.5 KG was applied in parallel with the C-axis of the compounds, in the Hall effect measurements.

3. Results and Discussion

3.1 Electrical Resistivity

The electrical resistivities (ρ) of $C_{24}K$ are plotted as a function of temperature (T). The data points can be fit to the equation $\rho = 2.54 \times 10^{-3} T^2 + 4.70 \times 10^{-2} T + 26.76$ in the temperature range between 4.2 and 300K except at the order-disorder temperature (98K). The linear term of T was dominant below 150K and the square term was above 150K. As far as the temperature-dependence is concerned the present result is consistent with the reported data [4, 5] and is accounted for on the basis of the intra- and inter-pocket scattering scheme [6].

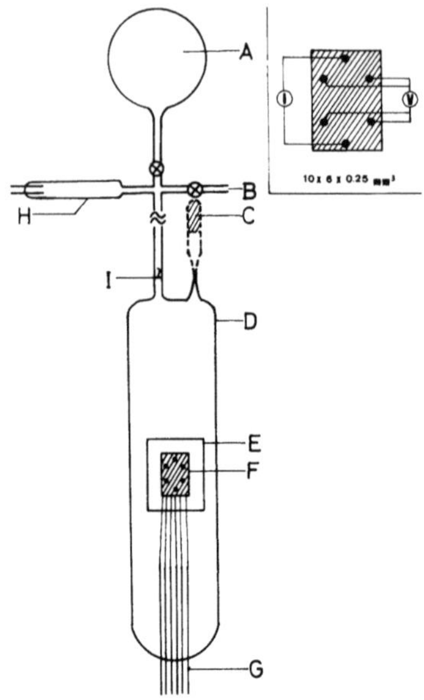

Fig.1 Schematic diagram of the sample cell. The inset shows the dimension of the sample
A: Gas reservoir B: Evacuation
C: Potassium D: Sample case
E: Sample box F: Sample
G: Leads H: Pirani guage
I: Breakable seal

Figs.2 and 3 show the temperature-dependences of the electrical resistivities of $C_{24}K(H_2)_2$ and $C_{24}K(D_2)_2$, respectively. The linear term of T was dominant except in the vicinity of 42K where the resistivity tends to level off, probably because of the contribution of residual resistivity. The ratio of the slopes of the lines were 1 : 1.23 : 1.92 for $C_{24}K$, $C_{24}K(H_2)_2$ and $C_{24}K(D_2)_2$. Hydrogen and deuterium gases were desorbed at 100-110K and 120-125K, respectively so that the resistivity decreased sharply and resumed to that of $C_{24}K$. It should be noted that the order-disorder transition (98K) of $C_{24}K$ disappears upon the hydrogen sorption, probably because the intercalant layers become more closely packed. The finding that the linear term dominates is likely to show that the contribution of the inter-pocket scattering is reduced according to the decrease of the carrier density upon hydrogen sorption. Inoshita et al. showed that the coefficient of the linear term is a function of $\omega_q \cdot |g_{kk'}|^2$ where ω_q and

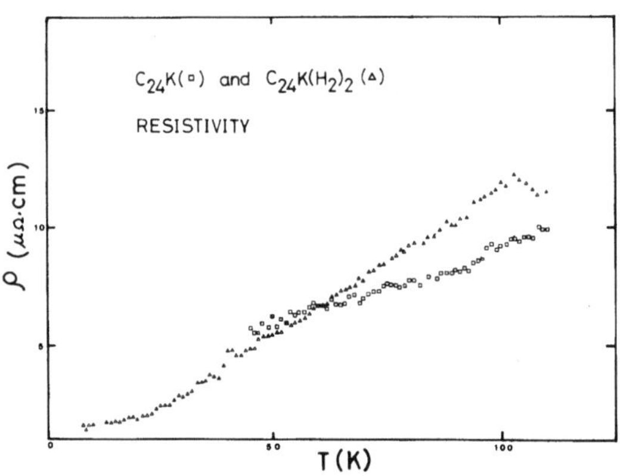

Fig.2 The electrical resistivities of $C_{24}K(H_2)_2$ are plotted against temperatures between 4.2 and 110K

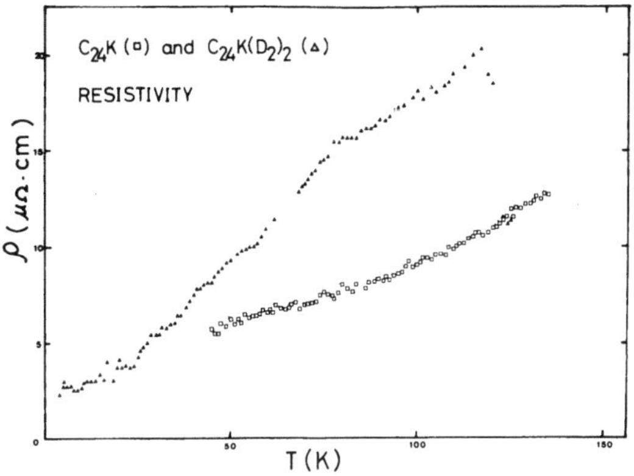

Fig.3 The electrical resistivities of $C_{24}K(D_2)_2$ are shown as a function of temperature in the range from 4.2 to 125K

$|g_{k'k}|^2$ represent the phonon frequency and the electron phonon matrix element related [6]. The increase of the slope upon hydrogen sorption is related, accordingly, to enhancement of the product $\omega_q \cdot |g_{kk'}|^2$ associated with the charge carrier transport. If a phonon mode coupled is the same in the H_2 and D_2 compounds, the coupling constant of the D_2 compound is **larger by 50% than that of the H_2 compound when ω_q is assumed** to be inversely proportional to the square root of the H_2 or D_2 mass.

3.2 Hall coefficient

Fig.4 shows the temperature dependence of the Hall coefficients (R_H) of $C_{24}K$ in the temperature range from 77 to 300K. R_H was almost independent of temperature below 100K so that a single carrier model is valid below this temperature. The carrier density was determined from R_H at 77K to be about $8.62 \times 10^{20}/cm^3$, which is comparable to the reported value $1.27 \times 10^{21}/cm^3$ [4]. In the case of $C_{24}K(H_2)_2$ and $C_{24}K(D_2)_2$, the Hall coefficients did not depend on temperature between 77 and 100K. Thus the single carrier picture is also applicable to these compounds. The values of R_H and ρ at 77K for $C_{24}K$, $C_{24}K(H_2)_2$ and $C_{24}K(D_2)_2$ are listed in Table 1. The carrier densi-

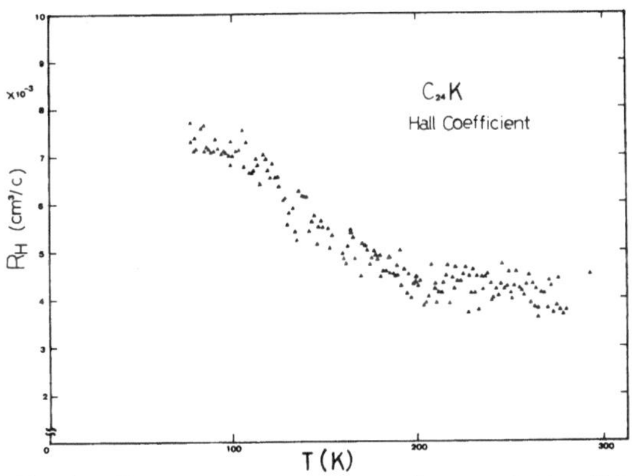

Fig.4 The temperature dependence of the Hall coefficients of $C_{24}K$

Table 1 Hall coefficients and electrical resistivities at 77K

	R_H [cm³/c] $\times 10^{-3}$	ρ [μΩ·cm]
$C_{24}K$	7.25 ± 0.22	7.42 ± 0.3
$C_{24}K(H_2)_2$	11.62 ± 0.18	8.53 ± 0.2
$C_{24}K(D_2)_2$	18.73 ± 0.20	12.04 ± 0.1

ties and Hall mobilities were evaluated by using these values shown in Table 1 and are summerized in Table 2. The valence electrons in the carbon layers were back-donated by 40 and 60% upon sorption of H_2 and D_2, respectively.

Table 2 Carrier densities and Hall mobilities in comparison with the energy of sorption

	$n \times 10^{20}$ [cm^{-3}]	μ [cm^3/v.s.]	The energy of sorption [kcal/mol] calc.	obs.[2]
$C_{24}K$	8.62 ± 0.26	977 ± 69	——	——
$C_{24}K(H_2)_2$	5.38 ± 0.08	1350 ± 50	2.17	2.14
$C_{24}K(D_2)_2$	3.34 ± 0.04	1570 ± 22	3.49	2.62

The mobility increase upon the hydrogen sorption can result from less contribution of the inter-pocket scattering due to the decrease of the carrier density. This marked isotope effect in the rate of the back-donation may be closely related to the isotope effect in the hydrogen sorption. The energy of hydrogen sorption was estimated according to the method by Salzano and Aronson [8] on the assumption that the carrier density is reduced by 40 and 60% in the $C_{24}K(H_2)_2$ and $C_{24}K(D_2)_2$, respectively, and hydrogen between the layers does not contribute. The values thus estimated are comparable to the reported heat of sorption [1, 2]. The isotope effect in the heat of sorption may arise from isotopic difference in the zero point energies of H_2 and D_2.

References

1. K. Watanabe, T. Kondow, M. Soma, T. Onishi, K. Tamaru: Proc. Roy. Soc. London A333, 51 (1973)
2. K. Watanabe, T. Kondow, M. Soma, T. Onishi, K. Tamaru: Chem. Letters 1972, 477 (1972)
3. D.E. Nixon, G.S. Parry: Brit. J. Appl. Phys., Ser. 2, 291 (1968)
4. H. Suematsu, K. Higuchi, S. Tanuma: J. Phys. Soc. Jpn. 48, 1541 (1980)
5. E. McRae, D. Billaud, J.F. Marêché, A. Hérold: Physica 99B+C, 489 (1980)
6. T. Inoshita, H. Kamimura: Synthetic Metals 3, 223 (1981)
7. M. Sano, H. Inokuchi: Chem. Letters 1979, 405 (1979)
8. F.J. Salzano, S. Aronson: J. Chem. Phys. 44, 4320 (1966)

Critical Currents in Layered Superconductors

L. Dobrosavljevič and Z. Radovič
Institute of Physics
Y-11001 Belgrade

Layered superconductors provide an excellent model system for the study of the flux pinning by inhomogeneities. We consider superconducting films consisting of stacked layers ... N/S/N/S/N ..., where S is a superconducting and N a normal metal. The increase of the critical current density j_c, which is observed experimentally when the external magnetic field H is parallel to the layers |1| , can be explained by the pinning of the flux lines in the N layers or at interfaces.

The behaviour of the system is strongly influenced by the proximity effects between N and S layers and varies very much with H and with temperature.

Experimentally, three types of variations in critical current density with magnetic field are found at a fixed temperature |1| |2|:

1. A monotonic decrease of j_c with increasing H
2. A constant j_c in low fields followed by a further monotonic decrease in higher fields (plateau regime)
3. A non-monotonic variation of j_c with H, i.e. characteristic peak structure (peak regime).

In the following we discuss the physical origin of the pinning mechanisms causing the observed $j_c(H)$ variations and give theoretical estimates of the critical current densities.

1. Monotonic Behaviour

This variation law is frequently observed in bulk superconductors; in layered superconductors it is always observed when the applied field is perpendicular to the layers. In a parallel field this type of variation is expected when the pinning occurs at the boundary between the N and S layers. This happens, e.g., in the lamellar eutectics |2|, where the lamellae are very large compared with the corresponding distances and S layers are decoupled.

A significant difference between the critical currents in parallel and perpendicular geometry is expected also when the thicknesses d_S and d_N of S and N layers are small compared to the corresponding coherence lengths $\xi_S(T)$ and $K^{-1}(T)$. In this case the system behaves as a single anisotropic superconductor, and the pinning in both parallel and perpendicular fields occurs at the boundary superconductor/vacuum. For an isotropic (thick) film of the thickness D, the corresponding critical current density is |3| |4|

$$j_c = \frac{3^{1/4}\, c\, \phi_o^{1/2}\, |M|}{2^{1/2}\, \lambda\, D\, B^{1/2}} \tag{1}$$

where $M=(B-H)/4\pi$ is the reversible magnetization. In the anisotropic case the flux penetration depends very much on the magnetic field orientation, as can be easily seen via a simple effective mass model |5|. Thus j_c is determined not only by the intensity of the applied field, but also by its orientation and by the amount of anisotropy. In both parallel and perpendicular geometry $j_c(H)$ is a monotonically decreasing function.

2. Plateau Regime

We explain the absence of the j_c variation in low fields (which occurs in parallel geometry only) by a special way of flux penetration in the specimen. When the N layers are thick, $d_N \gg K^{-1}(T)$, the normal material behaves as a type II superconductor, even if N is a type I superconductor at low temperature |6|. In each N layer a position dependent Ginzburg-Landau parameter k can be defined. As long as its value k_N at the N/S interface is smaller than 1, the flux penetrates in the middle of the N regions in form of quantized fluxoids |6|. This penetration begins at a lower critical field

$$H_{C1} = \frac{\phi_o}{8\, k_N} \cdot \frac{K^{3/2}}{(\frac{d_N}{2} - \rho)^{1/2}} \cdot (\mathrm{ch}\, K \frac{d_N}{2})^{-1} \tag{2}$$

where

$$\rho = K^{-1} \left(\ln \frac{1}{k_N} - 0.116\right)$$

and the fluxoids penetrate in a band of the thickness $d_N - 2\rho$. In the most experimentally studied cases |7| H_{C1} has a very low value, $H_{C1} \lesssim 10$ Oe.

In low fields ρ is field independent, up to a field $H_{min} = \phi_o K^2/2\pi$; with increasing induction it is only the distance between fluxoids along each N layer axis which diminishes. A slow variation of ρ with H occurs between H_{min} and $H_{max} = H_{min}/k_N$ at which the flux penetrates up to the N/S interface. In the field domain $H < H_{max}$ there is no flux penetration in S layers and the critical current is determined by the pinning of fluxoids in the N regions. The pinning force per fluxoid is determined as the ratio of its self energy and the coherence length, $f_L = \phi_o H_{C1}/4\pi\, K^{-1}$, whereas the pinning force per unit volume is $F_L = (B/\phi_o) f_L$. This force must be balanced by the Lorentz force (per unit volume) $(1/c) j_c B$. Thus the mean critical current density j_c is induction (and field) independent

$$j_c = \frac{c}{\phi_o} f_L = \frac{c}{4\pi} \frac{H_{C1}}{K^{-1}} \quad . \tag{3}$$

Its temperature dependence (at T not too close to T_{CS}, where the necessary condition $k_N << 1$ is satisfied |6|) appears via H_{C1} and K^{-1}, giving a monotonic decrease of j_c with T. If the N layers consist of a metal which is superconducting at $T < T_{CN}$, the condition $d_N >> K^{-1}$ is violated when one approaches T_{CN}. The N layers broaden and when $K\, d_N/2 << 1$ the normal regions become superconducting as well experimentally, this is manifested by the rounding of the plateau at the curve $j_C(H)$ at low T. In the experiment with the Pb-In/Sn-In layered sample with $T_{CN} \approx 3.8$ K, the plateau regime occurs at 4.8 K where $d_N/2 > K^{-1}$, whereas at 4.2 K a peak is observed |1|. This indicates another way of flux penetration which is theoretically expected as well, since at this temperature $d_N/2 \lesssim K^{-1}$. In the experiment with the Pb-In/Ag sample |1|, where $T_{CN} = 0$, the rounding of the $j_C(H)$ curve begins at $T \lesssim 1.5$ K.

In higher field regions ($H > H_{max}$) the flux penetration in the S layer starts. A monotonic decrease of j_C with H is expected because of the pinning at N/S interfaces, followed by a further drop of j_C due to the interaction of fluxoids.

3. Peak regime

If the thickness of the N layers is small, the whole specimen behaves similarly to a modulated superconductor as emphasized by RAFFY |1|. When the residual pinning effects are subtracted, the critical current density Δj_C is the spatial average of the supercurrent which flows when the rows of fluxoids are displaced from their equilibrium positions in the N layers |8|. This "resonant" current increases with H up to a maximum, which corresponds to the stability limit of the commensurate configuration. In higher fields the appearance of defects leads to a decrease of Δj_C with H.

Using the analogy with periodically modulated superconductors |7|, where the penetration depth λ varies periodically along the modulation direction |8|

$$\lambda^2 = \bar{\lambda}^2 \left(1 + \delta \cos \frac{2\pi}{d} x\right)$$

we calculate Δj_C in the commensurate phase taking $d = d_S + d_N$. If $1 \lesssim d/\bar{\lambda} \lesssim 4$, Δj_C has a single maximum in the vicinity of the principal matching where $B \cong \sqrt{3}\phi_0/2d^2$ |1|. The maximum amplitude is |8|

$$(\Delta j_C)_{max} \approx 0.1 \quad \delta' \frac{c\phi_0}{d^3} \quad \frac{\ell n\, \bar{k}}{\bar{k}^2} \qquad (4)$$

where $\bar{k}$ is an effective Ginzburg-Landau parameter and δ'/δ is a measure of the effective pinning potential for a given modulation δ. It can be shown |9| that $\delta'/\delta \approx 1$ when $d/\bar{\lambda}(T) >> 1$ and goes rapidly to zero when $d/\bar{\lambda}(T)$ decreases. Qualitatively variation (4) of $(\Delta j_C)_{max}$ with d and T is in a fairly good agreement with the experimental results |1|.

References

1. H. Raffy: Thesis, Paris University (1980) and the references therein
2. C.R. Spencer, P. Martinoli, E.D. Gibson, J.D. Verhoeven, and D.K. Finnemore: Phys.Rev. B 18, 1216 (1978)
3. V.V. Shmidt: Soviet Physics J.E.T.P. 34, 211 (1972)
4. A.M. Campbell and J.E. Evetts: Advances in Physics 21, 199 (1972)
5. D.R. Tilley: Proc. Phys.Soc. 85, 1177 (1965); and 86 289 (1965); and 86, 678 (1965)
6. L. Dobrosavljević and P.G.de Gennes, Solid State Commun. 5, 177 (1967)
7. C. Petipas: Thesis, Paris University (1969)
8. Z. Radović and L. Dobrosavljević: In *Recent Developments in Condensed Matter Physics* (Plenum, New York, 1981) 5, 405
9. Z. Radović and L. Dobrosavljević: to be published

Index of Contributors

MIX
Papier aus verantwortungsvollen Quellen
Paper from responsible sources
FSC® C105338

If you have any concerns about our products,
you can contact us on
ProductSafety@springernature.com

In case Publisher is established outside the EU,
the EU authorized representative is:
Springer Nature Customer Service Center GmbH
Europaplatz 3, 69115 Heidelberg, Germany

Printed by Libri Plureos GmbH
in Hamburg, Germany